Vertical External Cavity Surface Emitting Lasers

Vertical External Cavity Surface Emitting Lasers

VECSEL Technology and Applications

Edited by
Michael Jetter
Peter Michler

Editors

Dr. Michael Jetter
University of Stuttgart
Institut für Halbleiteroptik
Allmandring 3
70569 Stuttgart
Germany

Prof. Dr. Peter Michler
University of Stuttgart
Institut für Halbleiteroptik
Allmandring 3
70569 Stuttgart
Germany

Cover: © fotojog/Getty Images

All books published by **WILEY-VCH** are carefully produced. Nevertheless, authors, editors, and publisher do not warrant the information contained in these books, including this book, to be free of errors. Readers are advised to keep in mind that statements, data, illustrations, procedural details or other items may inadvertently be inaccurate.

Library of Congress Card No.: applied for

British Library Cataloguing-in-Publication Data A catalogue record for this book is available from the British Library.

Bibliographic information published by the Deutsche Nationalbibliothek The Deutsche Nationalbibliothek lists this publication in the Deutsche Nationalbibliografie; detailed bibliographic data are available on the Internet at <http://dnb.d-nb.de>.

© 2022 WILEY-VCH GmbH, Boschstr. 12, 69469 Weinheim, Germany

All rights reserved (including those of translation into other languages). No part of this book may be reproduced in any form – by photoprinting, microfilm, or any other means – nor transmitted or translated into a machine language without written permission from the publishers. Registered names, trademarks, etc. used in this book, even when not specifically marked as such, are not to be considered unprotected by law.

Print ISBN: 978-3-527-41362-1
ePDF ISBN: 978-3-527-80794-9
ePub ISBN: 978-3-527-80797-0
oBook ISBN: 978-3-527-80795-6

Typesetting   Straive, Chennai, India
Printing and Binding CPI Group (UK) Ltd, Croydon, CR0 4YY

Printed on acid-free paper

C068890_290821

Contents

Preface *xiii*

Part I Continuous wave VECSEL *1*

1 History of Optically Pumped Semiconductor Lasers – VECSELs *3*
Mark E. Kuznetsov
1.1   Introduction   *3*
1.2   OPS-VECSELs: Concept and History   *4*
1.3   Micracor   *8*
1.4   OPSL Development at Micracor: First Steps   *11*
1.5   OPS Development at Micracor: Pushing Forward   *14*
1.6   OPS Development at Micracor: Final Chapter   *16*
1.7   VECSELs beyond Micracor   *20*
         References   *22*

2 VECSELs in the Wavelength Range 1.18–1.55 μm *27*
Antti Rantamäki and Mircea Guina
2.1   Introduction   *27*
2.2   Overview of GaAs-based Gain Mirror Technologies for Long-wavelength Infrared VECSELs   *28*
2.2.1   InGaAs QWs   *28*
2.2.2   GaInNAs QWs   *28*
2.2.3   InAs QDs   *30*
2.2.4   GaAsSb QWs   *31*
2.3   Overview of InP-based Gain Mirror Technologies for Long-wavelength Infrared VECSELs   *32*
2.3.1   Monolithic InP-based DBRs   *32*
2.3.2   Dielectric and Metamorphic DBRs   *33*
2.3.3   Semiconductor-dielectric-metal Compound Mirrors   *34*
2.3.4   Wafer-bonded GaAs-based DBRs   *37*
2.3.4.1   Direct Wafer Bonding   *39*

2.3.4.2	Low Temperature Bonding *44*
2.3.5	Gain Structures in Transmission *47*
2.4	Conclusion *50*
	References *50*

3 Single-frequency and High Power Operation of 2–3 Micron VECSEL *63*

Marcel Rattunde, Peter Holl, and Joachim Wagner

3.1	Introduction *63*
3.2	Semiconductor Lasers for the MIR Range *64*
3.3	III-Sb Material System *66*
3.4	2–3 µm VECSEL Design *68*
3.4.1	Standard Barrier Pumped Structures *68*
3.4.2	In-well Pumping *69*
3.4.3	Low Quantum Deficit Barrier Pumping *70*
3.5	Mounting Technologies *72*
3.5.1	Intracavity Heatspreader *74*
3.5.2	Thin Device *76*
3.5.3	Double-sided Heatspreader *77*
3.6	Single-frequency Operation (SFO) of 2–3 µm VECSEL *78*
3.6.1	Key Parameters for Single-Frequency Operation *79*
3.6.2	SFO with Intracavity Heatspreader *81*
3.6.2.1	Laser Cavity Setup *82*
3.6.2.2	Wavelength Tuning *83*
3.6.2.3	Emission Linewidth *85*
3.6.2.4	Active Stabilization and Influence of Sampling Time *88*
3.6.2.5	Conclusion *90*
3.6.3	SFO with Wedged Heatspreader *91*
3.6.4	SFO with Microcavity VECSELs *92*
3.6.5	SFO without Intracavity Heatspreader *94*
3.7	Conclusion *99*
	References *101*

4 Highly Coherent Single-Frequency Tunable VeCSELs: Concept, Technology, and Physical Study *109*

Mikhael Myara

4.1	Introduction: Lasers for Applications *109*
4.2	The "Ideal" Laser *111*
4.3	Toward Single-Mode Operation *113*
4.4	Toward High Coherence *118*
4.5	The VeCSEL in the State of the Art *121*
4.6	Highly Coherent, Tunable VeCSEL Design *122*
4.7	Limits and Solutions *125*
4.8	Highly Coherent, Tunable VeCSEL: Main Characteristics *127*

4.9	Ultrahigh-Purity Single-mode Operation	*129*
4.10	Spatial Coherence	*131*
4.11	Time Domain Coherence and Noise	*131*
4.11.1	Noise in Photonics: Basics	*131*
4.11.2	Intensity Noise of a VeCSEL	*135*
4.11.3	Phase Noise, Frequency Noise, and Linewidth of a VeCSEL	*136*
4.12	Conclusion	*139*
	Acknowledgements	*140*
	References	*140*

5 Terahertz Metasurface Quantum Cascade VECSELs *145*
Benjamin S. Williams and Luyao Xu

5.1	Introduction	*145*
5.1.1	Waveguides for THz QC-Lasers	*146*
5.1.2	Overview of Metasurface QC-VECSEL Concept	*148*
5.2	Metasurface Design	*149*
5.3	QC-VECSEL Model	*152*
5.3.1	Confinement Factor	*156*
5.3.2	Metasurface and Cavity Optimization	*157*
5.4	THz QC-VECSEL Performance: Power, Efficiency, and Beam Quality	*159*
5.4.1	Effect of Metasurface on Spectrum	*160*
5.4.2	Effect of Output Coupler	*161*
5.4.3	Focusing Metasurface VECSEL	*162*
5.4.4	Intra-cryostat Cavity QC-VECSEL	*165*
5.5	Polarization Control in QC-VECSELs	*166*
5.6	Conclusion	*169*
	References	*170*

6 DBR-free Optically Pumped Semiconductor Disk Lasers *175*
Alexander R. Albrecht, Zhou Yang, and Mansoor Sheik-Bahae

6.1	Introduction	*175*
6.2	DBR-free Semiconductor Disk Lasers	*176*
6.2.1	Opportunities and Advantages	*177*
6.2.2	Thermal Analysis	*178*
6.2.3	Longitudinal Mode Structure and Broadband Tunability	*180*
6.3	Device Fabrication	*182*
6.4	DBR-free SDL Implementation	*185*
6.4.1	High Power Operation	*185*
6.4.2	Broad Tunability	*187*
6.4.3	Wafer-scale Processing	*189*
6.5	Novel Concepts	*189*
6.6	Conclusions	*192*
	References	*193*

7	**Optically Pumped Red-Emitting AlGaInP-VECSELs and the MECSEL Concept** *197*	
	Hermann Kahle, Michael Jetter, and Peter Michler	
7.1	Introduction *197*	
7.2	Direct Red-Emitting AlGaInP-VECSELs and Second-Harmonic Generation *199*	
7.2.1	GaInP Quantum Wells and the AlGaInP Material System *199*	
7.2.2	GaInP Quantum Well VECSELs: A Comparison *201*	
7.2.2.1	Architecture of the Semiconductor Structures *202*	
7.2.2.2	Experimental Setup *203*	
7.2.2.3	Characterization Results *204*	
7.2.2.4	Internal Efficiency *204*	
7.2.3	Power Scaling via Quantum Well and Multi-Pass Pumping *208*	
7.2.3.1	Quantum Well Pumping *208*	
7.2.3.2	Multi-Pass Pumping *210*	
7.2.4	Second-Harmonic Generation into the UV-A Spectral Range *211*	
7.3	The Membrane External-Cavity Surface-Emitting Laser (MECSEL) *212*	
7.3.1	The Semiconductor Active Region Membrane *213*	
7.3.2	MECSEL Setup *215*	
7.3.3	MECSEL Characterization *216*	
7.3.3.1	Output Power Measurements *216*	
7.3.3.2	Beam Profile and Beam Quality Factor *218*	
7.3.3.3	Spectra *218*	
7.4	Conclusions *221*	
	References *221*	

Part II Mode-Locked VECSEL *229*

8	**Recent Advances in Mode-Locked Vertical-External-Cavity Surface-Emitting Lasers** *231*	
	Anne C. Tropper	
8.1	Introduction *231*	
8.1.1	Ultrafast Lasers *232*	
8.1.2	Ultrafast Semiconductor Lasers; Diodes, VECSELs, and MIXSELs *233*	
8.2	Ultrafast Pulse Formation in a Surface-Emitting Semiconductor Laser *235*	
8.2.1	Surface-Emitting Gain Chip Design *235*	
8.2.2	Gain Filtering *238*	
8.2.3	Gain Saturation and Recovery *239*	
8.2.4	Saturable Absorbers for ML-VECSELs and MIXSELs *241*	
8.3	Performance of Passively Mode-Locked Semiconductor Lasers *244*	
8.3.1	Pulse Duration *244*	
8.3.2	Pulse Repetition Rate *246*	
8.3.3	Mode-Locked VECSELs: Visible to Mid-Infrared *248*

8.3.4	Simulation and Modeling	*249*
8.3.5	Noise	*251*
8.4	Applications	*252*
8.4.1	Biological Imaging	*252*
8.4.2	Quantum Optics	*253*
8.4.3	Supercontinuum Generation and Frequency Combs	*253*
8.4.4	Terahertz Imaging and Spectroscopy	*254*
8.5	Summary and Outlook	*255*
	References	*256*

9 Ultrafast Nonequilibrium Carrier Dynamics in Semiconductor Laser Mode-Locking *267*

I. Kilen, J. Hader, S.W. Koch, and J.V. Moloney

9.1	Introduction	*267*
9.2	Background Theory	*269*
9.2.1	Pulse Propagation	*269*
9.2.2	Microscopic Theory	*273*
9.3	Domain Setup/Modeling	*277*
9.3.1	The VECSEL Cavity	*277*
9.3.2	The Gain Region	*278*
9.3.3	The Relaxation Rates and the Round Trip Time	*280*
9.3.4	Noise Buildup to Pulse	*281*
9.4	Numerical Results	*282*
9.4.1	Single-Pass Investigation of QWs and SAMs on the Order of Second Born–Markov Approximation	*282*
9.4.1.1	Inverted Quantum Well	*282*
9.4.1.2	Saturable Absorber	*285*
9.4.2	Mode-Locked VECSELs	*288*
9.4.2.1	Gain, Absorption, and Dispersion	*288*
9.4.2.2	Pulse Buildup and Initial Conditions	*290*
9.4.2.3	Self-Phase Modulation from QWs	*290*
9.4.2.4	Mode-Locked Pulse Family	*291*
9.4.2.5	Influence of Loss on the Mode-Locked Pulse	*294*
9.4.2.6	Limits on the Shortest Possible Pulse and the Hysteresis Effect	*296*
9.5	Outlook	*299*
	References	*300*

10 Mode-Locked AlGaInP VECSEL for the Red and UV Spectral Range *305*

Roman Bek, Michael Jetter, and Peter Michler

10.1	Introduction	*305*
10.2	Epitaxial Layer Design of AlGaInP-SESAM Structures	*306*
10.2.1	Quantum Well SESAMs	*306*
10.2.2	Quantum Dot SESAMs	*307*
10.3	Temporal Response of AlGaInP SESAMs	*307*

10.4	Cavity Designs	*309*
10.5	Characterization Methods	*310*
10.6	Mode-Locking Results	*311*
10.6.1	Quantum Well Mode-Locked AlGaInP VECSELs	*311*
10.6.1.1	High Output Power	*311*
10.6.1.2	Femtosecond Operation	*312*
10.6.2	Quantum Dot Mode-Locked AlGaInP VECSELs	*314*
10.7	Second Harmonic Generation into the UV Spectral Range	*315*
10.8	Summary and Outlook	*317*
	References	*318*

11 Colliding Pulse Mode-locked VECSEL *321*
Alexandre Laurain

11.1	Introduction	*321*
11.2	Principle of Colliding Pulse Modelocking	*322*
11.3	Requirements for Stable Colliding Pulse Modelocking	*324*
11.3.1	Pulse Timing	*324*
11.3.2	Gain Recovery and Pumping Rate	*324*
11.3.3	Polarization	*326*
11.3.4	Mode Waist and Saturation Fluence	*326*
11.4	Design of an Ultrafast CPM VECSEL	*327*
11.4.1	The Optical Cavity	*327*
11.4.2	The Gain Structure	*328*
11.4.3	The SESAM	*333*
11.5	Modelocking Results	*335*
11.5.1	Robustness of the Modelocking Regime	*335*
11.5.2	Cross Correlation of the Output Beams	*336*
11.5.3	Pulse Duration Optimization	*338*
11.5.4	Multipulse Regime	*340*
11.6	Pulse Interactions in the Saturable Absorber	*341*
11.6.1	Field Intensity Distribution	*341*
11.6.2	Saturable Absorption Model	*343*
11.6.3	Dynamics of the Carrier Density Distribution	*345*
11.6.4	Absorption Losses and Pulse Shaping	*347*
11.6.5	Saturation Fluence of the Absorber	*349*
11.6.6	Power Balance in CPM Operation	*350*
11.7	Summary and Outlook	*352*
	Acknowledgments	*353*
	References	*353*

12 Self-Mode-Locked Semiconductor Disk Lasers *357*
Arash Rahimi-Iman

12.1	Introduction	*357*
12.2	Mode-Locking Techniques for Optically Pumped SDLs at a Glance	*358*
12.3	History of Saturable-Absorber-Free Pulsed VECSELs	*360*

12.3.1	Self-Mode-Locked Optically Pumped VECSELs *360*	
12.3.1.1	Once Upon a Time – Beyond Magic *361*	
12.3.1.2	Mode Competition – A Struggle for Acceptance *363*	
12.3.1.3	More Than a Flash in the Pan – Triggered Wave of Results *364*	
12.3.2	Harmonic Self-Mode-Locking *366*	
12.3.3	Self-Mode-Locking Quantum-Dot VECSEL *368*	
12.3.4	SML Cavity Configurations *369*	
12.3.5	SML VECSEL at Other Wavelengths *371*	
12.4	Overview on SESAM-Free Mode-Locking Achievements *373*	
12.4.1	Spotlight on SML VECSELs *373*	
12.4.1.1	Pulse Duration *373*	
12.4.1.2	Peak Power *374*	
12.4.1.3	Repetition Rate *375*	
12.4.2	SESAM-Free Alternatives to SML VECSEL *375*	
12.4.2.1	Graphene or Carbon Nanotube Saturable Absorber Mode-Locked VECSELs *375*	
12.4.2.2	SESAM-Free VECSEL Design with Intracavity Kerr Medium *375*	
12.5	Investigations into the Mechanisms and Outlook *376*	
12.5.1	First Studies Concerning the Mechanisms Behind SML *376*	
12.5.2	Z-Scan Measurements of the Nonlinear Refractive Index in a VECSEL Chip *377*	
12.5.3	Applications and Expected Advances *380*	
	Acknowledgments *381*	
	References *382*	

Index *387*

Preface

The first reports in 1997 of vertical external-cavity surface-emitting laser (VECSEL), also known as optically pumped semiconductor disk laser (SDL), attracted a lot of attention in the research and laser community. This relatively new type of laser combines the versatility in wavelength from the semiconductor-active region with superior emission properties such as multi-watt output powers in cylindrically symmetric TEM_{00} mode and nearly diffraction-limited beam quality. The extended cavities allow an efficient intracavity frequency conversion and enable, by incorporating semiconductor saturable absorber mirrors (SESAM), short-pulse mode locking operation.

Since the first summarizing book of the achievements in this field in 2010 from Oleg G. Okhotnikov, a tremendous progress in demonstrating the expected VECSEL properties took place. The direct emitting wavelength of these lasers expands meanwhile from the blue to the mid-infrared range. With frequency doubling, even the UV spectral range is reached with high optical powers. SESAM mode-locked lasers demonstrated pulse widths below 100 fs and pulse repetition rates over 100 GHz. However, not only the predicted features were confirmed but also new varieties of VECSELs were invented in the meantime. The most prominent ones are the mode-locked integrated external-cavity surface-emitting laser (MIXSEL), combining the SESAM and gain mirror in one semiconductor stack, and the membrane external-cavity surface-emitting laser (MECSEL), basically a heat spreader–semiconductor sandwich in an external cavity, enabling wavelength ranges which were not reachable due to semiconductor growth restrictions. Next to these nice efforts and lively research, these devices as well took the step to commercialization. Several companies offer nowadays laser equipment including SDL either in cw or mode-locked operation.

It is our pleasure and honor to present with this book some of the most recent developments in VECSEL research. It is written by internationally renowned experts who actively advance this field of laser research. The book is structured into two sections, a continuous wave and a pulsed laser part. It includes the development of the VECSELs, recent advances, technology aspects, and some applications of VECSEL. It can be useful for engineers and scientists working in this field as well as for graduate students interested in the technology of these laser devices.

We like to thank all the authors for their valuable contributions to this book and their patience. Furthermore, we like to thank the editorial staff of Wiley-VCH GmbH for keeping this book on track for publication.

Stuttgart, March 2021

Michael Jetter
Peter Michler

Part I

Continuous wave VECSEL

1

History of Optically Pumped Semiconductor Lasers – VECSELs

Mark E. Kuznetsov

Axsun/Excelitas Technologies, Billerica, MA, USA

1.1 Introduction

Optically pumped semiconductor lasers (OPSLs) are an old concept that originated in the early days of semiconductor lasers in the 1960s, and that remained a scientific curiosity until the mid-1990s, when its potential capabilities for high power with excellent beam quality were first fully demonstrated, spurring subsequent rapid development of the science and technology of these versatile lasers. Distinguishing features of OPSLs today are light emission normal to the plane of the semiconductor chip, laser cavity external to the chip to stabilize fundamental transverse laser mode and to enable insertion of intracavity functional optical elements, and the use of optical pumping for efficient and high output power operation. A wide range of applications is enabled by additional remarkable properties of this laser family, such as wavelength operation from ultraviolet (UV) to mid-infrared (IR) and even terahertz range, and passively modelocked operation with output pulses shorter than 100 fs. These lasers are also widely known by two other names, descriptive of their geometry: vertical external-cavity surface-emitting lasers (VECSELs) and semiconductor disk lasers (SDLs). Alternatively, VECSELs can also be electrically pumped, but achievable laser output powers are then typically much lower than for the optically pumped version.

OPSL development in the 1990s was spearheaded by Aram Mooradian in Micracor, a small start-up company that spun out technology from Aram's group in the MIT Lincoln Laboratory. I worked with Aram in Micracor to carry out this development. In 2011, the first annual VECSEL conference was held at SPIE Photonics West, with the VECSELs-XI conference scheduled for 2022. The first book about these lasers, *Semiconductor Disk Lasers: Physics and Technology* [1], was published in 2010; it was edited by Oleg Okhotnikov from the Tampere University of Technology in Finland and described the then state of the art in chapters contributed by researchers from around the world. Since the publication of this book, science, technology, and applications of VECSELs have made a significant step forward, and hence the present book to bring VECSELs overview up to date. This chapter

Vertical External Cavity Surface Emitting Lasers: VECSEL Technology and Applications, First Edition.
Edited by Michael Jetter and Peter Michler.
© 2022 WILEY-VCH GmbH. Published 2022 by WILEY-VCH GmbH.

describes the history of OPSLs, the people that took part in their development, and it's also a personal story of the OPSL development by our team at Micracor. Sadly, both Aram Mooradian and Oleg Okhotnikov, who have contributed so much to the early development of these lasers, have passed away since the publication of the first book. This historical chapter is dedicated to their memory.

1.2 OPS-VECSELs: Concept and History

The first laser invented in 1960 was a flashlamp-pumped solid-state ruby laser. Other laser gain media and pumping schemes soon followed, and in 1962 a semiconductor diode laser pumped by current injection in a semiconductor p-n junction was demonstrated [2–6]. Semiconductors offered the possibility of operating at different wavelengths, depending on the material composition – already in 1962, together with the near-IR operation of binary GaAs lasers, ternary alloy GaAsP semiconductor diode laser in the visible was also demonstrated [5]. Electron-hole pairs for laser excitation in semiconductors can be created by various means. Current injection pumping of diode lasers, although requiring more complex device fabrication, is appealing for its simplicity of use and the possibility of direct laser output modulation by current modulation. Yet other schemes for semiconductor laser pumping were also investigated, including optical and electron beam pumping. Semiconductor diode lasers have seen tremendous development from the 1960s to the 1990s, driven primarily by applications in optical fiber communication, CD and DVD optical disk readout, and pumping of solid-state and fiber lasers and amplifiers. One challenge had remained, however. While diode lasers could produce very large, from watts to hundreds of watts, powers, this power was produced with poor beam quality: in highly transverse multimode, high aspect ratio output beams, or from large arrays of lasers. Single transverse mode output, and especially circular Gaussian output beam, could be produced at only much smaller power levels of at most a few hundred milliwatts. Optical pumping had remained an essentially experimental tool to demonstrate lasing capability of the semiconductor gain medium or laser structure, on the way to the ultimately useful diode-current-injection electrical pumping.

Why would semiconductor lasers with high multiwatt output power and beam quality be useful and important? The alternative high-power laser technologies, e.g. solid state, gas, and atomic vapor lasers, rely on discrete atomic transitions and thus are restricted to discrete unique operating wavelengths. Semiconductor lasers, via material composition and bandgap-engineered quantum-confined structures, can produce a very wide range of operating wavelengths, from UV to mid-IR, both directly and using nonlinear optical, including intracavity, conversion. This allows, by design, an essentially continuous coverage of this spectrum and even dual wavelength laser operation. High-power good beam quality semiconductor lasers can offer unique operating parameters not accessible by other types of lasers. If you add to this femtosecond pulse capability with high peak powers, potentially compact size and low fabrication cost, this makes such semiconductor lasers useful, and sometimes possibly unique, for a wide range of applications.

Early semiconductor lasers were edge emitting and emitted light in the plane of the wafer, so that enough gain could be accumulated over the length of the device. Such edge-emitting geometry, both gain and index guided, limits transverse profile beam quality for high powers. Vertical cavity surface-emitting lasers (VCSELs) were first described by Kenichi Iga at the Tokyo Institute of Technology in 1979 [7, 8] and further developed to efficient low threshold operation by Jack Jewell at Bell Laboratories, Holmdel, in 1989 [9–11]. VCSELs surface-emitting geometry, with light emitted normal to the plane of the wafer, because of low gain of the short gain region, requires very high reflectivity semiconductor or dielectric multilayer mirrors. Such vertical cavity geometry allows single transverse mode circular Gaussian beam output, but typically only for milliwatt level powers, limited by the difficulty of heat dissipation from the small mode area of a few microns in diameter.

Optical pumping of semiconductor lasers has a long history; it has been used for various purposes, such as characterization of novel semiconductor laser materials and structures, generation of higher output powers, or short pulse generation. As early as 1965, an OPSL has been demonstrated by Robert Phelan and Robert Rediker at MIT Lincoln Laboratory [12], where an edge-emitting InSb laser was pumped by an edge-emitting GaAs diode laser. Remarkably, both concepts, optical pumping of semiconductors and the use of diode lasers as pumps, are already in use this early in the history of lasers. In 1966, Nikolay Basov's group at the Lebedev Physical Institute in Moscow introduced the concept of a "radiating mirror" [13], Figure 1.1a: a thin semiconductor gain layer placed on top of a mirror and a heatsink, with an external output coupling mirror completing the laser cavity. Both optical and electron beam pumping were envisaged and demonstrated as the possible excitation sources. Large lateral extent of the gain medium, greater than its thickness, would ensure effective heat removal and thus the possibility of high output power. This is essentially the concept of a "disk laser" geometry, which would prove so effective many years later in both solid-state [14, 15] and semiconductor [1] laser configurations. Basov's 1966 "radiating mirror" concept, Figure 1.1a, is remarkably similar to the 1996 Micracor OPSL configuration, Figure 1.1b. In his paper, Basov reported operation of a "radiating mirror" laser with optical pumping of CdSe using two-photon absorption of a Q-switched Nd-doped glass pump laser. So as a concept, SDL had been already introduced and demonstrated in 1966; however, its full potential was yet to be explored and developed.

In the late 1960s, Nick Holonyak's group at the University of Illinois, Urbana, reported several studies of optically pumped CdS, GaAs, and GaAsP semiconductor thin platelet lasers [16], some using GaAsP diode laser pumping, and considered both edge- and surface-emitting laser geometry. Transparent sapphire heatsinking windows had been used to remove heat and help improve power performance of the devices, foreshadowing the future use of such transparent heatspreaders. Optical pumping here is mainly used to explore lasing in different semiconductor materials. In a 1973 publication from Aram Mooradian's MIT Lincoln Lab group, Stephen Chinn demonstrated pulsed operation of optically pumped edge-emitting bulk GaAs semiconductor lasers [17], with the goal of efficient high power generation. Later in 1981, Julian Stone, Jay Wiesenfeld, Andrew Dentai, and coworkers from

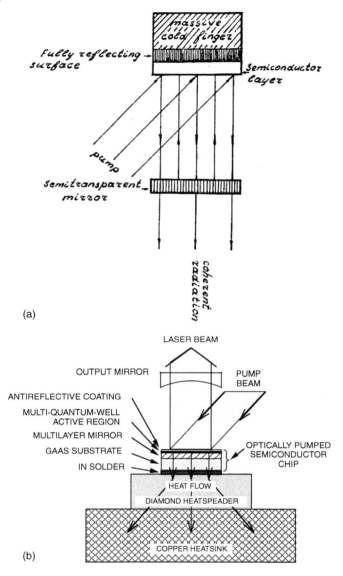

Figure 1.1 (a) Semiconductor laser with radiating mirrors. *Source*: Reprinted, with permission, from Basov et al. [13], © 1966 IEEE. (b) Optically pumped semiconductor vertical external-cavity surface-emitting laser (OPS-VECSEL), Micracor, 1996.

the Bell Laboratories Crawford Hill Lab, used surface-emitting thin-film ultrashort cavity InGaAsP lasers [18] to generate gain-switched picosecond pulses in the 0.83–1.59 mm wavelength range using dye laser pumping. Wavelength versatility of semiconductor materials is explored here, with short pulse generation as the primary goal. Using an external optical cavity for pulse repetition rate and transverse mode control, optically pumped mode locking was demonstrated with a CdS thin platelet laser by Charles Roxlo and Michael Salour at MIT in 1981 [19]. In 1991, Mooradian's group observed high peak power in an external-cavity GaAs platelet laser pumped by a Ti:sapphire laser [20]. Note that the above OPSLs largely used bulk semiconductor material without internal heterostructures. As a result, the performance of these lasers was limited.

When semiconductor heterostructures and quantum wells had been developed, they also had been used for OPSL experiments, e.g. in Holonyak's group [21]. When the first efficient VCSEL semiconductor lasers were demonstrated by Jack Jewell in 1989, early experiments used optical pumping of bulk GaAs [9] and later InGaAs and GaAs quantum well laser structures [10, 11]. On-wafer high reflectivity semiconductor multilayer distributed Bragg reflector (DBR) mirrors were used for these vertical cavity lasers. At this point, the full power of the semiconductor bandstructure engineering was beginning to be applied to OPSLs. But again, optical pumping was used only as an initial characterization of laser structures on the way to eventual diode current-injection pumping [11].

In the 1990s, a series of papers explored optical pumping of semiconductor lasers with the goal of obtaining high power with good beam quality, as well as short pulse generation. A wide variety of laser configurations were explored, both edge and surface emitting, with a range of pump laser options: diode, solid-state, and gas lasers. Using diode laser pumping, low-power 10 mW continuous wave (CW) operation was demonstrated with GaAs VCSEL by Steve Brueck's group at the University of New Mexico [22]; in external cavity, however, such lasers emitted only 20 μW [23, 24]. External resonator was also used with an electrically pumped VCSEL by a group from the University of California, Berkeley, in an attempt to increase its single-transverse-mode output power; however, only 2.4 mW was achieved in CW operation [25]. A diode-laser-pumped surface-emitting optical amplifier was demonstrated at 1.5 μm using InGaAs–InGaAlAs multi-quantum-well structures by Shojiro Kawakami's group at Tohoku University [26]. Using 77 K low temperature operation and a Nd:YAG pump laser, 190 mW continuous output power was obtained from an external-cavity InGaAs–InP surface-emitting laser by Wenbin Jiang in the Yoshihisa Yamamoto's group at the NTT Basic Research Laboratories [27]. The same group used a similar configuration to demonstrate an external-cavity GaAs VCSEL at 77 K with CW output power of 700 mW using a 1.8 W krypton–ion pump laser [28]. Modelocked femtosecond pulse operation was demonstrated with a periodic gain vertical cavity laser in an external cavity by Wenbin Jiang in the John Bowers group at the University of California, Santa Barbara [29]; with synchronous pumping by a modelocked Ti:sapphire laser, 6 mW average output power was obtained. Specially designed edge-emitting InGaAs–GaAs laser structures were used with diode laser pumping to generate as much as 4 W average power by Han Le at MIT Lincoln Laboratory [30, 31]; however, the beams were strongly elongated with aspect ratios between 10 and 50 to 1. To summarize the state of the art by the mid-1990s, semiconductor lasers could emit watt-class power only with poor beam quality from edge-emitting structures; vertical cavity lasers operated with good beams but only with milliwatt class output, external-cavity operation of surface-emitting lasers hadn't produced particularly high powers, and optical pumping remained just a tool for scientific exploration of novel laser configurations.

1.3 Micracor

Micracor (1992–1997) was a small company started by Aram Mooradian, see Figure 1.2, in Acton, a suburb of Boston, to commercialize several technologies from the Quantum Electronics Group at MIT Lincoln Laboratory, where Aram was the group leader for more than 20 years. Micracor's core technologies and nascent products were diode-pumped solid-state microchip lasers, work led by Kevin Wall, and tunable external cavity diode lasers, work led by Ken German. OPSLs were a concept explored previously in Aram's group at MIT Lincoln Lab. Micracor made a quixotic attempt to take this, at the time vague, concept, demonstrate it, and develop and commercialize such devices. The key initial target application was 980 nm pumping of Er-doped fiber amplifiers. Micracor's core technologies weren't very successful commercially, while OPSL development after three and a half years showed tremendous promise. But investors eventually lost patience, and

(a)

(b)

Figure 1.2 Aram Mooradian and the Micracor logo.

at the end of 1996 the company was shut down. Luis Spinelli from Coherent Inc. recognized the potential of OPSL technology and drove the purchase of Micracor's assets by Coherent. At this point our group at Micracor published two papers on OPS-VECSELs [32, 33], where the term VECSEL was first introduced. The results of Micracor's more than three-year effort on OPSLs were finally made public. These publications triggered subsequent development and exploration of VECSELs by the scientific community around the world. VECSEL technology was successfully commercialized by Coherent for applications as diverse as entertainment, forensics, life sciences, and medical. If it weren't for OPSLs, Micracor would be forgotten today. A recent Google search on Micracor yielded a puzzled response – "Did you mean: microcar?"

How did I come to Micracor and what was my role there in the development of OPSLs? After Micracor was founded, in 1993 Aram got a Small Business Innovation Research (SBIR) grant from the US Department of Defense to develop high-power OPSLs. However, there was nobody at the company who could actually carry out this work. Phase I money was being spent, but no progress was made. I was hired in August of 1993 to carry the development of what would become OPSLs. Micracor was funded by Rothschild Ventures, and I remember visiting their offices in New York, with Rothschild family portraits on the walls. I came to Micracor after graduate school at MIT with Hermann Haus and Erich Ippen and seven years at Crawford Hill Lab of Bell Laboratories in the department of Ivan Kaminow. Looking back, I was well prepared and had the right background and experience to embark on the risky and challenging development of OPSLs. At Bell Labs, I had worked on electronically tunable quantum well diode lasers, I had both theoretical and experimental experience with semiconductor lasers, and my work has involved extensive modeling and design of edge-emitting semiconductor laser structures, semiconductor device processing, laser fabrication and characterization, as well as device performance analysis. I had also been exposed to earlier work on OPSLs: at MIT, Michael Salour's lab was next door and I had attended multiple talks from the group; at Bell Labs, Julian Stone, Jay Wiesenfeld, and Jack Jewell were colleagues whose work I closely followed. Micracor was a small company of about 20 people working on a variety of projects; to develop OPSLs, I had to rely on myself to get the job done – no grand team to attack the problem; we had limited resources, equipment, and money. Compare this to Novalux, founded by Aram Mooradian several years after closing of Micracor, where $193 million (!) was spent developing electrically pumped VECSELs. Several people at Micracor played crucial roles in the OPSL development: Bob Sutherland (thin semiconductor wafer polishing), Bob Sprague (AR coatings), and Farhad Hakimi (pushing powers higher at the later stages of development). Aram Mooradian was the visionary who initiated the program, guided the program along, and with whom we discussed all aspects and nuances of the work to overcome a continuous string of challenges. Aram had the physical intuition to see OPS laser operation, even when he couldn't tell exactly how to get there. I felt we were "father and mother" team with Aram, nursing and raising our "baby" OPSLs.

Why do I think we succeeded at Micracor? Various groups had worked on OPSLs for several decades, but a common thread was that when they got results good enough for a publication, they were satisfied and stopped at the publication. We were a small company and had to push OPS concept to the commercial performance level to get funding for our work and for the company, both from government program sponsors and from investors. We were fighting for survival as a company, and just a publication was not good enough. And we had persevered, overcoming challenge, after challenge, after challenge, finding a path forward at every step. First, we got some initial miniscule amount of light from the laser. At that point, we had something to work with, and we just kept optimizing and improving, and we never stopped, until the company went out of business. In this process, physical understanding was critical, and fabrication ability was critical. Our work went through multiple iteration cycles: physical modeling and design – device fabrication – characterization of the materials and devices – analysis of the device characteristics – and then finding ways to improve in the next iteration cycle.

Another reason we succeeded at Micracor is the tremendous progress that had been made in the semiconductor technology in the preceding years and that was now available to us. Semiconductor lasers had progressed from simple edge-emitting homojunction GaAs devices grown by liquid-phase epitaxy (LPE) in the 1960s, to the MBE (molecular-beam epitaxy) and MOCVD/MOVPE (metal-organic chemical vapor deposition/vapor-phase epitaxy) grown semiconductor structures with bandgap engineering to manipulate their electronic and optical properties, to strain-engineered and strain-compensated quantum-confined structures, to vertical cavity laser structures with grown multilayer semiconductor DBR mirrors. Such semiconductor materials and structures were now also understood well enough to be grown commercially in companies such as Epitaxial Products International (EPI) PLC (now IQE PLC). These developments enabled diode VCSELs in the late 1980s; we applied these technologies to the optically pumped vertical external-cavity configuration in the early 1990s. Another key enabler was the availability of new multiwatt multitransverse-mode semiconductor diode pump lasers. Such pump lasers had been developed in the 1980s for pumping solid-state lasers, e.g. 808 nm pumps for Nd:YAG lasers.

What were the resources and facilities available to us at Micracor? We had an electron beam evaporator for optical dielectric coatings and a thermal evaporator for chip metallization; we had a polishing facility for thinning the wafers; we also had a spectrophotometer for optical characterization of wafers and coatings. However, we did not have an optical spectrum analyzer (OSA) – commercial OSA was just too expensive. So I made a homemade OSA, converting a grating monochromator to a rudimentary calibrated OSA by rigging a fiber input and a detector output to the input and output slits, driving the grating stepper motor from a computer D/A output and reading the detector output into computer A/D. I periodically calibrated this OSA using a He-Ne laser. To my surprise, coming from Bell Labs, I found that it is possible to do interesting science in a small company environment and with limited resources, of course given the right circumstances.

1.4 OPSL Development at Micracor: First Steps

In August of 1993 when I came to Micracor, it had an SBIR program running, Phase I, on the topic of "High Power Multi-Segmented Semiconductor Lasers." Figure 1.3 shows the device that was promised ultimately to the sponsors, a high-power multiple-bounce optically pumped surface-emitting semiconductor laser. The project was funded by the Department of Defense, U.S. Army Space and Strategic Defense Command, colloquially known then as the "Star Wars" initiative. With not much to show for the program accomplishments at this point, I had two months to finish the project, write the final report, and convince program managers to give us more money for Phase II of the project.

I wrote and submitted the final report for Phase I at the end of September 1993: we had seen no laser light but had done enough experiments to project high hopes for the future; now we just had to sell these hopes to the program managers. We had two semiconductor samples, epigrown DBR stack plus quantum-well gain region on top, to work with, grown and available through Aram's numerous connections: an 850 nm AlGaAs/GaAs wafer from Art D'Asaro at Bell Labs and a 900 nm InGaAs/GaAs wafer grown by Stephen Hersee at the University of New Mexico. Both wafers had major design shortcomings. What had been accomplished in Phase I? I had learned to align pump optics and laser cavity using Si cameras, I imaged and characterized pump and sample spontaneous emission beams, I had demonstrated 1.2 W pump power into ~100 μm diameter spot from a 785 nm 3 W/500 μm-wide stripe diode laser, and I observed strong amplified spontaneous emission (ASE) at 900 nm into the laser mode – but no lasing. I had sketched out a basic OPS laser model, e.g. see Figure 1.4, and determined a set of parameters to make a "single-bounce" OPS surface-emitting laser. Proposed designs had a semiconductor wafer structure with a DBR mirror and a gain region with ~10

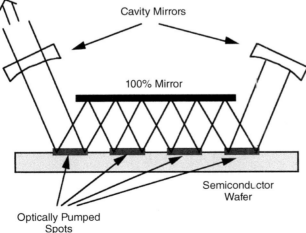

Figure 1.3 Multi-bounce OPS laser Micracor promised in 1993 to make ultimately in its SBIR program "High Power Multi-Segmented Semiconductor Lasers."

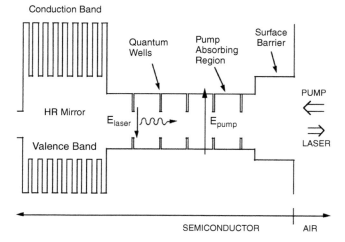

Figure 1.4 Band diagram of the proposed optically pumped surface-emitting semiconductor laser structure in the September 1993 final report for SBIR Phase I. Quantum wells are placed in a resonant periodic gain structure [34].

quantum wells in a resonant periodic gain structure [34], with two specific wafer designs proposed, an external spherical mirror with ~100 mm radius of curvature and a reflectivity $R > 97\%$, and a stable fundamental transverse mode laser cavity with ~1 W threshold pump power for a 100 μm diameter mode. And I had a basic pump optics design with cylindrical lenses. Based on all this, while lacking an actual laser demonstration, we had claimed in the Phase I final report that we had demonstrated the "feasibility" of our optically pumped semiconductor OPS laser approach.

By the time I wrote Phase II SBIR project proposal in April 1994, we had a first major milestone accomplishment – we had seen first laser light at 900 nm with antireflection (AR) coated samples. Only a tiny amount of pulsed light was observed, but we already had something to work with. I wrote a very optimistic proposal, trying to overcome rudimentary results of the first phase, and I described in detail our approach to getting high power operation. Figure 1.5 shows our proposed initial single-bounce OPS laser configuration. We got funding for Phase II.

When we started Phase II of the project in June 1994, we were already characterizing low power pulsed laser light. Our 900 nm InGaAs/GaAs samples had four quantum wells; the chips were AR coated, thinned to 100 μm, metallized, and soldered onto Cu heatsink. With $R > 99\%$ output mirror, and the laser driven with 100 μs pump pulses at 1 kHz, we saw very low power laser light but with an excellent beam profile, see Figure 1.6; chip heating turned the laser off for longer pulses. Thermal impedance of the chip was high; thermal path from active region to heatsink included DBR region and a 100 μm thick residual substrate. But the important thing was that we already had some light to work with, and we started to characterize and optimize our laser.

With money now available to continue the project, in the summer of 1994 we started by designing and growing a new optimized wafer structure based on our

1.4 OPSL Development at Micracor: First Steps

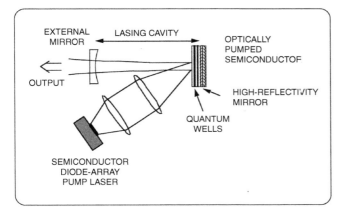

Figure 1.5 Single-bounce OPS laser configuration in the April 1994 SBIR Phase II proposal.

Figure 1.6 Excellent beam profile of the pulsed optically pumped 900 nm InGaAs/GaAs laser, June 1994.

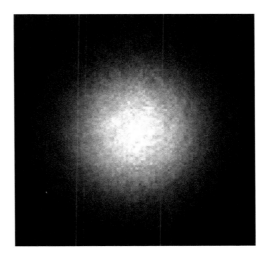

prior analysis and experiments. We targeted operating wavelength of 980 nm for application to pumping Er-doped fiber amplifiers; wafer structures included 10 strained quantum wells in the mature InGaAs/GaAs material system. The structures were designed to be pumped at 808 nm and even included pump-reflecting mirror in the DBR for more efficient double pass pump absorption. The wafers were grown successfully by MOVPE at EPI PLC in Cardiff, United Kingdom. These were fairly complex designs with many DBR and quantum well semiconductor layers; it's impressive that a commercial company at the time could already grow successfully such sophisticated structures. By then, we had developed techniques to characterize the grown OPS wafers with spectrophotometer wafer reflectance spectra, as well as surface and edge photoluminescence; laser power and spectral behavior were also extensively characterized, as well as their temperature dependence. We continued detailed modeling to explain the observed wafer and laser characteristics, both optical and thermal, and had designed several different pump

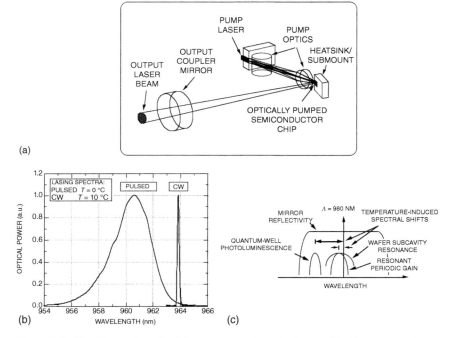

Figure 1.7 Two figures from the Micracor second quarterly report to the sponsor, January 1995: (a) laser and pump configuration, (b) pulsed and CW laser spectra. (c) Figure from the Phase II final report, July 1996: thermal offset design for OPS vertical cavity laser.

optics configurations. In our first Phase II quarterly report in September 1994, we described our accomplishments: the new wafers lased only in the pulsed mode; using $R = 97\%$ output reflecting mirror, threshold was ~400 mW and maximum pulsed output power was only 30 mW, limited by heating.

So a year after the program started, we had learned a lot, but got only 30 mW pulsed – pretty far away from the promised watts of CW power. This was time to get philosophical in the face of challenges, to collect thoughts and to see how to proceed forward. At the time, I was reminded of the quote from Harold Edgerton, MIT professor of the flash photography fame, "We worked and worked, but didn't get anywhere. That's how you know you're doing research." Well, we definitely were doing research. I was also thinking of my work then as a friendly wrestling match with Nature; Nature is a tough opponent, but she is not malicious. I was learning to speak the language of Nature, learning to listen carefully and understand what she says, and trying to speak back to her in her own language; and if I were successful, she would listen and understand me, and do what I asked.

1.5 OPS Development at Micracor: Pushing Forward

In the next quarter we pushed forward, and here are the important advances from the second quarterly report in January of 1995, see Figure 1.7. Laser chips were

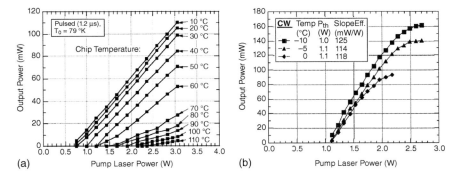

Figure 1.8 (a) Pulsed laser power up to 110 °C with weak temperature dependence, third quarterly report, March 1995. (b) 160 mW CW output power, program extension proposal, May 1995.

thinned to 20–30 μm, to reduce thermal impedance between the quantum well active layer and heatsink, and soldered to Cu submounts – we got 130 mW pulsed output power and, crossing an important barrier, we had ~1 mW CW! (at 0 °C). We had concluded that thermal effects limited our CW output power, and we focused on several approaches to improve further laser power performance. (i) We lowered thermal impedance of on-chip mirror and substrate in order to decrease detrimental temperature rise of the active region. (ii) We made laser threshold less sensitive to temperature. To this end, we improved quantum well design with stronger carrier confinement to prevent carrier escape from the wells at higher temperatures. And finally, (iii) we placed the lasing mode resonance on the long wavelength side of the material gain peak at room temperature. This would compensate the larger temperature shift of the gain spectrum compared to the smaller temperature shift of the resonant wavelength and would align material gain spectral peak and lasing mode wavelength at the increased operating temperature of the laser active region. The key to laser power performance was addressing these thermal issues.

Our understanding and improvement steps proved to be correct. By the time I wrote our Phase II third quarterly report in March of 1995, we had made significant progress. Our second iteration redesigned 980 nm wafer structure was grown – the measured thermal impedance improved by a factor of two to 45 °C/W, in part due to elimination of the pump-reflecting mirror; improved quantum well (QW) carrier confinement resulted in weaker dependence of threshold power on temperature, see Figure 1.8a; CW lasing was now observed at up to 30 °C; and we measured 57 mW CW output power at −5 °C. By May 1995 we had improved performance even further – the chip was now only 17 μm thick and soldered onto diamond heatspreader, which was in turn soldered to a Cu heatsink; CW output power was now 160 mW! See Figure 1.8b. And, based on these results, we got infusion of additional money from the sponsor in the program extension. We could keep going with OPS laser development.

In August of 1995, I had a chance for the first time to attend the Topical Meeting on Semiconductor Lasers in Keystone, Colorado – attending conferences was a luxury not always available in a small company. There I discovered that we had competition – in a poster session, John Sandusky described their work with Steve Brueck at the University of New Mexico on "A CW external-cavity surface-emitting laser" [23]. Their configuration was generically similar to ours, and they reported 20 µW of CW output power using a ring dye laser pump. This was frustrating – at the time we were already measuring 200 mW CW and with a diode laser pump, four orders of magnitude higher power, but we couldn't report anything at the conference, our work was commercial and proprietary, not for publication. So all I could do is talk to the presenter and comment on their nice work. We did apply for several patents on Micracor work. Once I even had to travel to Washington DC to argue for one of our patent applications in front of a patent examiner – I argued successfully against the objections, and we got that patent. I remember the patent office building, with stacks full of paper copies of old patents, famous patents on display; paper was king then.

As we were increasing OPS output powers, we had discovered serious laser power degradation over time and had found the culprit – under pump illumination, dark line defects were growing in our chips, caused by crystal dislocations due to excessive accumulated strain –thickness product of our multiple-quantum-well (MQW) structures, see Figure 1.9. Moving to a new spot on the chip restored the power, but then it would degrade slowly again. While a single quantum well in our structure was within the strain-thickness stability limits of Matthews and Blakeslee [35], multiple quantum wells violated that limit. In an attempt to address this, by September 1995 in our third iteration wafer design we had reduced the cumulative strain-thickness product and had hoped that large 100 nm separation between wells would also help. We got higher CW output powers – now 200 mW in Transverse Electro-Magnetic (TEM) TEM01 and 140 mW in TEM00 Gaussian beam modes using a 1.2 W fiber-coupled diode laser pump, Figure 1.10. But power degradation and dark line defects were still present.

At this point we were at the threshold of commercially relevant output powers. Aram was already dreaming of optical frequency doubling into the visible. The initial promising application was pumping of Er-doped fiber amplifiers, which required several hundred milliwatts at 980 nm in a single mode fiber. We had contacted several potential customers to inquire how much power was desired. A memorable answer came from Bell Labs: "Too much power is almost enough!" How true. But we still needed to solve the dark line power degradation problem – our devices were useless, if all that power decayed on the scale of hours. We were yet again facing a critical challenge – a fundamental semiconductor materials problem: crystal layer strain in our structures was too high, and attempts to reduce it did not help.

1.6 OPS Development at Micracor: Final Chapter

Reviewing the literature, I had found that strain compensation should solve our problem: thin tensile strain-compensating GaAsP layers to balance the compressively strained InGaAs quantum wells on GaAs. Strain compensation was a

Figure 1.9 Photoluminescence image of the strained multi-quantum-well InGaAs/GaAs OPS chip, 808 nm pump, 980 nm photoluminescence. Dark line defects in the image and a dark spot at the lasing location, caused by dislocation defect lines in the crystal due to excessive crystal strain. Third iteration wafer design, Micracor fifth quarterly report, February 1996.

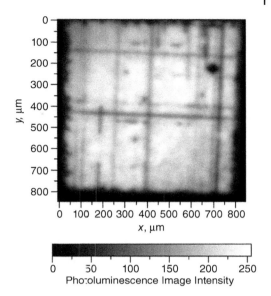

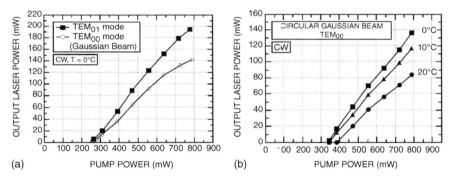

Figure 1.10 (a) OPS CW output power in TEM01 and TEM00 (Gaussian beam) modes, and (b) temperature dependence of TEM00 mode, second iteration wafer design, Micracor fourth quarterly report, September 1995.

relatively new technique at the time, and we hoped that EPI could grow such structures. I had designed the required structure and even calculated its theoretical crystal X-ray diffraction pattern, so that the grower could measure and verify strain balancing. I also calculated tolerance limits of strain compensation, to make sure composition and thickness errors during growth won't destroy crystal strain structural stability. And I verified band edge alignment of the strain-compensating layers in our structure to make sure pump-generated carriers are not blocked on their way to the quantum wells. EPI had accepted the wafer growth order and was learning to do strain compensation.

While EPI was growing our fourth-generation wafer design, now incorporating strain-compensated semiconductor multi-quantum-well structure, I had further

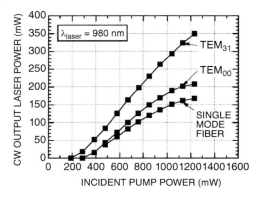

Figure 1.11 OPS CW output power in TEM31, TEM00 (Gaussian beam) modes, and power coupled into single mode fiber, third iteration wafer design, Micracor fifth quarterly report, February 1996.

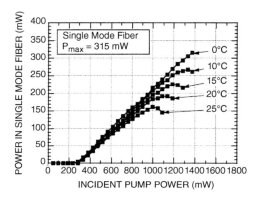

Figure 1.12 OPS power coupled into single mode fiber, fourth generation wafer design, with strain compensation, Phase II Final Report, July 1996. OPS output power now did not degrade with time, wafer was free of photoluminescence dark line defects.

improved our OPS laser performance. I learned how to better thin our chips, experimented with Au/Sn, and thin evaporated indium solders, diamond heatspreaders, etc. By our fifth quarterly report in February 1996, Figure 1.11, using our third iteration wafer design, we had already achieved 350 mW output power, albeit in a higher order transverse mode; fundamental Gaussian beam output was 200 mW; and 160 mW was coupled into single mode fiber.

Wafer growth at EPI was a success, and fourth iteration design with strain compensation worked well – there were no dark lines, and output optical power was not decaying! And we also improved laser performance. In the Phase II program Final Report in July of 1996, OPS power was now 500 mW in TEM01 mode, 425 mW in TEM00 mode, and 315 mW coupled into single mode fiber, see Figure 1.12. At the time, state-of-the-art high power edge-emitting lasers had only ~200 mW in single mode fiber. We started to look into ways of packaging our OPS lasers. We also thought of making our laser tunable and explored using an external mirror on an electrostatically deflectable membrane. The process of device characterization and performance optimization and improvement continued.

In April of 1996 we applied for, and later received, a Small Business Technology Transfer (STTR) program from the Defense Advanced Research Projects Agency (DARPA) of the US Department of Defense to work together with Sandia National Laboratory on the topic of "High-Power Optically-Pumped Vertical Cavity

Figure 1.13 OPS power in TEM11 mode, TEM00 fundamental circular Gaussian mode, and coupled into single mode fiber, fifth generation design with inverted wafer structure, DARPA Final Report, December 1996. Output power here is pump power limited.

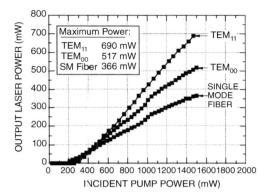

Semiconductor Lasers with Diffraction-Limited Beams." By then we knew and had announced to the potential sponsor that Micracor had demonstrated a new watt-class high-power optically pumped semiconductor OPS laser technology; and we wanted to push OPS powers even higher to 1 W and beyond. Chip thermal impedance still limited our performance, but we couldn't thin substrate between the DBR and heatspreader any further by polishing. So we applied the "flip-chip" approach to get active region closer to the heatsink. We designed an inverted wafer structure, where the MQW active region was grown on the substrate first, with the DBR region grown on top of it. We then flipped the chip and soldered the DBR mirror directly onto the diamond heatspreader. The substrate, which was now on top of the active region, was then selectively etched away. We had several inverted wafer designs grown at EPI and also by Hong Hou at Sandia Lab. Using this flip-chip approach, we got our OPS thermal impedance down to ~20°K/W, a factor of 2–3 lower than for the non-inverted structure. Laser performance had indeed improved. In our final report to DARPA in December of 1996, OPS laser power was 0.69 W in TEM11 mode, 0.51 W in a fundamental Gaussian TEM00 mode, and 0.36 W coupled into single mode fiber, see Figure 1.13. Output powers for the inverted OPS laser structure were now pump power limited, unlike our earlier lasers, where output power was thermally limited and was saturating and rolling off at the higher pump powers. If we had a more powerful pump, or added a second pump to the laser, laser output power would have been even higher. Indeed, several years later in 2003 a group from Osram Opto Semiconductors used this wafer design, with higher power pump, to demonstrate an OPS laser with 8 W output [36].

So almost three and a half years after the start of the OPS program, we had demonstrated at Micracor the powerful capabilities of the OPS laser approach: watt-class output powers in a circular Gaussian beam, with stable and reliable wafer structure, and multi-transverse-mode diode laser pumps. A path was now open to future developments: yet higher output powers, intracavity optical frequency doubling and visible light generation, modelocking for short pulse generation, new semiconductor materials and new wavelengths, and other yet unknown possibilities. Figure 1.14 shows evolution of Micracor OPS laser power over time: it took us a year to get some pulsed light out of the structures, but after that, continuous improvements in wafer design and chip processing drove the power relentlessly up from the milliwatt scale of the VCSELs to the watt scale of the OPS VECSELs. One thing this chart does not reflect is all the challenges and frustrations that had to be overcome along the way.

But at this point Micracor reached its finale. At the end of 1996 Micracor ran out of money, and investors closed the company. Largely due to the foresight of Luis Spinelli from Coherent Inc., who visited Micracor and realized the potential of OPS technology, Coherent bought the assets of what remained of Micracor, which related to materials and intellectual property of solid-state microchip and OPS lasers. I transferred the OPS laser technology to Coherent and helped to set it up there; Juan Chilla took over the OPS project at Coherent. I was anxious to publish at least some portion of our work, the results of more than three years of our labor. Fortunately, Coherent agreed, we got permission to publish, and in March of 1997 I submitted a short paper on OPS-VECSELs to *IEEE Photonics Technology Letters* (PTL) [32]. At Micracor, from the very beginning we called our technology "OPS lasers," but for the paper Aram came up with the term VECSEL, vertical *external*-cavity surface-emitting laser, to distinguish more clearly the external cavity of our approach from the conventional monolithic short-cavity VCSEL. At this time, the term OPS-VECSEL was born. More than a year later, in November of 1998, when I was already working at MIT Lincoln Laboratory, I wrote a longer and more detailed paper on OPS-VECSELs and our work at Micracor and submitted it to *IEEE Journal of Selected Topics in Quantum Electronics* (JSTQE) [33].

1.7 VECSELs beyond Micracor

I was excited by the results of our work and their publication, but I did not foresee the impact this had on the laser field and the subsequent broad development of VECSEL science and technology in academia and in the commercial world. Following publication of the Micracor OPS papers, after a brief period, VECSEL-related papers started coming out; and I got to review many of them. In 1998 Aram Mooradian started a new company in California, Novalux, to develop electrically pumped VECSELs. Soon, Novalux Extended-Cavity Surface-Emitting Laser (NECSEL) was demonstrated, first in the IR, and then in 2001 Victor Lazarev demonstrated

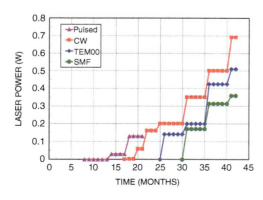

Figure 1.14 Evolution of Micracor OPS laser power over time since the start of the program in August 1993.

intracavity frequency doubling of NECSELs into the blue. Coherent Inc. released commercial products based on OPSL technology, first in the IR (at Optical Fiber Communication (OFC) conference 1999) and later in the visible (Sapphire™ blue laser, 2001). It's interesting that OPS-VECSEL, and NECSEL for that matter, were first developed with 980 nm pumping of Er-fiber amplifiers in mind, and that's one application that never took, either for optically or electrically pumped version. In 1999, Sandia reported OPS-VECSEL intracavity doubled into the blue [37]. In 2000, the first SESAM, semiconductor saturable absorber mirror, passively modelocked picosecond OPS-VECSEL was reported in a joint work between Ann Tropper's group in the University of Southampton in the UK and Ursula Keller's group in the Swiss Federal Institute of Technology ETH in Zurich [38]. Later, VECSEL pulses entered high power femtosecond regime [39] at GHz repetition rates, with promising applications to multiphoton imaging demonstrated by the Ursula Keller's group [40]. The same group developed modelocked integrated external-cavity surface-emitting lasers (MIXSELs), where the gain region and saturable absorber are integrated on a common substrate, with promising applications such as dual-comb spectroscopy [41]. In the following years, CW VECSELs have been demonstrated in a variety of materials and wavelengths, from UV to near-IR, mid-IR, and terahertz (THz) range [1]. Alexei Sirbu and Eli Kapon at École Polytechnique Fédérale de Lausanne (EPFL) in Switzerland, together with the Oleg Okhotnikov group at Tampere University, Finland, demonstrated wafer-fused VECSELs at 1.3 and 1.5 µm [42]. Mid-IR VECSELs were demonstrated by the Hans Zogg group in ETH Zurich [43]. Many originally unexpected applications and features of VECSELs were also demonstrated. Highly sensitive spectroscopic applications of VECSELs, intracavity laser absorption spectroscopy (ICLAS) [44] and cavity ringdown spectroscopy (CRDS) [45], took OPS ideas in new directions, as demonstrated by a collaboration of researchers from several French universities, including Alain Campargue at Université Grenoble and Arnaud Garnache at Université Montpellier. Arnaud Garnache and his group demonstrated a number of VECSEL properties and novel applications, such as spatial vortex beam generation [46]. VECSELs were found to operate in the low noise class-A laser regime and found applications for radio frequency (RF) optical analog links by a group from several French institutions [47]. David Wineland's group at the National Institute of Standards and Technology (NIST) in Boulder, Colorado, in collaboration with Mircea Guina's group from Tampere University in Finland, utilized wavelength versatility, narrow linewidth, power scaling, and intensity stability of VECSELs to demonstrate applications in atomic physics for cooling and trapping of atoms [48] with potential future applications in quantum computing. High power yellow VECSELs in Mircea Guina's group found application for sodium laser guide stars [49]. Jacob Khurgin from the Johns Hopkins University [50] and Mansoor Sheik-Bahae from the University of New Mexico [51] in the US have described the use of VECSELs for optical refrigeration - laser cooling of solids. Benjamin Williams from the University of California Los Angeles has developed terahertz quantum-cascade metasurface-based watt-level and tunable VECSELs [52]. Along the way, the term semiconductor disk laser (SDL) was introduced, by analogy with the similar solid-state disk lasers [14, 15]. *Semiconductor*

Disk Lasers book was published by Oleg Okhotnikov in 2010 [1], and 10 annual conferences focused on VECSELs have been held at SPIE Photonics West in San Francisco. Many researchers, research groups, and commercial companies around the world have contributed to the advancement of VECSELs over the years and have made VECSELs what they are today; they are too numerous for all to be listed here.

From the initial concepts of Nikolay Basov in 1966, and after three decades of advances by many groups working on OPSLs, in 1997 Micracor used modern bandgap-engineered semiconductor structures to demonstrate OPS-VECSELs with watt-class output powers and excellent circular Gaussian beams. By demonstrating feasibility, power, and implementation details of the OPS approach, Micracor work opened the gates for the rapid and broad development of this class of lasers. OPSL – VECSEL – SDL approach combined strengths of semiconductor lasers and diode-pumped solid-state lasers, such combination proving to have unique advantages. Optical pumping and external cavity of VECSELs, which initially appeared as their weaknesses, proved to be instrumental for the rich versatility of these lasers. VECSELs frequently outperform and displace other types of lasers because of their unique combination of properties, such as >100 W output power, femtosecond pulse operation, or output wavelengths from UV to THz. The future of VECSELs lies in further understanding of their physical properties, improvement of their various operational parameters, and broadening the range of their applications, which today range from entertainment to solid-state laser pumping, spectroscopy and atomic physics, fluorescence excitation in biomedicine, retinal photocoagulation therapy, and multiphoton imaging. Yearly VECSEL conferences at SPIE Photonics West bring VECSEL scientists together to exchange their latest results and ideas. This second book on VECSELs reviews their state of the art today. Thanks to the contributions of researchers prior to Micracor, the Micracor work itself, and all the contributions from researchers and engineers since, VECSELs have established their place among the important laser technologies today, with expected rich and multifaceted future.

References

1 Okhotnikov, O.G. (ed.) (2010). *Semiconductor Disk Lasers: Physics and Technology*. Weinheim: Wiley-VCH.
2 Hall, R.N., Fenner, G.E., Kingsley, J.D. et al. (1962). Coherent light emission from GaAs junctions. *Phys. Rev. Lett.* 9: 366–368.
3 Nathan, M.I., Dumke, W.P., Burns, G. et al. (1962). Stimulated emission of radiation from GaAs p-n junctions. *Appl. Phys. Lett.* 1: 62–64.
4 Quist, T.M., Rediker, R.H., Keyes, R.J. et al. (1962). Semiconductor maser of GaAs. *Appl. Phys. Lett.* 1: 91–92.
5 Holonyak, N. Jr., and Bevacqua, S.F. (1962). Coherent (visible) light emission from Ga(As1-xPx) junctions. *Appl. Phys. Lett.* 1: 82–83.
6 Nathan, M.I. (1987). Invention of the injection laser at IBM. *IEEE J. Quantum Electron.* QE-23: 679–683.

7 Soda, H., Iga, K., Kitahara, C., and Suematsu, Y. (1979). GaInAsP/InP surface emitting injection lasers. *Jpn. J. Appl. Phys.* 18: 2329–2330.
8 Iga, K. (2000). Surface-emitting laser – its birth and generation of new optoelectronics field. *IEEE J. Sel. Topics Quantum Electron.* 6: 1201–1215.
9 Jewell, J.L., McCall, S.L., Lee, Y.H. et al. (1989). Lasing characteristics of GaAs microresonators. *Appl. Phys. Lett.* 54: 1400–1402.
10 Jewell, J.L., Huang, K.F., Tai, K. et al. (1989). Vertical cavity single quantum well laser. *Appl. Phys. Lett.* 55: 424–426.
11 Tell, B., Lee, Y.H., Brown-Goebeler, K.F. et al. (1990). High-power cw vertical-cavity top surface-emitting GaAs quantum well lasers. *Appl. Phys. Lett.* 57: 1855–1857.
12 Phelan, R.J. Jr., and Rediker, R.H. (1965). Optically pumped semiconductor laser. *Appl. Phys. Lett.* 6: 70–71.
13 Basov, N.G., Bogdankevich, O.V., and Grasyuk, A.Z. (1966). Semiconductor lasers with radiating mirrors. *IEEE J. Quantum Electron.* QE-2: 594–597.
14 Giesen, A., Hügel, H., Voss, A. et al. (1994). Scalable concept for diode-pumped high-power solid-state lasers. *Appl. Phys. B* 58: 365–372.
15 Giesen, A. and Speiser, J. (2007). Fifteen years of work on thin-disk lasers: results and scaling laws. *IEEE J. Sel. Topics Quantum Electron.* 13: 598–609.
16 Johnson, M.R. and Holonyak, N. Jr., (1968). Optically pumped thin-platelet semiconductor lasers. *J. Appl. Phys.* 39: 3977–3985.
17 Chinn, S.R., Rossi, J.A., Wolfe, C.M., and Mooradian, A. (1973). Optically pumped room-temperature GaAs lasers. *IEEE J. Quantum Electron.* QE-9: 294–300.
18 Stone, J., Wiesenfeld, J.M., Dentai, A.G. et al. (1981). Optically pumped ultrashort cavity In1-xGaxAsyP1-y lasers: picosecond operation between 0.83 and 1.59 mm. *Opt. Lett.* 6: 534–536.
19 Roxlo, C.B. and Salour, M.M. (1981). Synchronously pumped mode-locked CdS platelet laser. *Appl. Phys. Lett.* 38: 738–740.
20 Le, H.Q., Di Cecca, S., and Mooradian, A. (1991). Scalable high-power optically pumped GaAs laser. *Appl. Phys. Lett.* 58: 1967–1969.
21 Holonyak, N. Jr., Kolbas, R.M., Dupuis, R.D., and Dapkus, P.D. (1978). Room-temperature continuous operation of photopumped MO-CVD AlxGa1-xAs-GaAs-AlxGa1-xAs quantum-well lasers. *Appl. Phys. Lett.* 33: 73–75.
22 McDaniel, D.L. Jr., McInerney, J.G., Raja, M.Y.A. et al. (1990). Vertical cavity surface emitting semiconductor laser with cw injection laser pumping. *IEEE Photon. Technol. Lett.* 2: 156–158.
23 Sandusky, J.V., Mukherjee, A., and Brueck, S.R.J. (1995). A cw external cavity surface-emitting laser. *Poster TuE11 in OSA/IEEE Topical Meeting on Semiconductor Lasers*, Keystone, Colorado (August 1995).
24 Sandusky, J.V. and Brueck, S.R.J. (1996). A CW external-cavity surface-emitting laser. *IEEE Photon. Technol. Lett.* 8: 313–315.
25 Hadley, M.A., Wilson, G.C., Lau, K.Y., and Smith, J.S. (1993). High single-transverse-mode output from external-cavity surface-emitting laser diodes. *Appl. Phys. Lett.* 63: 1607–1609.

26 Hanaizumi, O., Jeong, K.T., Kashiwada, S.-Y. et al. (1996). Observation of gain in an optically pumped surface-normal multiple-quantum-well optical amplifier. *Opt. Lett.* 21: 269–271.

27 Jiang, W.B., Friberg, S.R., Iwamura, H., and Yamamoto, Y. (1991). High powers and subpicosecond pulses from an external-cavity surface-emitting InGaAs/InP multiple quantum well laser. *Appl. Phys. Lett.* 58: 807–809.

28 Sun, D.C., Friberg, S.R., Watanabe, K. et al. (1992). High power and high efficiency vertical cavity surface emitting GaAs laser. *Appl. Phys. Lett.* 61: 1502–1503.

29 Jiang, W., Shimizu, M., Mirin, R.P. et al. (1993). Femtosecond periodic gain vertical-cavity lasers. *IEEE Photon. Technol. Lett.* 5: 23–24.

30 Le, H.Q., Goodhue, W.D., and Di Cecca, S. (1992). High-brightness diode-laser-pumped semiconductor heterostructure lasers. *Appl. Phys. Lett.* 60: 1280–1282.

31 Le, H.Q., Goodhue, W.D., Maki, P.A., and Di Cecca, S. (1993). Diode-laser-pumped InGaAs/GaAs/AlGaAs heterostructure lasers with low internal loss and 4-W average power. *Appl. Phys. Lett.* 63: 1465–1467.

32 Kuznetsov, M., Hakimi, F., Sprague, R., and Mooradian, A. (1997). High-power (>0.5-W CW) diode-pumped vertical external-cavity surface-emitting semiconductor lasers with circular TEM00 beams. *IEEE Photon. Technol. Lett.* 9: 1063–1065.

33 Kuznetsov, M., Hakimi, F., Sprague, R., and Mooradian, A. (1999). Design and characteristics of high-power (>0.5-W CW) diode-pumped vertical external-cavity surface-emitting semiconductor lasers with circular TEM00 beams. *IEEE J. Sel. Topics Quantum Electron.* 5: 561–573.

34 Raja, M.Y.A., Brueck, S.R.J., Osinsky, M. et al. (1989). Resonant periodic gain surface-emitting semiconductor lasers. *IEEE J. Quantum Electron.* 25: 1500–1512.

35 Matthews, J.W. and Blakeslee, A.E. (1974). Defects in epitaxial multilayers: I. Misfit dislocations. *J. Crystal Growth* 27: 118–125.

36 Lutgen, S., Albrecht, T., Brick, P. et al. (2003). 8-W high-efficiency continuous-wave semiconductor disk laser at 1000 nm. *Appl. Phys. Lett.* 82: 3620–3622.

37 Raymond, T.D., Alford, W.J., Crawford, M.H., and Allerman, A.A. (1999). Intracavity frequency doubling of a diode-pumped external-cavity surface-emitting semiconductor laser. *Opt. Lett.* 24: 1127–1129.

38 Hoogland, S., Dhanjal, S., Tropper, A.C. et al. (2000). Passively mode-locked diode-pumped surface-emitting semiconductor laser. *IEEE Photon. Technol. Lett.* 12: 1135–1137.

39 Waldburger, D., Link, S.M., Mangold, M. et al. (2016). High-power 100 fs semiconductor disk lasers. *Optica* 3: 844–852.

40 Voigt, F., Emaury, F., Bethge, P. et al. (2017). Multiphoton in vivo imaging with a femtosecond semiconductor disk laser. *Biomed. Opt. Express* 8: 3213–3231.

41 Link, S.M., Maas, D.J.H.C., Waldburger, D., and Keller, U. (2017). Dual-comb spectroscopy of water vapor with a free-running semiconductor disk laser. *Science* 356: 1164–1168.

42 Sirbu, A., Rantamäki, A., Saarinen, E.J. et al. (2014). High performance wafer-fused semiconductor disk lasers emitting in the 1300 nm waveband. *Opt. Express* 22: 29398–29403.

43 Rahim, M., Khiar, A., Felder, F. et al. (2009). 4.5 μm wavelength vertical external cavity surface emitting laser operating above room temperature. *Appl. Phys. Lett.* 94: 201112.

44 Garnache, A., Liu, A., Cerutti, L., and Campargue, A. (2005). Intracavity laser absorption spectroscopy with a vertical external cavity surface emitting laser at 2.3μm: application to water and carbon dioxide. *Chem. Phys. Lett.* 416: 22–27.

45 Cermak, P., Chomet, B., Ferrieres, L. et al. (2016). CRDS with a VECSEL for broad-band high sensitivity spectroscopy in the 2.3 μm window. *Rev. Sci. Instr.* 87: 83109.

46 Seghilani, M.S., Myara, M., Sellahi, M. et al. (2016). Vortex laser based on III-V semiconductor metasurface: direct generation of coherent Laguerre-Gauss modes carrying controlled orbital angular momentum. *Nat. Sci. Rep.* 6: 38156.

47 Baili, G., Morvan, L., Pillet, G. et al. (2014). Ultralow noise and high-power VECSEL for high dynamic range and broadband RF/optical links. *J. Lightw. Technol.* 32: 3489–3494.

48 Burd, S.C., Allcock, D.T.C., Leinonen, T. et al. (2016). VECSEL systems for the generation and manipulation of trapped magnesium ions. *Optica* 3: 1294–1299.

49 Kantola, E., Leinonen, T., Ranta, S. et al. (2014). High-efficiency 20 W yellow VECSEL. *Opt. Express* 22: 6372–6380.

50 Khurgin, J.B. (2020). Radiation-balanced tandem semiconductor/Yb3+:YLF lasers: feasibility study. *J. Opt. Soc. Am. B* 37: 1836–1895.

51 Yang, Z., Meng, J., Albrecht, A.R. et al. (2019). Radiation-balanced Yb:YAG disk laser. *Opt. Express* 27: 1392–1400.

52 Curwen, C.A., Reno, J.L., and Williams, B.S. (2019). Broadband continuous single-mode tuning of a short-cavity quantum-cascade VECSEL. *Nature Photonics* 13: 855–859.

2

VECSELs in the Wavelength Range 1.18–1.55 μm

Antti Rantamäki[1] and Mircea Guina[2]

[1]*Formerly with Optoelectronics Research Centre, Tampere University of Technology, Tampere, Finland*
[2]*Optoelectronics Research Centre, Tampere University of Technology, Tampere, Finland*

2.1 Introduction

This chapter provides an overview of vertical external-cavity surface-emitting lasers (VECSELs) with fundamental emission in the wavelength range of 1.18–1.55 μm. The main challenges relate to the properties of the available semiconductor materials. On one hand, InP-based gain structures are the preferred choice for emission at wavelengths >1.3 μm, but the III–V alloys lattice matched to InP are not suitable for fabricating a high-performance distributed Bragg reflector (DBR) section. On the other hand, GaAs-based materials can offer high-performance DBRs, but the fabrication of GaAs-based active regions is challenging at wavelengths longer than 1.1 μm. This challenge of GaAs-based materials has been tackled by using strain compensation, GaInNAs and GaAsSb quantum wells (QWs), and InAs quantum dots (QDs). These techniques are detailed within the first part of this chapter. The focus is then shifted to InP-based VECSELs and the measures that have been taken to overcome the limitations of InP-based DBRs. The most impressive results have been obtained with wafer bonding, which enables combining the better parts of both worlds, i.e. the integration of InP-based gain sections with GaAs-based DBRs. For this reason, an entire section is devoted to the topic of wafer bonding. Further advances in the DBR performance are also discussed, including the possibility of omitting the DBR section altogether. Finally, the first demonstration of InP-based VECSEL at 1.5 μm with flip-chip geometry and the first demonstration of a 1.5 μm VECSELs exploiting QD gain structures are reviewed.

Vertical External Cavity Surface Emitting Lasers: VECSEL Technology and Applications, First Edition.
Edited by Michael Jetter and Peter Michler.
© 2022 WILEY-VCH GmbH. Published 2022 by WILEY-VCH GmbH.

2.2 Overview of GaAs-based Gain Mirror Technologies for Long-wavelength Infrared VECSELs

2.2.1 InGaAs QWs

Impressive performance has been demonstrated from GaAs-based VECSELs at wavelengths around 1 μm with output powers reaching >100 W in continuous-wave operation [1], multiple watts in modelocked operation [2, 3], and tens of watts at the visible wavelengths via intracavity frequency doubling [4]. This level of performance is made possible by the favorable properties of the semiconductor materials. The active regions benefit from InGaAs QWs that provide high gain and temperature-insensitive operation, while the DBR can be fabricated using Al(Ga)As/GaAs layers that provide sufficient thermal conductivity and high refractive index contrast.

Unfortunately, the performance of these GaAs-based VECSELs is limited at wavelengths >1.2 μm. The limitation arises because emission at these longer wavelengths requires a higher In content in the InGaAs QWs, which is associated with higher compressive strain. To some extent, this issue can be alleviated using tensile-strained GaAsP layers between the QWs for strain compensation [5, 6]. One can also use lower growth temperatures, which prevent the highly strained InGaAs layers from evolving into 3D growth. However, the lower growth temperatures can also increase the number of nonradiative centers, point defects, and the surface roughness in the structure [6]. The growth temperature is therefore a trade-off between these factors. VECSELs with an output power as high as 72 W emitting around 1.18 μm have been demonstrated using strained InGaAs QWs but reaching comparable power levels beyond 1.2 μm has remained so far elusive [7]. The corresponding output power characteristics are shown in Figure 2.1.

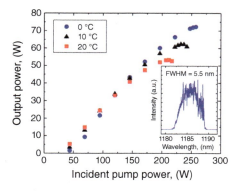

Figure 2.1 The output characteristic of the 1185 nm VECSEL. The maximum output power of 72 W is obtained at a heatsink temperature of 0 °C. Source: The figure is reproduced from [7] with permission from IET.

2.2.2 GaInNAs QWs

Due to the favorable properties of GaAs-based materials, several approaches have been undertaken to extend their emission wavelength beyond 1.2 μm. One of the most prominent approaches is the incorporation of small amounts of nitrogen into the InGaAs QWs. The incorporation of nitrogen is beneficial, because it reduces the

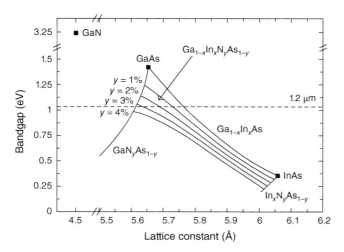

Figure 2.2 A representation of the bandgap and lattice constants for dilute nitride GaInNAs QWs with varying N content. Source: The figure is reproduced from [14] with permission.

bandgap and shrinks the lattice constant of InGaAs QWs [8, 9]. GaInNAs QWs therefore provide longer emission wavelengths and reduced strain when compared to InGaAs QWs. This bowing of the InGaAs bandgap at different nitrogen concentrations is illustrated Figure 2.2. The nitrogen content in GaInNAs QWs is usually kept below 3–4%, which is sufficient for reducing the bandgap to 0.8 eV that corresponds to 1550 nm in wavelength [10]. However, more typically the nitrogen content is kept below 2% [11–13] in order to decrease the detrimental influence of N-related defects on the device lifetime.

When compared to their InP-based counterparts at wavelengths <1.3 μm, GaInNAs QWs benefit from better carrier confinement and therefore improved operation at elevated temperatures [9, 15]. In addition, the shortcomings of InP-based DBRs are mitigated using GaAs-based materials. Specifically, there is a lack of InP-based lattice-matched materials having a high refractive index contrast, which is an essential requirement for the DBR. InP-based DBRs therefore have narrow reflection bands and require a very high number of DBR pairs (i.e. in the range of 50 layer pairs) to obtain high reflectivity, which increases their electrical and thermal resistance [16].

On the other hand, the introduction of nitrogen into InGaAs QWs requires rather specialized expertise [17] and is associated with the formation of structural defects that can act as nonradiative recombination traps. In practice, there is a trade-off in the In/N ratio: a higher N content reduces the compressive strain, but at the same time degrades the quality of the structure [11]. As an example of this, the maximum output power of a GaInNAs VECSEL is 11 W at 1.18 μm [13], while the maximum output power at a longer wavelength of 1.3 μm is considerably lower with 0.6 W [18]. The output power is then further decreased to below 100 mW at 1.5 μm [19].

Overall, it seems that the most promising operation range for dilute nitride VECSELs is in the wavelength range 1.2–1.3 μm, where GaInNAs VECSELs have been demonstrated with frequency-doubled output powers of >10 W from a 1230 nm VECSEL [20] and >6 W from 1250 nm VECSEL [21]. At their fundamental emission wavelength, GaInNAs VECSELs have been used as pump sources for Raman fiber lasers that highly benefit from the combination of low noise and high beam quality provided by VECSELs [22, 23]. GaInNAs VECSELs have also been demonstrated in modelocked operation generating 5 ps pulses and an average output power of 2.75 mW at the wavelength of 1.2 μm [24]. Finally, it is noteworthy that a GaInNAsSb VECSEL has been demonstrated at a longer wavelength of 1.55 μm with 80 mW of output power, owing to introduction of Sb [19]. The purpose of the Sb was to act as a surfactant that improves the structural quality, widens the growth parameter window, and decreases the bandgap of highly strained GaInNAs structures [15, 25].

2.2.3 InAs QDs

With GaAs-based materials, the wavelength range >1.2 μm can also be accessed using InAs QDs. QDs are essentially nanostructured semiconductors that are placed into semiconductor material having a larger bandgap. The special properties of QDs include high characteristic temperatures, low threshold currents, and high material gain [26, 27]. These features arise from their discrete density of electronic states that result from carrier confinement in all three dimensions [28].

The methods for fabricating QD-based lasers include submonolayer (SML) and Stranski–Krastanow (S–K) growth [29, 30]. The QD lasers emitting at wavelengths around 1 μm are usually grown using the SML growth, where the QDs are formed by alternating depositions InAs and GaAs with an average layer thickness below one monolayer. In the S–K growth mode, the QDs are formed via relaxation of the grown layer after some critical thickness has been reached. The critical thickness, the size of the QDs, and the density of the QDs all depend on the growth conditions and the lattice mismatch between the substrate and the deposited layers [31]. The S–K growth is generally more suitable at the longer wavelengths >1.2 μm [29], though 6 W of output power at 1040 nm has also been demonstrated from such a structure [32].

Similar to dilute nitride structures, QD-based lasers have certain drawbacks despite their many potential advantages. For instance, even though the material gain is very high in QDs, the relatively small gain volume and small optical confinement factor (i.e overlap between the optical field and the active material) lead to smaller modal gains when compared to their QW-based counterparts [29, 31]. The peak gain can also be relatively low in QD-based active media due to inhomogeneous broadening that arises from size and composition variation in the QDs [31]. Therefore, QD lasers are preferably fabricated with several QD layers [33, 34] and high areal dot density [35] in order to enhance the modal gain and increase the saturation intensity.

QD-based lasers also face additional challenges at wavelengths >1.3 µm. These issues arise because the QDs emitting at longer wavelengths need to be larger, which leads to increased strain and reduced dot densities [29]. To some extent, the issues with increasing strain can be mitigated by covering the QDs with a strain-reducing layer, such as InGaAs [36]. Further improvement can be obtained by embedding the InAs QDs into InGaAs QWs in a so-called dots-in-a-well (DWELL) structure. These DWELL structures offer attractive features such as narrow emission linewidths, high QD densities, and efficient carrier capture into the QDs [37, 38]. However, at wavelengths >1.3 µm the performance of DWELL structures can be limited by reduced carrier confinement and an excessive amount of defects in the InGaAs QWs [39, 40]. Alternative methods are therefore used in extending the wavelength of QD-based lasers beyond 1.3 µm [31, 41, 42], such as metamorphic growth and bilayer structures [43, 44].

As for the performance of QD-based VECSELs, a comparative study of two different gain structures at the wavelength of 1.25 µm was performed by Albrecht et al. in [45]. The authors fabricated two similar gain structures that each had 12 layers of QDs but with different distributions over the antinodes of the optical field in the resonant periodic gain structure. In the first structure, the QD layers were placed at the antinodes of the optical field with a 12×1 distribution, while the second structure had a 4×3 distribution over four antinodes. The 12×1 structure exhibited superior performance with 3.25 W of output power, whereas the 4×3 structure produced power levels four times lower. The authors attributed this difference in performance to the superior strain relief in the first structure having only one QD layer per antinode, but also to the increased pump absorption in the thicker active region. In any case, the demonstrations of monolithic QD-based VECSELs have been limited to wavelengths below 1.3 µm. In the wavelength range 1.1–1.3 µm, the best results in terms of output power include 6–7 W at 1.18 µm [32, 46], 2 W at 1.2 µm [47], 4.65 W at 1.25 µm [48], and 200 mW at 1.3 µm [49].

2.2.4 GaAsSb QWs

The third approach in extending the emission wavelength of GaAs-based VECSELs to the 1.2 µm range is the use of GaAsSb-based active regions [50–52]. This approach has been implemented in VECSELs leading to the demonstration of close to 100 mW of output power for a gain element temperature of −15 °C [53]. Unfortunately, these structures are not suitable for high power operation, because their low conduction band offset leads to weak electron confinement and therefore high temperature sensitivity [30, 54]. The performance can be improved using type-II QWs with "W"-type GaInAs/GaAsSb/GaInAs structures, where the transitions occur indirectly across material interfaces as illustrated in Figure 2.3. In such a structure, the electron and hole confinements are spatially separated, so that the conduction and valence band states can be engineered independently [55]. These type-II "W" QWs can also enable reduced Auger recombination [55–57] and provide high gain due to a good overlap between the electron and hole probability density

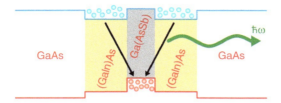

Figure 2.3 An illustrative example of a type-II "W"-type QW. The electrons are confined in the GaInAs QWs and the holes in the GaAsSb QW. The radiative recombination occurs via indirect transitions through the material boundary as shown by the black arrows. As a result, a photon illustrated with the green arrow is emitted. Source: The figure is reproduced from [55] by courtesy of Prof. Dr. Wolfram Heimbrodt with the permission of AIP Publishing LLC.

functions [58, 59]. In contrast to type-I QWs, type-II "W" QW structures exhibit a blue shift in material gain with increasing carrier density [60, 61].

To date, the highest reported output powers from a type-II GaInAs/GaAsSb/GaInAs VECSEL are 4 W and 1.6 W in multi-transverse mode operation at gain element temperatures of −15 and 15 °C, respectively [61]. In addition, single transverse mode operation has been reported with a lower output power of 400 mW at a temperature of 15 °C [62]. All of these demonstrations were performed using structures emitting at 1.2 μm, but further extension to the wavelength range 1.2–1.47 μm could be possible to reach using appropriate growth conditions [59]. Even further extension to the 1.55 μm range could be possible using GaInNAs/GaAsSb/GaInNAs type-II QWs, which overcome the issues of high strain in the GaInAs/GaAsSb/GaInAs structures at these wavelengths [63, 64].

2.3 Overview of InP-based Gain Mirror Technologies for Long-wavelength Infrared VECSELs

InP-based heterostructures have been well established as laser gain materials emitting at >1.3 μm. They have played an important role in the emergence of telecom laser diode technology and vertical-cavity surface-emitting lasers [65], even if the material system suffers from poor performance DBRs. Starting from the ideal choice of developing monolithic structures, the development of InP-based gain mirrors has evolved toward complementary technologies that aim to mitigate the poor performance of InP-based DBRs. Figure 2.4 presents a generic description of various gain mirror technologies, which are detailed in this section with a focus on InP-based materials.

2.3.1 Monolithic InP-based DBRs

The initial demonstrations of InP-based long-wavelength VECSELs include a monolithically grown structure emitting at 1.55 μm [67]. The active region comprised 20 InGaAsP QWs and a 48 layer pair InP/InGaAsP DBR with a calculated reflectivity >99.8%. The thermal management of the structure was performed by thinning the InP substrate to about 50 μm, after which the bottom of the substrate was bonded

2.3 Overview of InP-based Gain Mirror Technologies for Long-wavelength Infrared VECSELs

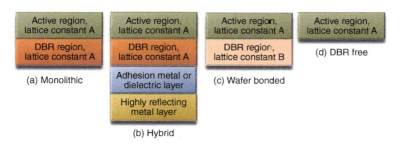

Figure 2.4 Main approaches for integration of gain and reflector regions. (a) Monolithic, (b) hybrid, (c) wafer bonded, and (d) DBR free. Source: The figure is reproduced from [66]. © IOP Publishing. Reproduced with permission. All rights reserved.

onto copper submount using indium. The VECSEL was pumped with a 1250 nm Raman fiber laser and produced 70 mW of output power in multi-transverse mode operation at a temperature of −40 °C. The performance was soon improved with a more effective thermal management strategy, where an intracavity silicon heatspreader was bonded onto the top emitting surface of the VECSEL [68]. When tested in a similar setting, the VECSEL produced 250 mW of output power with multi-transverse mode operation at a temperature of −33 °C. Further improvement was achieved using an intracavity diamond heatspreader with 780 mW of output power at a temperature of −33 °C [69]. This VECSEL was also the first InP-based VECSEL with a reasonable output power of 100 mW at room temperature. A similar configuration with a 50 μm thick chemical vapour deposition (CVD) diamond [70] was later utilized in demonstrating modelocked operation with 3.2 ps pulses at an average output power of 120 mW [71] and single-frequency operation with an output power of 470 mW in [72].

2.3.2 Dielectric and Metamorphic DBRs

In order to increase the thermal conductance and the reflectivity of InP-based DBRs, Symonds et al. demonstrated a 1.55 μm flip-chip VECSEL with a mirror section comprising two pairs of SiN_x/Si and a highly reflecting Au layer [73]. The authors estimated that the thermal conductance of this dielectric-metal mirror was three times higher when compared to an InP-based DBR. The improved thermal characteristics also permitted the demonstration of 45 mW of average output power at a gain element temperature of 0 °C. Further improvement was achieved using a GaAs-based DBR that was metamorphically grown onto the InP-based active region [74]. As with previous demonstration, the reflectivity of the DBR was enhanced by finishing it with a highly reflective Au layer. This flip-chip VECSEL produced up to 80 mW of average output power at a temperature of 20 °C at 1.55 μm. Later, the thermal properties of metamorphically grown GaAs-based DBRs were confirmed to be superior to the use of dielectric mirrors, both experimentally and theoretically [75].

Based on using metamorphic GaAs-based DBRs, long-wavelength VECSELs have been reported in various configurations at room temperature. The first demonstration of single-frequency operation with 77 mW of output power at 1.55 μm was reported in [74, 75]. This work was further advanced in [76], where Baili et al. demonstrated a single-frequency VECSEL with 110 mW of output power at 1.55 μm with an emphasis on the class A operation of VECSELs, i.e. the absence of relaxation oscillations due to the photon lifetime exceeding the carrier lifetime [77]. This VECSEL was also implemented in a wideband photonic link and was shown to outperform a low-noise distributed feedback (DFB) laser. In addition, single-frequency operation has been demonstrated at a slightly longer wavelength of 1.61 μm with 8 mW of output power [78], though in this case the gain element comprised InAs quantum dashes.

InP-based VECSELs with metamorphic GaAs-based DBRs have also been demonstrated in dual-frequency operation. These VECSELs possess two orthogonal frequency components that are partially separated at the gain element using a birefringent intracavity element but still pumped using the same pump spot. One of the novelties of this implementation arise from the fact that the two orthogonal beams share the same cavity fluctuations, i.e. the intensity and phase fluctuations are strongly correlated, so that a high-purity beat note with a linewidth in the 10 kHz range can be obtained in free-running operation [79]. Such a VECSEL was demonstrated in [80] at the wavelength of 1.55 μm with about 50 mW of power in each of the two orthogonal modes. The work was then further advanced in [81] with a detailed experimental and theoretical study on the correlation between the intensity noises of the two orthogonal laser modes and on the correlation between the phase noise of the generated radio frequency (RF) beat note and the intensity noises of the two orthogonal laser modes. Moreover, such a VECSEL has been proposed for use in distributed optical fiber sensors (OFS) that are based on Brillouin scattering [82].

Finally, GaAs-based metamorphic DBRs on InP-based QWs have been utilized in ultrashort pulse generation via passive modelocking. Near transform-limited pulses were demonstrated in [83] with a 1.7 ps pulse duration and an average power of 15 mW at the repetition rate of 2 GHz. The laser utilized a GaInNAs semiconductor saturable absorber mirror (SESAM) and had a free-running RF linewidth less than 1 kHz. Further reduction in pulse duration was reported in [84] with transform-limited 900 fs pulses at an average output power of 10 mW. The time-bandwidth product of the pulses was 0.36, which is 1.14 times the Fourier transform limit.

2.3.3 Semiconductor-dielectric-metal Compound Mirrors

Typically, the DBR section has been made thinner by reducing the number of DBR pairs and then compensating for the consequent reflectivity loss with a highly reflecting metal layer. Inherently, this scheme also increases the residual pump reflection due to the high reflectance of the final metal layer, provided that the DBR is transparent for the pump radiation. The highly reflective metal layer is usually gold, but other metals such as aluminum, silver, or copper can also be used. In these mirror structures, the node of the optical field is placed close to the semiconductor–metal interface for obtaining phase matching between the semiconductor DBR and the highly reflecting metal layer [85]. However, the drawback of these structures is the highly absorbing adhesion metal that is often needed between the semiconductor DBR and the highly reflecting metal layer. This adhesive metal layer can hinder the reflective properties of the semiconductor-metal mirrors [86–89].

Fortunately, the absorbing adhesion metal layer can also be avoided using a thin dielectric layer between the semiconductor DBR and the highly reflecting metal layer. This concept is based on having a transparent dielectric layer that is thin enough, so that its thermal conductance is high even though the thermal conductivity of the material itself is relatively low [90, 91]. When compared to semiconductor-metal mirrors, the semiconductor-dielectric-metal compound mirrors enable the fabrication of even thinner mirror structures with reduced growth times, but also lower overall strain leading to higher lifetimes [92]. They can also ensure higher reflectivity for the pump radiation, which is beneficial in terms of overall efficiency due to the recirculation of the unabsorbed pump light. Moreover, these goals can be accomplished without compromising the reflectivity for the VECSEL signal wavelength, the thermal conductance of the mirror section, or the reliability of overall structure.

To date, semiconductor-dielectric-metal mirrors have been demonstrated in edge-emitting semiconductor lasers [93] and SESAMs [94–97]. In the simplest case, only one dielectric layer is placed between the metal reflector and the semiconductor structure. The thickness of this dielectric layer should be slightly lower than the conventional $0.25\text{-}\lambda$ due to the phase shift that occurs in the metal reflector. The equation for calculating this optimal thickness is provided in [94], with an optimal value of about $0.222\text{-}\lambda$. The benefits of these compound mirrors have been realized in broadband SESAMs, which have enabled several demonstrations of modelocked solid-state lasers with pulse durations in the fs range at 0.7–1.6 μm. These SESAMs included mirror structures such as InAlAs-SiO_2-Au [94, 96], InAlAs-SiO_2-TiO_2-SiO_2-Au [95], and AlGaAs-Al_2O_3-Ag [97].

However, care should be taken when choosing the materials for a semiconductor-dielectric-metal compound mirror. The materials should provide high adhesion between the adjacent layers but also limited material diffusion for long-term stability [98, 99]. Both of these requirements are fulfilled for semiconductor-Al_2O_3-Al structures that have been demonstrated for VECSELs on several occasions [90, 100, 101]. Further improvement could be obtained using Au as the last metal layer, since Au provides higher reflectivity than Al (at wavelengths > 600 nm). However, the use of Au would also require exchanging the Al_2O_3 layer to a more suitable dielectric material that provides better adhesion to Au. This requirement is fulfilled by fluoride-dielectrics [102] that could also provide higher thermal conductivities than oxide-dielectrics [103]. In fact, such a configuration has been depicted in [104] with semiconductor-BaF_2-Au mirror structures. These structures targeted VECSELs emitting at <700 nm, where the properties of semiconductor DBRs are limited.

The optical properties of semiconductor-dielectric-metal mirrors are illustrated in Figure 2.5. The simulated data points show the minimum number of DBR layer pairs that are required at a given dielectric layer thickness for >99.8% reflectivity. The figure is meant to show the trade-off between the number of DBR layer pairs and the thickness of the dielectric layer, when the thickness of the dielectric layer is varied from a few nm up to $0.222\text{-}\lambda$. For comparison, the simulated data is provided

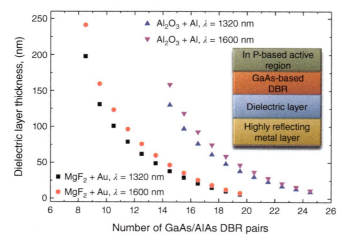

Figure 2.5 The number of DBR pairs required to attain a reflectivity of >99.8% as a function of Al_2O_3 and MgF_2 layer thicknesses (simulation performed using MacLeod thin film tool). Inset: schematic of the flip-chip structure used in the simulation. Source: The figure is reproduced from [66]. © IOP Publishing. Reproduced with permission. All rights reserved.

for VECSEL structures at 1.32 and 1.6 µm for mirror structures having Al_2O_3-Al and MgF_2-Au as the final layers. The data for the 1.32 µm VECSEL is obtained by simulating the reflectivity of the structure described in [105], while the data for the 1.6 µm VECSEL is obtained assuming a similar structure with a correspondingly thicker structure. In both cases, it is clear that the thickness of the dielectric layer does not have to be as high as 0.222-λ in order to enable a significant reduction in the required number of DBR layer pairs. Specifically, a significant reduction in the required DBR layer pairs is already obtained using a dielectric layer of 100 nm in thickness. In such a case, a reflectivity >99.8% is obtained with 15 GaAs/AlAs layer pairs using the GaAs/AlAs-Al_2O_3-Al mirror and with 11 GaAs/AlAs layer pairs using the GaAs/AlAs-MgF_2-Au mirror. This is an important feature from a thermal point of view, because a dielectric layer with 100 nm in thickness is expected to provide a negligible increase in the thermal resistance of such structures [90, 91]. Finally, it should also be noted that another variation of a compound mirror has also been proposed in [106]. In this design, the flip-chip VECSEL gain element (with a reduced number of DBR layers) is bonded onto a CVD diamond heatspreader with a transparent bonding layer, and a secondary Bragg mirror is then placed at the bottom of the diamond heatspreader to obtain >99.8% reflectivity for the signal wavelength.

When considering long-wavelength VECSELs at 1.3–1.6 µm, the combination of wafer bonding (covered in Section 2.3.4) and GaAs/AlAs-dielectric-metal compound mirrors (covered in Section 2.3.3) could bring several benefits. For one, GaAs-based DBRs are transparent to the 980 nm pump radiation, so they are suitable for residual pump recirculation. Consequently, due to the high reflectance of the final metal layer, semiconductor-dielectric-metal mirrors can easily provide well over 90% reflection for the residual pump radiation without resorting to thick double-band

Figure 2.6 The output characteristics of a flip-chip VECSEL with a mirror section comprising 18-layer GaAs/AlAs pairs, 100 nm Al_2O_3, and 150 nm of Al. The optical spectrum of the VECSEL at the highest output power is shown in the inset. Source: ©2015 IEEE. Reprinted with permission from [101].

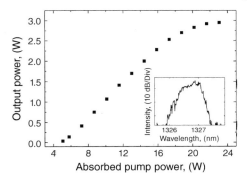

DBR structures [107, 108]. Second, reducing the number of DBR layers is expected to provide increasing benefits at longer wavelengths with respect to higher thermal conductance, less material use, less strain, and lower number of misfit dislocations [92, 101], because the $\lambda/4$ layer thickness increases with wavelength.

The output characteristic of a 1.32 μm VECSEL with a GaAs/AlAs-Al_2O_3-Al compound mirror is shown in Figure 2.6 [101]. The gain mirror comprised an InP-based active region that was wafer bonded onto a GaAs-based DBR. The details of the VECSEL structure are covered in [109], while the utilized SiO_2–SiO_2 wafer-bonding procedure is covered in the Section 2.3.4.2. The VECSEL was pumped with a fiber-coupled 980 nm pump diode that was focused onto the gain element with a spot diameter of 200 μm. The maximum output power reached about 3 W at a gain element temperature of 10 °C. This power level is slightly less than the 4 W that was obtained in the intracavity diamond configuration in Figure 2.17 (utilizing the same semiconductor wafers), but this would be expected using such a relatively small pump spot size [110]. At this point, it should also be pointed out that a similar comparison between flip-chip and intracavity diamond configurations was reported in [111] for wafer-fused structures emitting at 1.3 μm. The output power of the flip-chip structure reached 5.6 W, while the intracavity diamond structure provided an output power of 7.1 W.

2.3.4 Wafer-bonded GaAs-based DBRs

In general, monolithic InP-based DBRs, dielectric DBRs, and metamorphically grown (GaAs-based) DBRs have only resulted in limited success for long-wavelength InP-based VECSELs with output powers in the hundreds of mW range. Further extension to multiwatt output powers has required the implementation of wafer bonding that enables the integration GaAs-based DBRs onto InP-based active regions [112, 113]. In this method, the DBR and the active region can be both grown on their preferred substrates, after which they are integrated in a single device. However, while conceptually simple, historically the term wafer bonding has been "a very general term applied to various techniques using different physical/chemical principles," and there has been "a high degree of confusion surrounding this field due to the total lack of standards" [114]. Nevertheless, wafer bonding of III–V semiconductors has played a significant role for 1.3–1.6 μm VECSELs.

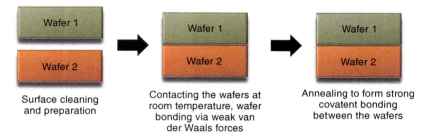

Figure 2.7 A generic illustration of wafer bonding.

In essence, wafer bonding enables the integration of materials that cannot be grown monolithically without introducing an excessive amount of defects due to lattice mismatch or mismatched coefficients of thermal expansion (CTEs) [115]. The generic wafer-bonding process comprises surface cleaning, surface preparation, wafer contacting, and thermal annealing. The process is roughly illustrated in Figure 2.7. A generic procedure could include:

1. Surface polishing to obtain surface roughness <1 nm [116]
2. Particle contamination removal via megasonic cleaning [117–119]
3. Organic contamination removal, e.g. via oxygen plasma [120] or UV/ozone treatment [121, 122]
4. Oxide removal [123, 124]
5. Surface activation with chemicals and/or energetic particles [123, 125, 126]
6. Surface rinsing and drying [117, 127]
7. Wafer contacting and annealing

It should be emphasized, however, that not all these steps are necessarily required for a given bonding process, and sometimes multiple points can be covered using a single processing step. For instance, sputter cleaning in ultrahigh vacuum (UHV) conditions can simultaneously provide oxide removal and surface activation via dangling bond creation [123]. Similarly, chemical–mechanical polishing (CMP) can itself create surfaces ready for wafer bonding [128].

In any case, the wafer surfaces should be sufficiently smooth for successful bonding. The roughness should preferably be <1 nm over an area of $2 \times 2\,\mu m^2$ [114, 129]; this can be assessed using an atomic force microscopy (AFM). The wafer surfaces also need to be clean of particles and organic contaminants. The particles are conventionally removed using megasonic cleaning, because it is less likely to cause surface damage than ultrasonic cleaning. The organic contaminants are often removed using oxygen plasma. Otherwise, organic contamination can lead to bubble creation at the bonding interface, because hydrocarbons provide agglomeration sites [130] for the gaseous side products that are generated when covalent bonds are formed between the wafers [114, 129, 130]. When the wafer surfaces are clean, they are usually activated using appropriate chemicals and/or sputtering procedures. The wafers are then contacted at room temperature [127] by placing them face-to-face and applying point pressure in the middle of the wafer stack. Such a procedure initiates a "bonding wave" at the center of the wafers [131], so that the wafers become

2.3 Overview of InP-based Gain Mirror Technologies for Long-wavelength Infrared VECSELs

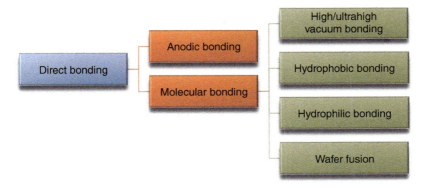

Figure 2.8 Generic summary of direct wafer-bonding methods. Source: The figure is adapted from [66]. © IOP Publishing. Reproduced with permission. All rights reserved.

contacted via van der Waals forces. Alternatively, the initial contacting can also be performed via capillary bonding using a suitable liquid [132]. Finally, the bonding is completed by transforming the van der Waals bonding into covalent bonding by thermal annealing under uniform pressure.

The following sections introduce the most general wafer-bonding methods related to VECSELs. The discussion is started by introducing the basic principles of direct wafer bonding. The section will also include the results that have been obtained with so-called wafer-fused VECSELs. The focus is then shifted into low temperature bonding methods and their application to III–V semiconductors.

2.3.4.1 Direct Wafer Bonding

The direct wafer-bonding methods can be generally summarized as shown in Figure 2.8 [66, 114, 133]. By definition, these bonding processes are performed without resorting to the use of intermediate layers. They can be roughly divided into anodic and molecular bonding. Anodic bonding is conventionally used to bond silicon wafers onto glass wafers having high alkali oxide content. It comprises thermal annealing and the application of an electric field to establish bonding at the silicon–glass interface. On the other hand, molecular bonding is based on joining the wafers via chemical bonds between the surfaces. In the simplest case, the bonding is performed between two clean wafers that are free of adsorbents. Such surfaces can be obtained in UHV conditions e.g. via thermal annealing and/or atomic hydrogen exposure [124, 134]. However, the commonly used wafer-bonding processes often comprise surface conditioning procedures that render the surfaces either hydrophobic or hydrophilic.

In hydrophobic bonding, the surface oxides are removed and the wafer surfaces terminated with hydrogen and/or fluoride, often using dilute HF. The HF treatment can also leave a few monolayers of residual HF on the wafer surfaces, which can provide bonding at room temperature via capillary bonding [135]. The initial van der Waals bonding at room temperature is obtained via hydrogen bridge bonds. Once the wafer stack is annealed at 300–700 °C, the hydrogen desorbs from the wafer surfaces and diffuses away from the bonding interface. In the case of silicon wafers, covalent bonding between the wafers takes place via the reaction [136]

$$\equiv Si - H + H - Si \equiv \Leftrightarrow \equiv Si - Si \equiv + H_2 \qquad (2.1)$$

In hydrophilic bonding, the wafer surfaces are rendered attractive to water. This is conventionally done by terminating the surfaces with reactive oxides and/or amine groups. A few monolayers of water are often left on the surfaces that help with the initial capillary bonding. As before, the initial van der Waals bonding is obtained via hydrogen bridge bonds. Upon annealing >150 °C, the residual water diffuses away from the bonding interface and/or reacts with the surrounding material. In the case of silicon wafers, the wafer bonding takes place via the reactions [126, 137]

$$\equiv Si - OH + OH - Si \equiv \Leftrightarrow \equiv Si - O - Si \equiv + H_2O \qquad (2.2)$$

$$\equiv Si - NH_2 + H_2N - Si \equiv \Leftrightarrow \equiv Si - N - N - Si \equiv + H_2 \qquad (2.3)$$

Finally, direct wafer bonding can also be obtained via wafer fusion [138], which also goes by the name of bonding by atomic rearrangement [139]. It should be noted, however, that in some cases these terms have also been used as synonyms to the broader category of "direct bonding" and "molecular bonding" [114]. Nevertheless, wafer fusion is placed in its own category here, because it is based on slight plastic deformation and material diffusion to create uniform bonding. The following description of the wafer fusion process roughly follows the one given in [140–142]:

1. Surface patterning to create outgassing channels for the residual gasses
2. Thinning one of the wafers to reduce the thermal stress after bonding
3. Removing organic contaminants with oxygen plasma or UV/ozone treatments
4. Surface oxide removal
5. Contacting the wafers and the application of uniform pressure
6. Annealing at about 550 °C
7. Cooling to room temperature

In wafer fusion, one of the wafers is usually patterned with grooves to provide outgassing channels to the species that are released at the bonding interface during high temperature annealing. Without the presence of these grooves, the bonding can become unreliable and large interfacial bubbles can appear at the bonding interface [141]. Thinning one of the substrates is also essential for the wafer fusion process, because it reduces the thermal stress at the bonding interface by making one of the substrates more compliant [143]. Otherwise, high temperature bonding of two rigid substrates with differing CTEs would lead to high stress at the bonding interface upon cooling to room temperature. The substrate thinning is followed by organic contaminant removal, after which the oxides are removed, e.g. using chemicals such as HF for InP and HCl for GaAs. However, the oxide removal can also be performed in oxygen-reduced environments such as nitrogen and/or hydrogen. In any case, it is important the wafers are not reoxidized prior to the bonding. The wafer fusion process therefore resembles conventional hydrophobic wafer bonding, even if the bond

2.3 Overview of InP-based Gain Mirror Technologies for Long-wavelength Infrared VECSELs

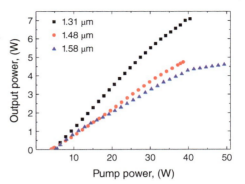

Figure 2.9 Some of the initial high-power demonstrations of 1.3–1.6 μm wafer-fused VECSELs. Source: The figure is adapted for the 1.31 and 1.58 μm VECSELs with permission from [111, 149] © The Optical Society. The data for the 1.48 μm is previously unpublished (courtesy of Jussi Rautiainen), but it was obtained using the gain structure described in [148].

formation is associated with material diffusion in the bonding interface. The bonding occurs when the wafers are contacted with a uniform pressure of 3 kPa–3 MPa and the temperature of the assembly is increased to >500 °C for about 30 minutes. At such high pressures and temperatures, it is said that indium and phosphorus are the mobile species that diffuse along the bonding interface and create uniform contacting of the wafers. The wafer stack is then cooled to room temperature, after which the assembly bows due to the differing CTEs of the wafers. The planarity is restored once the (thinned) substrate is removed [144].

The wafer fusion process was initially demonstrated for long-wavelength VCSELs in the early nineties [145, 146], but it was not applied to VECSELs until 2008 [112, 113]. The initial high power attempts with wafer-fused VECSELs comprised several demonstrations at the wavelength range 1.3–1.6 μm [147–149], as summarized in Figure 2.9. The thermal management for all these VECSELs utilized a 300 μm thick CVD intracavity diamond heatspreader. The optical pumping was performed with a 980 nm fiber-coupled diode laser that was focused onto the gain chips with a spot diameter of 300 μm. The data for the 1.31 and 1.58 μm VECSELs were measured at gain element temperature of 8 °C using V-cavities, while the data for the 1.48 μm VECSEL was measured at 15 °C in a linear cavity. The average output powers reached 7.1, 4.8, and 4.6 W for the 1.31, 1.48, and 1.58 μm VECSELs, respectively.

Further power scaling of wafer-fused long-wavelength VECSELs was later achieved using flip-chip structures emitting at 1.27 μm. Such flip-chip structures are advantageous in avoiding unwanted cavity reflections that are associated with the use of intracavity heatspreaders. The initial work comprised demonstrations with output powers reaching 6.1 W at 7 °C [150] and 8.5 W at 5 °C [89]. Later, the output power was scaled up to multiple tens of watts with the demonstration of 33 W at a temperature of −5 °C [151]. The corresponding output characteristics are shown in Figure 2.10. The VECSEL was pumped using two different 980 nm fiber-coupled diode lasers that were focused on the gain chip with a spot diameter of about 860 μm. The 75 mm long linear optical cavity comprised a curved output coupler with a radius of curvature of 150 mm and an output coupling ratio of 2.5%. A slight kink in the output power was observed when the second pump laser was

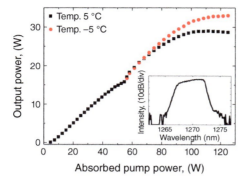

Figure 2.10 The output characteristics of a wafer-fused flip-chip VECSEL generating over 30 W of output power. The corresponding optical spectrum is shown in the inset. Source: The figure is reproduced with permission from [151] © The Optical Society.

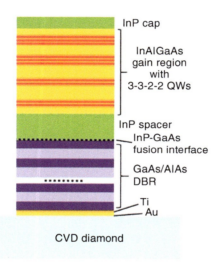

Figure 2.11 The gain mirror structure used for the first flip-chip VECSEL at ~1.54 μm. Source: ©2015 IEEE. Reprinted with permission from [152].

introduced. This kink was presumed to originate from imperfect overlap of the two pump spots.

Recently, a wafer-fused flip-chip VECSEL was also demonstrated at a longer wavelength of 1.54 μm [152]. The corresponding gain mirror is shown in Figure 2.11. The laser was tested in a V-cavity configuration with a 2.2% output coupler. The optical pumping was performed with a 980-mm fiber-coupled diode laser that was focused onto the gain element with a spot diameter of about 470 μm. A maximum output power of 3.65 W was achieved in fundamental mode operation at a heatsink temperature of 11 °C. These results are put in perspective with the other demonstrations of flip-chip VECSELs at the infrared wavelengths with the summary shown in Figure 2.12.

Finally, the first QD-based wafer-fused VECSEL operating at 1.53 μm was demonstrated in [153]. The gain structure comprised 20 layers of high-density InAs QDs. They were grown on a 2-inch InP(311)B substrate using the S–K method with gas source molecular beam epitaxy. The InAs QDs were formed at 480 °C after the deposition of a few monolayers of InAs onto an InP-lattice-matched $Ga_{0.2}In_{0.8}As_{0.44}P_{0.56}$

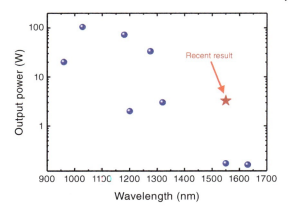

Figure 2.12 A summary of output power levels achieved using flip-chip VECSELs (spheres for early results; star for recent demonstration). Source: ©2015 IEEE. Reprinted with permission from [152].

alloy. Independent control of QD density and wavelength emission was achieved by applying a double-cap growth sequence [154]. The optimization of the spacer layers thicknesses and strain compensation were applied to enable the nucleation of a relatively high number (i.e. 5) of QD layers per each antinode within the resonant periodic gain structure. The gain structure was then wafer fused with a GaAs-based DBR that was grown by metal-organic vapor-phase epitaxy (MOVPE). The InP substrate was then removed by wet etching and a 2.7 × 2.7 mm² gain chip was capillary bonded onto an intracavity diamond heatspreader. The gain chip was pumped with 980 nm fiber-coupled diode laser that was focused onto the gain element with a spot of 190 μm in diameter. The maximum output power reached 2.2 W at gain element temperature of 15 °C. In addition, using a 0.5 mm thick birefringent filter, the operation wavelength could be tuned over 60 nm.

The ability to generate multiwatt emission in the wavelength range 1.25–1.6 μm via wafer fusion has also spurred interest in accessing the wavelength range 650–800 nm via intracavity frequency doubling. This interest arises due to the limited performance of directly emitting VECSEL structures at these wavelengths. In particular, direct-emitting VECSELs at 650–800 nm suffer from issues such as low carrier confinement in the QWs that leads to increased temperature sensitivity, material degradation, AlGaAs/AlAs DBRs with reduced refractive index contrast and thermal conductivity, and the requirement for pump lasers emitting at the visible wavelengths [30, 155–157]. Consequently, the output powers of directly emitting red VECSELs have conventionally been limited to <1 W [158], though higher output powers have been recently demonstrated and further advances are expected [159, 160]. In any case, wafer-fused VECSELs provide an interesting alternative to reach these wavelengths via intracavity frequency doubling with demonstrated output powers of 3 W at 650 nm [147], 1.5 W at 750 nm [161], and 1 W at 785 nm [149]. In addition, a wafer-fused 1.56 μm VECSEL has been demonstrated with 1 W of output power in single-frequency operation at a gain element temperature of 15 °C [162].

Wafer fusion has also contributed to modelocked VECSELs in the wavelength range 1.25–1.6 μm. These demonstrations include 6.4 ps pulses with an average output power of 100 mW at 1.3 μm [163] and 16 ps pulses with an average output power of 860 mW at 1.57 μm [164]. In the former case, the SESAM was a similar wafer-fused structure as the gain element, while in the latter case the SESAM was a dilute nitride GaInNAs structure. Later on, the output of a modelocked 1.57 μm VECSEL was amplified in an Er/Yb-doped fiber amplifier to reach an average output power of 4.5 W, which was then used for supercontinuum generation using a highly GeO_2-codoped silica fiber [165]. Finally, VECSELs have also been used as pump sources for modelocked Raman fiber lasers in several configurations [166, 167]. VECSELs are particularly suitable for this task due to their low noise and high beam quality that enable high coupling efficiencies into single mode fibers. These features have led to the demonstrations of VECSEL-pumped modelocked Raman fiber lasers at 1.38 μm [168] and 1.6 μm [169] using pump VECSELs emitting at 1.3 and 1.48 μm, respectively.

2.3.4.2 Low Temperature Bonding

The main limitation in wafer bonding is the thermal annealing that is required for transforming the initial van der Waals bonding into covalent bonding between the wafers [114]. Consequently, if the CTEs of the wafers differ, high bonding temperatures can lead to significant stress in the bonding interface, warping, and even wafer breakage [114, 143, 170]. Therefore, a push exists for low temperature bonding techniques for III–V semiconductor materials [115].

The most obvious route for low bonding temperature is the utilization of hydrophilic bonding, because it enables relatively low bonding temperatures. However, hydrophilic wafer bonding can be problematic for III–V semiconductors, because the surface oxides of III–V semiconductors are not stable (with respect to time and temperature) and the surfaces are prone to roughening during surface cleaning [171]. III–V semiconductor materials also react with the residual water in the bonding interface at elevated temperatures, which leads to gaseous side products and eventually interface bubbles [131]. Thus, alternative routes for low temperature bonding of III–V semiconductors have been devised. The most common routes include the use of various surface activation methods and the use of intermediate bonding layers.

Bonding via Surface Activation The surface activation methods rely on surface treatments that lower the amount of thermal energy that needs to be provided to the bonding interface for covalent bonding. The activation procedures can be generally divided into wet and dry processes. The wet activation methods include chemical–mechanical polishing/planarization (CMP) [128, 130] and the use of various chemicals [137], while the dry activation methods include plasma exposure [117, 172, 173], surface-activated bonding (SAB) or sputtering [125], and UV illumination [174, 175].

An illustration of wafer cleaning/activation via sputtering is shown in Figure 2.13 [125]. A very similar procedure is also provided by Electronic Visions Group (EVG) with the development of their ComBond process for III–V semiconductors [123], which relies on plasma activation/sputtering prior to the bonding. This process has also been demonstrated for III–V/Si multi-junction solar cells with 31.3% efficiency [176]. However, to date, such a process has not been demonstrated in long-wavelength VECSELs.

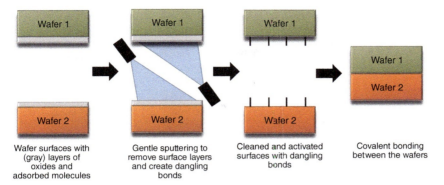

Figure 2.13 Wafer cleaning and activation via sputtering to generate dangling bonds on the wafer surfaces. The process needs to be performed in high vacuum to prevent surface oxidation after the surface cleaning/activation.

Bonding via Intermediate Layers The intermediate bonding layer approach relies on using intermediate materials that enable lower bonding temperatures than the original wafer materials. A summary of various intermediate bonding layer techniques is shown in Figure 2.14 [66, 114, 133, 177]. The division between different bonding methods is based on the intermediate materials that are used for the bonding process. However, the discussion here is limited to the intermediate layers that are suitable for (optically pumped) VECSELs, i.e. to the bonding methods that provide optically transparent interfaces with relatively thin bonding sections in the range of a few nm. This leaves thin dielectric layers and monolayers as viable options, while glass frit, metal, and adhesive bonding are omitted from the discussion.

The benefits of using thin intermediate layers include avoiding the issues related to the III–V semiconductor oxides, relatively modest bonding temperatures, reduced requirements for high vacuum conditions, and the possibility of utilizing bonding techniques that are well-established for silicon-based materials. The use of non-conducting intermediate layers is also a viable option for optically pumped VECSELs, since electrical conductivity is not required. However, the intermediate layer should be sufficiently thin to prevent any noticeable increase in the interface thermal resistance [90, 110].

The suitable materials for wafer-bonding III–V semiconductors include Si_3N_4 [178] and SiO_2 [126, 179], which both benefit from the well-established procedures for hydrophilic silicon bonding. This is beneficial because hydrophilic bonding is less prone to hydrocarbon contamination [128] and less sensitive to the wafer surface roughness [120] when compared to hydrophobic bonding. A schematic of such a process utilizing intermediate SiO_2 layers is illustrated on the left side in Figure 2.15. The process is started by dipping the SiO_2-covered wafers into NH_4OH and drying them with nitrogen. The wafers are then placed face-to-face and pressure is applied in the middle of the stack to initiate the bonding wave. The bonding is concluded by annealing the stack under uniform pressure of about 0.5 MPa at 200 °C. A scanning electron microscope (SEM) image of such a VECSEL structure

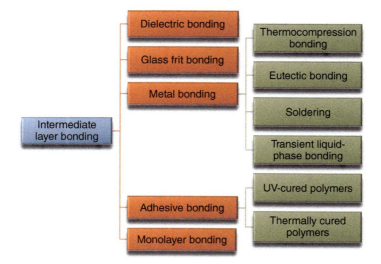

Figure 2.14 Generic summary of the intermediate layers that have been used in wafer bonding. Source: The figure is adapted from [66]. © IOP Publishing. Reproduced with permission. All rights reserved.

is shown on the right side of Figure 2.15. This process has been demonstrated for VECSELs with multiwatt-level output powers [100, 101, 105].

Another variation of intermediate layer bonding is the utilization of self-assembling monolayers (SAMs), where a molecule with at least two different functional groups is used for the bonding [180, 181]. The first functional group is used for attaching the molecule onto the first wafer, while the second functional group is utilized to obtain bonding with the opposing wafer. Such a procedure has been demonstrated for VECSELs using (3-mercaptopropyl)trimethoxysilane (MPTMS) [100, 101, 105, 109]. In addition to enabling relatively low bonding temperatures, such a bonding process can also relax the requirements for the surface roughness [182].

The MPTMS process is started by dipping the SiO_2-covered GaAs-based DBR into NH_4OH and drying it with nitrogen. The DBR structure is then placed into a low-vacuum chamber with an open container of MPTMS [183]. As the MPTMS molecules evaporate from their container, the hydrophilic ends of MPTMS attach to the hydrophilic NH_2 and OH groups on the SiO_2-covered GaAs surface, as illustrated on the left side of Figure 2.16. The process leaves the GaAs-based DBR structure terminated with hydrophobic SH groups, which can be used for wafer bonding with hydrophobic InP surface.

The output characteristic of an MPTMS-bonded VECSEL is shown in Figure 2.17 [109]. The VECSEL utilized a 300 μm thick CVD intracavity diamond heatspreader for thermal management. The gain element was pumped with a 980 nm laser diode that was focused onto the gain element with a spot diameter of 200 μm. The temperature of the gain element was set to 10 °C and the maximum output power reached 3.9 W. This output power is comparable to the output power that was obtained from similar semiconductor materials processed with wafer fusion

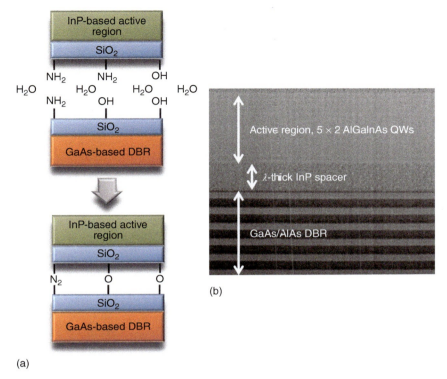

Figure 2.15 Wafer bonding of long-wavelength VECSELs via intermediate SiO$_2$ layers. Source: The figure reproduced from [66]. © IOP Publishing. Reproduced with permission. All rights reserved.

[184]. It is also noteworthy that the thermal simulations performed in [110] suggest that an intermediate layer with a thickness of a few nanometers and a thermal conductivity of $1\,\text{W}\,\text{m}^{-1}\,\text{K}^{-1}$ introduces negligible increase in the overall thermal resistance of the structure. The use of thin intermediate layers can therefore be considered as a viable approach for high power optically pumped VECSELs.

2.3.5 Gain Structures in Transmission

At this point, it is safe to state that significant measures have been taken to circumvent the limitations brought by the DBR section in InP-based VECSELs. In essence, all of these techniques target to overcome the low refractive index contrast and low thermal conductivity of InP-based DBRs with alternative mirror designs. However, the limitations of InP-based DBRs can also be overcome by omitting the DBR section altogether and using the InP-based active region in transmissive mode, i.e. replacing the semiconductor DBR with external dielectric mirrors. These structures are labeled DBR-free in this section.

A schematic of a DBR-free 1.36 μm VECSEL is illustrated in Figure 2.18 [185]. In contrast to the InP-based VECSELs with InGaAsP QWs in Section 2.3.4, the active

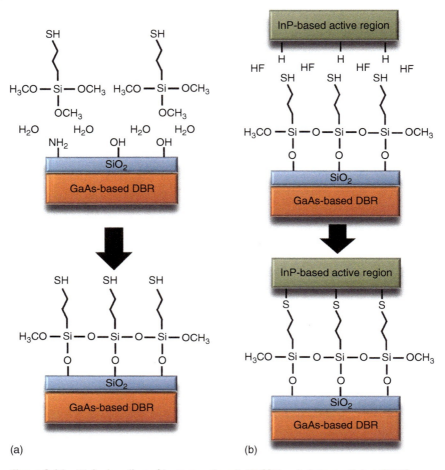

Figure 2.16 Wafer bonding of long-wavelength VECSELs via intermediate MPTMS monolayers.

region comprised 30 pairs of AlGaInAs QWs due to their higher conduction band offset and better carrier confinement [186]. The backside of the InP substrate was mechanically polished, after which antireflection coatings were applied on both sides of the gain chip. The pumping was performed with a 1.06 μm Q-switched solid-state laser that operated at repetition rates 5–50 kHz with pulse durations of 15–50 ns. At the peak pump power of 8.5 kW at 5 kHz, the VECSEL produced a maximum peak power of 1.5 kW. The output power was presumed to be limited by pump absorption saturation in the barriers layers of the active region. The corresponding pump saturation intensity was estimated to be 8.2 MW/cm^2, which is two to three orders of magnitude higher than conventional solid-state laser crystals due to the shorter fluorescence decay time [187]. On the other hand, at a higher pump repetition rate of 50 kHz, the output power of the uncooled gain element was considered to be limited by thermal rollover.

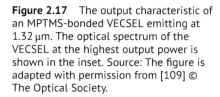
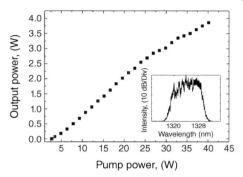

Figure 2.17 The output characteristic of an MPTMS-bonded VECSEL emitting at 1.32 μm. The optical spectrum of the VECSEL at the highest output power is shown in the inset. Source: The figure is adapted with permission from [109] © The Optical Society.

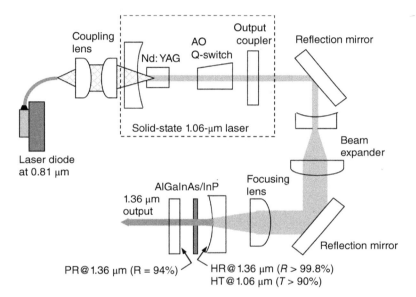

Figure 2.18 The cavity configuration of a pulsed VECSEL emitting at 1.36 μm. The active region comprised 30 pairs of AlGaInAs QWs on InP substrate. The VECSEL is operated in a transmissive mode without the DBR section in a 10 mm long optical cavity. Source: The figure is reproduced with permission from [185] © The Optical Society.

The work on long-wavelength DBR-free VECSELs was later extended to the eye-safe wavelength range of 1.57 μm [188]. The overall cavity and VECSEL designs were very similar to the 1.36 μm demonstration described above, but now both of the cavity mirrors were flat and the cavity was stabilized by a thermal lens in the VECSEL gain element. At a pump repetition rate of 20 kHz, the VECSEL produced peak output powers of 290 W at a peak pump power of 2.3 kW. In subsequent work, the barrier pumping with a 1.06 μm laser was exchanged for in-well pumping using a 1.32 μm Q-switched Nd:YVO$_4$ pump laser [189]. Using the same pump repetition rate of 20 kHz, the VECSEL produced peak powers of 0.52 kW at a pump peak power of 3.7 kW. The concept of in-well pumping was then extended to wavelengths 1.23 μm [190] and 1.53 μm [191] with peak powers of 0.76 and 0.56 kW, respectively.

Finally, the utilization of improved thermal management with intracavity diamond heatspreaders enabled high power demonstrations at much higher repetition rates of hundreds of kHz [191–193]. The peak powers in these demonstrations were in the range 400–600 W, while the average output powers reached multiwatt levels. Specifically, a maximum average output power >3 W was obtained at a repetition rate of 200 kHz [191].

2.4 Conclusion

The main technological approaches deployed in developing VECSELs operating at long infrared wavelengths (i.e. 1.18–1.55 µm) have been reviewed. As a reflection of the largest research effort, design versatility, and performance, a detailed presentation of wafer-bonding technologies was provided. Amongst the reviewed VECSELs, we note the demonstration of 33 W at 1.27 µm, the first demonstration of a multiwatt flip-chip VECSEL at 1.55 µm, and the demonstration of the first InP-based QD VECSEL at >1.5 µm.

References

1 Heinen, B., Wang, T.-L., Sparenberg, M. et al. (2012). 106 W continuous-wave output power from vertical-external-cavity surface-emitting laser. *Electronics Letters* 48 (9): 516–517.

2 Scheller, M., Wang, T., Kunert, B. et al. (2012). Passively modelocked VECSEL emitting 682 fs pulses with 5.1 W of average output power. *Electronics Letters* 48 (10): 1–2.

3 Rudin, B., Wittwer, V., Maas, D. et al. (2010). High-power MIXSEL: an integrated ultrafast semiconductor laser with 6.4 W average power. *Optics Express* 18 (26): 27582–27588.

4 Berger, J., Anthon, D., Caprara, A. et al. (2012). 20 Watt CW TEM$_{00}$ intracavity doubled optically pumped semiconductor laser at 532 nm. *Proceedings of SPIE* 8242: 824206.

5 Fan, L., Hessenius, C., Fallahi, M. et al. (2007). Highly strained InGaAs/GaAs multiwatt vertical-external-cavity surface-emitting laser emitting around 1170 nm. *Applied Physics Letters* 91 (13): 1114.

6 Ranta, S., Hakkarainen, T., Tavast, M. et al. (2011). Strain compensated 1120 nm GaInAs/GaAs vertical external-cavity surface-emitting laser grown by molecular beam epitaxy. *Journal of Crystal Growth* 335 (1): 4–9.

7 Kantola, E., Penttinen, J.-P., Ranta, S., and Guina, M. (2018). 72-W vertical-external-cavity surface-emitting laser with 1180-nm emission for laser guide star adaptive optics. *Electronics Letters* 54 (19): 1135–1137.

8 Weyers, M., Sato, M., and Ando, H. (1992). Red shift of photoluminescence and absorption in dilute GaAsN alloy layers. *Japanese Journal of Applied Physics* 31 (7A): L853–L855.

9 Kondow, M., Kitatani, T., Nakatsuka, S. et al. (1997). GaInNAs: a novel material for long-wavelength semiconductor lasers. *IEEE Journal of Selected Topics in Quantum Electronics* 3 (3): 719–730.
10 Jaschke, G., Averbeck, R., Geelhaar, L., and Riechert, H. (2005). Low threshold InGaAsN/GaAs lasers beyond 1500nm. *Journal of Crystal Growth* 278 (1): 224–228.
11 Guina, M., Leinonen, T., Härkönen, A., and Pessa, M. (2009). High-power disk lasers based on dilute nitride heterostructures. *New Journal of Physics* 11 (12): 125019.
12 Korpijärvi, V.-M., Guina, M., Puustinen, J. et al. (2009). MBE grown GaInNAs-based multi-Watt disk lasers. *Journal of Crystal Growth* 311 (7): 1868–1871.
13 Korpijärvi, V.-M., Leinonen, T., Puustinen, J. et al. (2010). 11 W single gain-chip dilute nitride disk laser emitting around 1180 nm. *Optics Express* 18 (25): 25633–25641.
14 Guina, M., Härkönen, A., Korpijärvi, V.-M. et al. (2012). Semiconductor disk lasers: recent advances in generation of yellow–orange and mid-IR radiation. *Advances in Optical Technologies* 2012: 1–19.
15 S. Bank, Bae, H., Goddard, L. et al. (2007). Recent progress on 1.55-µm dilute-nitride lasers. *IEEE Journal of Quantum Electronics* 43 (9): 773–785.
16 Karim, A., Björlin, S., Piprek, J., and Bowers, J. (2000). Long-wavelength vertical-cavity lasers and amplifiers. *IEEE Journal of Selected Topics in Quantum Electronics* 6 (6): 1244–1253.
17 Guina, M., Wang, S., and Aho, A. (2018). Molecular beam epitaxy of dilute nitride optoelectornic devices. In: *Molecular Beam Epitaxy* (ed. M. Henini), 73–94. Elsevier.
18 Hopkins, J.-M., Smith, S., Jeon, C.-W. et al. (2004). 0.6 W CW GaInNAs vertical external-cavity surface emitting laser operating at 1.32 µm. *Electronics Letters* 40 (1): 30–31.
19 Korpijarvi, V.-M., Kantola, E., Leinonen, T. et al. (2015). Monolithic GaInNAsSb/GaAs VECSEL operating at 1550 nm. *IEEE Journal of Selected Topics in Quantum Electronics* 21 (6): 480–484.
20 Kantola, E., Leinonen, T., Penttinen, J.-P. et al. (2015). 615 nm GaInNAs VECSEL with output power above 10 W. *Optics Express* 23 (16): 20280–20287.
21 Penttinen, J.-P., Leinonen, T., Rantamäki, A. et al. (2017). High power VECSEL prototype emitting at 625 nm. *Advanced Solid State Lasers*: ATu1A-8.
22 Chamorovskiy, A., Rautiainen, J., Rantamäki, A. et al. (2011). 1.3 µm Raman-bismuth fiber amplifier pumped by semiconductor disk laser. *Optics Express* 19 (7): 6433–6438.
23 Chamorovskiy, A., Rautiainen, J., Rantamäki, A., and Okhotnikov, O. (2011). Low-noise Raman fiber amplifier pumped by semiconductor disk laser. *Optics Express* 19 (7): 6414–6419.
24 Rautiainen, J., Korpijärvi, V.-M., Puustinen, J. et al. (2008). Passively mode-locked GaInNAs disk laser operating at 1220 nm. *Optics Express* 16 (20): 15964–15969.

25 Yuen, H., Bank, S.R., Bae, H. et al. (2009). The role of antimony on properties of widely varying GaInNAsSb compositions. *Journal of Applied Physics* 99 (9): 093504.

26 Ledentsov, N. (2002). Long-wavelength quantum-dot lasers on GaAs substrates: from media to device concepts. *IEEE Journal of Selected Topics in Quantum Electronics* 8 (5): 1015–1024.

27 Blood, P. (2009). Gain and recombination in quantum dot lasers. *IEEE Journal of Selected Topics in Quantum Electronics* 15 (3): 808–818.

28 Bimberg, D. and Pohl, U. (2011). Quantum dots: promises and accomplishments. *Materials Today* 14 (9): 388–397.

29 Pohl, U. and Bimberg, D. (2010). Semiconductor disk lasers based on quantum dots. In: *Semiconductor Disk Lasers: Physics and Technology* (ed. O. Okhotnikov), 187–211. Wiley.

30 Calvez, S., Hastie, J., Guina, M. et al. (2009). Semiconductor disk lasers for the generation of visible and ultraviolet radiation. *Laser & Photonics Reviews* 3 (5): 407–434.

31 Hogg, R. and Zhang, Z. (2013). Quantum dot technologies. In: *The Physics and Engineering of Compact Quantum Dot-based Lasers for Biophotonics* (ed. E.U. Rafailov), 7–41. Wiley.

32 Butkus, M., Rautiainen, J., Okhotnikov, O. et al. (2011). Quantum dot based semiconductor disk lasers for 1–1.3 μm. *IEEE Journal of Selected Topics in Quantum Electronics* 17 (6): 1763–1771.

33 Salhi, A., Martiradonna, L., Visimberga, G. et al. (2006). High-modal gain 1300-nm In(Ga)As-GaAs quantum-dot lasers. *IEEE Photonics Technology Letters* 18 (16): 1735–1737.

34 Salhi, A., Fortunato, L., Martiradonna, L. et al. (2006). Enhanced modal gain of multilayer InAs InGaAs/GaAs quantum dot lasers emitting at 1300 nm. *Journal of Applied Physics* 100 (12): 123111.

35 Mukai, K., Nakata, Y., Otsubo, K. et al. (1999). 1.3-μm CW lasing of InGaAs-GaAs quantum dots at room temperature with a threshold current of 8 mA. *IEEE Photonics Technology Letters* 11 (10): 1205–1207.

36 Nishi, K., Saito, H., Sugou, S., and Lee, J.-S. (1999). A narrow photoluminescence linewidth of 21 meV at 1.35 μm from strain-reduced InAs quantum dots covered by $In_{0.2}Ga_{0.8}As$ grown on GaAs substrates. *Applied Physics Letters* 74 (8): 1111–1113.

37 Liu, H., Hopkinson, M., Harrison, C. et al. (2003). Optimizing the growth of 1.3 μm InAs/InGaAs dots-in-a-well structure. *Journal of Applied Physics* 93 (5): 2931–2936.

38 Liu, H., Sellers, I., Gutierrez, M. et al. (2005). Optimizing the growth of 1.3-μm InAs/InGaAs dots-in-a-well structure: achievement of high-performance laser. *Materials Science and Engineering: C* 25 (5): 779–783.

39 Seravalli, L., Minelli, M., Frigeri, P. et al. (2007). Quantum dot strain engineering of InAs/InGaAs nanostructures. *Journal of Applied Physics* 101 (2): 024313.

40 Seravalli, L., Frigeri, P., Trevisi, G., and Franchi, S. (2008). 1.59 μm room temperature emission from metamorphic InAs/InGaAs quantum dots grown on GaAs substrates. *Applied Physics Letters* 92 (21): 213104.
41 Ledentsov, N., Shchukin, V., Kettler, T. et al. (2007). MBE-grown metamorphic lasers for applications at telecom wavelengths. *Journal of Crystal Growth* 301–302: 914–922.
42 Le Ru, E., Howe, P., Jones, T., and Murray, R. (2003). Strain-engineered InAs/GaAs quantum dots for long-wavelength emission. *Physical Review B* 67 (16): 165303.
43 Majid, M., Childs, D., Shahid, H. et al. (2011). Toward 1550-nm GaAs-based lasers using InAs/GaAs quantum dot bilayers. *IEEE Journal of Selected Topics in Quantum Electronics* 17 (5): 1334–1342.
44 Ledentsov, N., Kovsh, A., Zhukov, A. et al. (2003). High performance quantum dot lasers on GaAs substrates operating in 1.5 μm range. *Electronics Letters* 39 (15): 1126–1128.
45 Albrecht, A., Hains, C., Rotter, T. et al. (2011). High power 1.25 μm InAs quantum dot vertical external-cavity surface-emitting laser. *Journal of Vacuum Science & Technology B* 29 (3): 03C113.
46 Al Nakdali, D., Gaafar, M., Shakfa, M. et al. (2015). High-power operation of quantum-dot semiconductor disk laser at 1180 nm. *IEEE Photonics Technology Letters* 27 (10): 1128–1131.
47 Rantamäki, A., Rautiainen, J., Toikkanen, L. et al. (2012). Flip chip quantum-dot semiconductor disk laser at 1200 nm. *IEEE Photonics Technology Letters* 24 (15): 1292–1294.
48 Albrecht, A., Rotter, T., Hains, C. et al. (2011). High-power 1.25 μm InAs QD VECSEL based on resonant periodic gain structure. *Proceedings of SPIE* 7919: 791904.
49 Rantamäki, A., Sokolovskii, G., Blokhin, S. et al. (2015). Quantum dot semiconductor disk laser at 1.3 μm. *Optics Letters* 40 (14): 3400–3403.
50 Anan, T., Nishi, K., Sugou, S. et al. (1998). GaAsSb: a novel material for 1.3 μm VCSELs. *Electronics Letters* 34 (22): 2127–2129.
51 Yamada, M., Anan, T., Tokutome, K. et al. (2000). Low-threshold operation of 1.3-μm GaAsSb quantum-well lasers directly grown on GaAs substrates. *IEEE Photonics Technology Letters* 12 (7): 774–776.
52 Yamada, M., Anan, T., Kurihara, K. et al. (2000). Room temperature low-threshold CW operation of 1.23 μm GaAsSb VCSELs on GaAs substrates. *Electronics Letters* 36 (7): 637–638.
53 Gerster, E., Ecker, I., Lorch, S. et al. (2003). Orange-emitting frequency-doubled GaAsSb/GaAs semiconductor disk laser. *Journal of Applied Physics* 94 (12): 7397–7401.
54 Johnson, S., Guo, C., Chaparro, S. et al. (2003). GaAsSb/GaAs band alignment evaluation for long-wave photonic applications. *Journal of crystal growth* 251 (1): 521–525.

55 Gies, S., Kruska, C., Berger, C. et al. (2015). Excitonic transitions in highly efficient (GaIn)As/Ga(AsSb) type-II quantum-well structures. *Applied Physics Letters* 107 (18): 182104.

56 Berger, C., Möller, C., Hens, P. et al. (2015). Novel type-II material system for laser applications in the near-infrared regime. *American Institute of Physics advances* 5 (4): 047105.

57 Zegrya, G. and Andreev, A. (1995). Mechanism of suppression of Auger recombination processes in type-II heterostructures. *Applied Physics Letters* 67 (18): 2681–2683.

58 Lammers, C., Stein, M., Berger, C. et al. (2016). Gain spectroscopy of a type-II VECSEL chip. *Applied Physics Letters* 109 (23): 232107.

59 Fuchs, C., Beyer, A., Volz, K., and Stolz, W. (2017). MOVPE growth of (GaIn)As/Ga(AsSb)/(GaIn)As type-II heterostructures on GaAs substrate for near infrared laser applications. *Journal of Crystal Growth* 464: 201–205.

60 Fuchs, C., Berger, C., Möller, C. et al. (2016). Electrical injection type-II (GaIn)As/Ga(AsSb)/(GaIn) As single 'W'-quantum well laser at 1.2 µm. *Electronics Letters* 52 (22): 1875–1877.

61 Möller, C., Fuchs, C., Berger, C. et al. (2016). Type-II vertical-external-cavity surface-emitting laser with Watt level output powers at 1.2 µm. *Applied Physics Letters* 108 (7): 071102.

62 Möller, C., Zhang, F., Fuchs, C. et al. (2016). Fundamental transverse mode operation of a type-II vertical-external-cavity surface-emitting laser at 1.2 µm. *Electronics Letters* 53 (2): 93–94.

63 Tansu, N. and Mawst, L. (2003). Design analysis of 1550-nm GaAsSb-(In)GaAsN type-II quantum-well laser active regions. *IEEE Journal of Quantum Electronics* 39 (10): 1205–1210.

64 Yeh, J.-Y., Mawst, L., Khandekar, A. et al. (2006). Long wavelength emission of InGaAsN/GaAsSb type II "W" quantum wells. *Applied Physics Letters* 88 (5): 051115.

65 Nishiyama, N., Caneau, C., Hall, B. et al. (2005). Long-wavelength vertical-cavity surface-emitting lasers on InP with lattice matched AlGaInAs-InP DBR grown by MOCVD. *IEEE Journal of Selected Topics in Quantum Electronics* 11 (5): 990–998.

66 Guina, M., Rantamäki, A., and Härkönen, A. (2017). Optically pumped VECSELs: review of technology and progress. *Journal Physics D: Applied Physics* 50: 383001.

67 Lindberg, H., Strassner, M., Bengtsson, J., and Larsson, A. (2004). InP-based optically pumped VECSEL operating CW at 1550 nm. *IEEE Photonics Technology Letters* 16 (2): 362–364.

68 Lindberg, H., Strassner, M., Bengtsson, J., and Larsson, A. (2004). High-power optically pumped 1550-nm VECSEL with a bonded silicon heat spreader. *IEEE Photonics Technology Letters* 16 (5): 1233–1235.

69 Lindberg, H., Strassner, M., Gerster, E., and Larsson, A. (2004). 0.8 W optically pumped vertical external cavity surface emitting laser operating CW at 1550 nm. *Electronics Letters* 40 (10): 601–602.

70 Lindberg, H., Strassner, M., and Larsson, A. (2005). Improved spectral properties of an optically pumped semiconductor disk laser using a thin diamond heat spreader as an intracavity filter. *IEEE Photonics Technology Letters* 17 (7): 1363–1365.

71 Lindberg, H., Sadeghi, M., Westlund, M. et al. (2005). Mode locking a 1550 nm semiconductor disk laser by using a GaInNAs saturable absorber. *Optics Letters* 30 (20): 2793–2795.

72 Lindberg, H., Larsson, A., and Strassner, M. (2005). Single-frequency operation of a high-power, long-wavelength semiconductor disk laser. *Optics Letters* 30 (17): 2260–2262.

73 Symonds, C., Dion, J., Sagnes, I. et al. (2004). High performance 1.55 μm vertical external cavity surface emitting laser with broadband integrated dielectric-metal mirror. *Electronics Letters* 40 (12): 734–735.

74 Tourrenc, J., Bouchoule, S., Khadour, A. et al. (2007). High-power single-longitudinal-mode VECSEL at 1.55 μm with an hybrid metal-metamorphic Bragg mirror. *Electronics Letters* 43 (14): 754–755.

75 Tourrenc, J.-P., Bouchoule, S., Khadour, A. et al. (2008). Thermal optimization of 1.55 μm OP-VECSEL with hybrid metal-metamorphic mirror for single-mode high power operation. *Optical and Quantum Electronics* 40: 155–165.

76 Baili, G., Morvan, L., Pillet, G. et al. (2014). Ultralow noise and high-power VECSELfor high dynamic range and broadband RF/optical links. *Journal of Lightwave Technology* 32 (20): 3489–3494.

77 Baili, G., Alouini, M., Dolfi, D. et al. (2007). Shot-noise-limited operation of a monomode high-cavity-finesse semiconductor laser for microwave photonics applications. *Optics Letters* 32 (6): 650–652.

78 Pes, S., Paranthoen, C., Levallois, C. et al. (2017). Class-A operation of an optically-pumped 1.6 μm-emitting quantum dash-based vertical-external-cavity surface-emitting laser on InP. *Optics Express* 25 (10): 11760–11766.

79 Baili, G., Morvan, L., Alouini, M. et al. (2009). Experimental demonstration of a tunable dual-frequency semiconductor laser free of relaxation oscillations. *Optics Letters* 34 (21): 3421–3423.

80 De, S., Baili, G., Alouini, M. et al. (2014). Class-A dual-frequency VECSEL at telecom wavelength. *Optics Letters* 39 (19): 5586–5589.

81 De, S., Baili, G., Bouchoule, S. et al. (2015). Intensity-and phase-noise correlations in a dual-frequency vertical-external-cavity surface-emitting laser operating at telecom wavelength. *Physical Review A* 91 (5): 053828.

82 Chaccour, L., Aubin, G., Merghem, K. et al. (2016). Cross-polarized dual-frequency VECSEL at 1.5 μm for fiber-based sensing applications. *IEEE Photonics Journal* 8 (6): 1–10.

83 Khadour, A., Bouchoule, S., Aubin, G. et al. (2010). Ultrashort pulse generation from 1.56 μm mode-locked VECSEL at room temperature. *Optics Express* 18 (19): 19902–19913.

84 Zhao, Z., Bouchoule, S., Song, J. et al. (2011). Subpicosecond pulse generation from a 1.56 μm mode-locked VECSEL. *Optics Letters* 36 (22): 4377–4379.

85 Devautour, M., Michon, A., Beaudoin, G. et al. (2013). Thermal management for high-power single-frequency tunable diode-pumped VECSEL emitting in the near- and mid-IR. *IEEE Journal of Selected Topics in Quantum Electronics* 19 (4): 1701108.

86 Hader, J., Wang, T.-L., Yarborough, J. et al. (2011). VECSEL optimization using microscopic many-body physics. *IEEE Journal of Selected Topics in Quantum Electronics* 17 (6): 1753–1762.

87 Wang, T.-L., Kaneda, Y., Hader, J. et al. (2012). Strategies for power scaling VECSELs. *Proceedings of SPIE* 8242: 824209.

88 Gbele, K., Laurain, A., Hader, J. et al. (2016). Design and fabrication of hybrid metal semiconductor mirror for high-power VECSEL. *IEEE Photonics Technology Letters* 28 (7): 732–735.

89 Keller, S., Sirbu, A., Iakovlev, V. et al. (2015). 8.5 W VECSEL output at 1270 nm with conversion efficiency of 59%. *Optics Express* 23 (13): 17 437–17 442.

90 Rantamäki, A., Saarinen, E., Lyytikäinen, J. et al. (2014). High power semiconductor disk laser with a semiconductor-dielectric-metal compound mirror. *Applied Physics Letters* 104 (10): 101110.

91 Govorkov, S. and Austin, R. (2012). Optically-pumped external-cavity surface-emitting semiconductor lasers with front-cooled gain-structures. US Patent US8170073 B2, 1 May 2012.

92 Jasik, A., Wierzchowski, W., Muszalski, J. et al. (2009). The reduction of the misfit dislocation in non-doped AlAs/GaAs DBRs. *Journal of Crystal Growth* 311 (16): 3975–3977.

93 Bedford, R. and Fallahi, M. (2004). Analysis of high-reflectivity metal-dielectric mirrors for edge-emitting lasers. *Optics Letters* 29 (9): 1010–1012.

94 Zhang, Z., Torizuka, K., Itatani, T. et al. (1998). Broadband semiconductor saturable-absorber mirror for a self-starting mode-locked Cr:Forsterite laser. *Optics Letters* 23 (18): 1465–1467.

95 Zhang, Z., Nakagawa, T., Torizuka, K. et al. (1999). Self-starting mode-locked Cr^{4+}:YAG laser with a low-loss broadband semiconductor saturable-absorber mirror. *Optics Letters* 24 (23): 1768–1770.

96 Zhang, Z., Nakagawa, T., Torizuka, K. et al. (2000). Gold-reflector-based semiconductor saturable absorber mirror for femtosecond mode-locked Cr^{4+}:YAG lasers. *Applied Physics B* 70 (1): S59–S62.

97 Zhang, Z., Nakagawa, T., Takada, H. et al. (2000). Low-loss broadband semiconductor saturable absorber mirror for mode-locked Ti:sapphire lasers. *Optics Communications* 176 (1): 171–175.

98 Lloyd, J., Clemens, J., and Snede, R. (1999). Copper metallization reliability. *Microelectronics Reliability* 39 (11): 1595–1602.

99 Lloyd, R. (1998). Reliability of Copper Metallization. Technology Report 514. Lloyd Technololgy Associates, Inc., Katonah, NY, USA.

100 Rantamäki, A., Saarinen, E., Lyytikäinen, J. et al. (2015). Towards high power flip-chip long-wavelength semiconductor disk lasers. *Proceedings of SPIE* 9349: 934908.

101 Rantamäki, A., Saarinen, E., Lyytikäinen, J. et al. (2015). Thermal management in long-wavelength flip-chip semiconductor disk lasers. *IEEE Journal of Selected Topics in Quantum Electronics* 21 (6): 1501507.

102 Zydzik, G., Van Uitert, L., Singh, S., and Kyle, T. (1977). Strong adhesion of vacuum-evaporated gold to oxide or glass substrates. *Applied Physics Letters* 31 (10): 697–699.

103 Guenther, A. and McIver, J. (1988). The role of thermal conductivity in the pulsed laser damage sensitivity of optical thin films. *Thin Solid Films* 163: 203–214.

104 Caprara, A., Chilla, J., and Spinelli, L. (2010). High-power external-cavity optically-pumped semiconductor lasers. Patent US 7,643,530 B2, 5 January 2010.

105 Rantamäki, A., Lyytikäinen, J., Heikkinen, J. et al. (2013). Multi-watt semiconductor disk laser by low temperature wafer bonding. *IEEE Photonics Technology Letters* 25 (22): 2233–2235.

106 Govorkov, S. and Spinelli, L. (2013). Optically-pumped surface-emitting semiconductor laser with heat-spreading compound mirror-structure. US Patent 8,611,383, 17 December 2013.

107 Kim, K.-S., Yoo, J., Kim, G. et al. (2007). Enhancement of pumping efficiency in a vertical-external-cavity surface-emitting laser. *IEEE Photonics Technology Letters* 19 (23): 1925–1927.

108 Demaria, F., Lorch, S., Menzel, S. et al. (2009). Design of highly efficient high-power optically pumped semiconductor disk lasers. *IEEE Journal of Selected Topics in Quantum Electronics* 3 (15): 973–977.

109 Heikkinen, J., Gumenyuk, R., Rantamäki, A. et al. (2014). A 1.33 μm picosecond pulse generator based on semiconductor disk mode-locked laser and bismuth fiber amplifier. *Optics Express* 22 (10): 11446–11455.

110 Vetter, S. and Calvez, S. (2012). Thermal management of near-infrared semiconductor disk lasers with AlGaAs mirrors and lattice (mis)matched active regions. *IEEE Journal of Quantum Electronics* 48 (3): 345–352.

111 Sirbu, A., Rantamäki, A., Saarinen, E. et al. (2014). High performance wafer-fused semiconductor disk lasers emitting in the 1300 nm waveband. *Optics Express* 22 (24): 29398–29403.

112 Rautiainen, J., Lyytikäinen, J., Sirbu, A. et al. (2008). 2.6 W optically-pumped semiconductor disk laser operating at 1.57-μm using wafer fusion. *Optics Express* 16 (26): 21 881–21 886.

113 Lyytikäinen, J., Rautiainen, J., Toikkanen, L. et al. (2009). 1.3-μm optically-pumped semiconductor disk by wafer fusion. *Optics Express* 17 (11): 9047–9052.

114 Dragoi, V. (2006). From magic to technology: materials integration by wafer bonding. *SPIE Proceedings* 6123: 612314.

115 Moutanabbir, O. and Gösele, U. (2010). Heterogeneous integration of compound semiconductors. *Annual Review of Materials Research* 40: 469–500.

116 Brightup, S. and Goorsky, M. (2010). Chemical-mechanical polishing for III–V wafer bonding applications: polishing, roughness, and an abrasive-free polishing model. *ECS Transactions* 33 (4): 383–389.

117 Dragoi, V., Farrens, S., and Lindner, P. (2005). Plasma activated wafer bonding for MEMS. *Proceedings of SPIE* 5116: 179–187.

118 Dragoi, V., Mittendorfer, G., Thanner, C., and Lindner, P. (2007). Plasma-activated wafer bonding: the new low-temperature tool for MEMS fabrication. *Proceedings of SPIE* 6589: 65890T.

119 Kern, W. (1990). The evolution of silicon wafer cleaning technology. *Journal of the Electrochemical Society* 137 (6): 1887–1892.

120 Pasquariello, D. and Hjort, K. (2002). Plasma-assisted InP-to-Si low temperature wafer bonding. *IEEE Journal of Selected Topics in Quantum Electronics* 8 (1): 118–131.

121 Fecioru, A. (2006). Low temperature uhv bonding with laser pre-cleaning. PhD thesis. Martin-Luther-Univ. Halle-Wittenberg.

122 McIntyre, N., Davidson, R., Walzak, T. et al. (1991). Uses of ultraviolet/ozone for hydrocarbon removal: applications to surfaces of complex composition or geometry. *Journal of Vacuum Science & Technology A: Vacuum, Surfaces, and Films* 9 (3): 1355–1359.

123 Eibelhuber, M., Flötgen, C., and Lindner, P. (2015). Optimising devices with wafer bonding. *Compound Semiconductor* 21 (1): 34–38.

124 Akatsu, T., Plößl, A., Scholz, R. et al. (2001). Wafer bonding of different III–V compound semiconductors by atomic hydrogen surface cleaning. *Journal of Applied Physics* 90 (8): 3856–3862.

125 Takagi, H. and Maeda, R. (2006). Direct bonding of two crystal substrates at room temperature by Ar-beam surface activation. *Journal of Crystal Growth* 292 (2): 429–432.

126 Tong, Q.-Y., Fountain, G., and Enquist, P. (2006). Room temperature SiO_2/SiO_2 covalent bonding. *Applied Physics Letters* 89 (4): 042110.

127 Dragoi, V. and Farrens, S. (2005). Low temperature MEMS manufacturing processes: plasma activated wafer bonding. *Materials Research Society Symposium Proceedings* 872: J7.1.1.

128 Plößl, A. and Kräuter, G. (1999). Wafer direct bonding: tailoring adhesion between brittle materials. *Materials Science and Engineering: R: Reports* 25 (1): 1–88.

129 Matthias, T., Dragoi, V., and Lindner, P. (2005). Aligned fusion wafer bonding for wafer-level packaging and 3D integration. *Proceedings of International Symposium On Microelectronics (IMAPS)*: 715–725.

130 Christiansen, S., Singh, R., and Gösele, U. (2006). Wafer direct bonding: from advanced substrate engineering to future applications in micro/nanoelectronics. *Proceedings of the IEEE* 94 (12): 2060–2106.

131 Gösele, U. and Tong, Q.-Y. (1998). Semiconductor wafer bonding. *Annual Review of Materials Science* 28 (1): 215–241.

132 Liau, Z. (2000). Semiconductor wafer bonding via liquid capillarity. *Applied Physics Letters* 77 (5): 651–653.

133 Higurashi, E. and Suga, T. (2016). Review of low-temperature bonding technologies and their application in optoelectronic devices. *Electronics and Communications in Japan* 99 (3): 63–71.

134 Reznicek, A., Senz, S., Breitenstein, O. et al. (2002). Electrical and structural investigation of bonded silicon interfaces. *Electrochemical Society Proceedings. Semiconductor Wafer Bonding: Science, Technology, and Applications VI* 2001-27: 114–125.

135 Tong, Q.-Y., Lee, T.-H., Gösele, U. et al. (1997). The role of surface chemistry in bonding of standard silicon wafers. *Journal of The Electrochemical Society* 144 (1): 384–389.

136 Tong, Q.-Y., Schmidt, E., Gösele, U., and Reiche, M. (1994). Hydrophobic silicon wafer bonding. *Applied Physics Letters* 64 (5): 625–627.

137 Tong, Q.-Y. and Gösele, U. (1999). Wafer bonding and layer splitting for microsystems. *Advanced Materials* 11 (17): 1409–1425.

138 Liau, Z. and Mull, D. (1990). Wafer fusion: a novel technique for optoelectronic device fabrication and monolithic integration. *Applied Physics Letters* 56 (8): 737–739.

139 Lo, Y., Bhat, R., Hwang, D. et al. (1991). Bonding by atomic rearrangement of InP/InGaAsP 1.5 µm wavelength lasers on GaAs substrates. *Applied Physics Letters* 58 (18): 1961–1963.

140 Black, A., Hawkins, A., Margalit, N. et al. (1997). Wafer fusion: materials issues and device results. *IEEE Journal of Selected Topics in Quantum Electronics* 3 (3): 943–951.

141 Horng, R., Peng, W., Wuu, D. et al. (2002). Surface treatment and electrical properties of directly wafer-bonded InP epilayer on GaAs substrate. *Solid-State Electronics* 46 (8): 1103–1108.

142 Ram, R., Dudley, J., Bowers, J. et al. (1995). GaAs to InP wafer fusion. *Journal of Applied Physics* 78 (6): 4227–4237.

143 Ogawa, S., Imada, M., and Noda, S. (2003). Analysis of thermal stress in wafer bonding of dissimilar materials for the introduction of an InP-based light emitter into a GaAs-based three-dimensional photonic crystal. *Applied Physics Letters* 82 (20): 3406–3408.

144 Sirbu, A., Mereuta, A., Caliman, A. et al. (2011). High power optically pumped VECSELs emitting in the 1310 nm and 1550 nm wavebands. *Proceedings of SPIE* 7355: 791903.

145 Babic, D., Dudley, J., Streubel, K. et al. (1994). Optically pumped all-epitaxial wafer-fused 1.52 µm vertical-cavity lasers. *Electronics Letters* 30 (9): 704–706.

146 Dudley, J., Ishikawa, M., Babic, D. et al. (1992). 144 °C operation of 1.3 µm InGaAsP vertical cavity lasers on GaAs substrates. *Applied Physics Letters* 61 (26): 3095–3097.

147 Rantamäki, A., Sirbu, A., Mereuta, A. et al. (2010). 3 W of 650 nm red emission by frequency doubling of wafer-fused semiconductor disk laser. *Optics Express* 18 (21): 21 645–21 650.

148 Lyytikäinen, J., Rautiainen, J., Sirbu, A. et al. (2011). High-power 1.48-μm wafer-fused optically pumped semiconductor disk laser. *IEEE Photonics Technology Letters* 23 (13): 917–919.

149 Rantamäki, A., Rautiainen, J., Lyytikäinen, J. et al. (2012). 1 W at 785 nm from a frequency-doubled wafer-fused semiconductor disk laser. *Optics Express* 20 (8): 9046–9051.

150 Rantamäki, A., Sirbu, A., Saarinen, E. et al. (2014). High-power flip-chip semiconductor disk laser in the 1.3 μm wavelength band. *Optics Letters* 39 (16): 4855–4858.

151 Leinonen, T., Iakovlev, V., Sirbu, A. et al. (2017). 33 W continuous output power semiconductor disk laser emitting at 1275 nm. *Optics Express* 25 (6): 7008–7013.

152 Mereuta, A., Nechay, K., Caliman, A. et al. (2019). Flip-chip wafer-fused OP-VECSELs emitting 3.65 W at the 1.55-μm waveband. *IEEE Journal of Selected Topics in Quantum Electronics* 25 (6): 1–5.

153 Nechay, K., Mereuta, A., Paranthoen, C. et al. (2019). InAs/InP quantum dot VECSEL emitting at 1.5 μm. *Applied Physics Letters* 115: 171105.

154 Paranthoen, C., Bertru, N., Dehaese, O. et al. (2001). Height dispersion control of InAs/InP quantum dots emitting at 1.55 μm. *Applied Physics Letters* 78 (12): 1751–1753.

155 Johnson, K., Hibbs-Brenner, M., Hogan, W., and Dummer, M. (2012). Advances in red VCSEL technology. *Advances in Optical Technologies* 2012: 1–13.

156 Erbert, G., Bugge, F., Knauer, A. et al. (1999). High-power tensile-strained GaAsP-AlGaAs quantum-well lasers emitting between 715 and 790 nm. *IEEE Journal of Selected Topics in Quantum Electronics* 5 (3): 780–784.

157 Mawst, L., Rusli, S., Al-Muhanna, A., and Wade, K. (1999). Short-wavelength (0.7 μm < λ < 0.78 μm) high-power InGaAsP-active diode lasers. *IEEE Journal of Selected Topics in Quantum Electronics* 5 (3): 785–791.

158 Hastie, J., Calvez, S., Dawson, M. et al. (2005). High power CW red VECSEL with linearly polarized TEM_{00} output beam. *Optics Express* 13 (1): 77–81.

159 Mateo, C., Brauch, U., Kahle, H. et al. (2016). 2.5 W continuous wave output at 665 nm from a multipass and quantum-well-pumped AlGaInP vertical-external-cavity surface-emitting laser. *Optics Letters* 41 (6): 1245–1248.

160 Kahle, H., Mateo, C., Brauch, U. et al. (2016). Semiconductor membrane external-cavity surface-emitting laser (MECSEL). *Optica* 3 (12): 1506–1512.

161 Saarinen, E., Lyytikäinen, J., Ranta, S. et al. (2015). 750 nm 1.5 W frequency-doubled semiconductor disk laser with a 44 nm tuning range. *Optics Letters* 40 (19): 4380–4383.

162 Rantamäki, A., Rautiainen, J., Sirbu, A. et al. (2013). 1.56 μm 1 watt single frequency semiconductor disk laser. *Optics Express* 21 (2): 2355–2360.

163 Rautiainen, J., Lyytikäinen, J., Toikkanen, L. et al. (2010). 1.3-μm mode-locked disk laser with wafer fused gain and SESAM structures. *IEEE Photonics Technology Letters* 22 (11): 748–750.

164 Saarinen, E., Puustinen, J., Sirbu, A. et al. (2009). Power-scalable 1.57 μm mode-locked semiconductor disk laser using wafer fusion. *Optics Letters* 34 (20): 3139–3141.

165 Chamorovskiy, A., Kerttula, J., Rautiainen, J. and Okhotnikov, O. (2012). Supercontinuum generation with amplified 1 57 μm picosecond semiconductor disk laser. *Electronics Letters* 48 (16): 1010–1012.

166 Chamorovskiy, A., Rautiainen, J., Rantamäki, A., and Okhotnikov, O. (2011). Raman fiber oscillators and amplifiers pumped by semiconductor disk lasers. *IEEE Journal of Quantum Electronics* 47 (9): 1201–1207.

167 Chamorovskiy, A. and Okhotnikov, O. (2012). Nonlinear fibre-optic devices pumped by semiconductor disk lasers. *Quantum Electronics* 42 (11): 964.

168 Chamorovskiy, A., Rantamäki, A., Sirbu, A. et al. (2010). 1.38-μm mode-locked Raman fiber laser pumped by semiconductor disk laser. *Optics Express* 18 (23): 23872–23877.

169 Chamorovskiy, A., Rautiainen, J., Lyytikäinen J. et al. (2010). Raman fiber laser pumped by a semiconductor disk laser and mode locked by a semiconductor saturable absorber mirror. *Optics Letters* 35 (20): 3529–3531.

170 Alexe, M., Dragoi, V., Reiche, M., and Gosele, U. (2000). Low temperature GaAs/Si direct wafer bonding. *Electronics Letters* 36 (7): 677–678.

171 Radu, I. (2010). In memoriam Ulrich Gösele: wafer bonding á la carte. *Meeting Abstracts, the Electrochemical Society* 27: 1742–1742.

172 Dragoi, V. and Lindner, P. (2006). Plasma activated wafer bonding of silicon: in situ and ex situ processes. *ECS Transactions* 3 (6): 147–154.

173 Plach, T., Hingerl, K., Tollabimazraehno, S. et al. (2013). Mechanisms for room temperature direct wafer bonding. *Journal of Applied Physics* 113 (9): 094905.

174 Lin, X., Liao, G., Tang, Z., and Shi, T. (2009). UV surface exposure for low temperature hydrophilic silicon direct bonding. *Microsystem Technologies* 15 (2): 317–321.

175 Holl, S., Colinge, C., Hobart, K., and Kub, F. (2006). UV activation treatment for hydrophobic wafer bonding. *Journal of the Electrochemical Society* 153 (7): G613–G616.

176 Bellini, E. (2017). Fraunhofer ISE and EVG set 31.3% efficiency for silicon multi-junction solar cell. *PV Magazine* 24: 03.

177 Rantamäki, A., Lindfors, J., Silvennoinen, M. et al. (2013). Low temperature gold-to-gold bonded semiconductor disk laser. *IEEE Photonics Technology Letters* 25 (11): 1062–1065.

178 Bower, R., Ismail, M., and Roberds, B. (1993). Low temperature Si_3N_4 direct bonding. *Applied Physics Letters* 62 (26): 3485–3487.

179 Liang, D., Fang, A., Park, H. et al. (2008). Low-temperature, strong SiO_2–SiO_2 covalent wafer bonding for III–V compound semiconductors-to-silicon photonic integrated circuits. *Journal of Electronic Materials* 37 (10): 1552–1559.

180 Kräuter, G., Bluhm, Y., Batz-Sohn, C., and Gösele, U. (1999). The joining of parallel plates via organic monolayers: chemical reactions in a spatially confined system. *Advanced Materials* 11 (12): 1035–1038.

181 Bakish, I., Artel, V., Ilovitsh, T. et al. (2012). Self-assembled monolayer assisted bonding of Si and InP. *Optical Materials Express* 2 (8): 1141–1148.

182 Zhang, W.-Y., Labukas, J., Tatic-Lucic, S. et al. (2005). Novel room-temperature first-level packaging process for microscale devices. *Sensors and Actuators A: Physical* 123: 646–654.

183 Mahapatro, A., Scott, A., Manning, A., and Janes, D. (2006). Gold surface with sub-nm roughness realized by evaporation on a molecular adhesion monolayer. *Applied Physics Letters* 88 (15): 151917.

184 Sirbu, A., Pierscinski, K., Mereuta, A. et al. (2014). Wafer-fused VECSELs emitting in the 1310 nm waveband. *Proceedings of SPIE* 8966: 89660G.

185 Su, K., Huang, S., Li, A. et al. (2006). High-peak-power AlGaInAs quantum-well 1.3-μm laser pumped by a diode-pumped actively Q-switched solid-state laser. *Optics Letters* 31 (13): 2009–2011.

186 Zah, C.-E., Bhat, R., Pathak, B. et al. (1994). High-performance uncooled 1.3-μm $Al_xGa_yIn_{1-x-y}$/As/InP strained-layer quantum-well lasers for subscriber loop applications. *IEEE Journal of Quantum Electronics* 30 (2): 511–523.

187 Sanchez, F., Brunel, M., and Aït-Ameur, K. (1998). Pump-saturation effects in end-pumped solid-state lasers. *Journal of Optical Society of America B* 15 (9): 2390–2394.

188 Huang, S., Chang, H., Su, K. et al. (2009). AlGaInAs/InP eye-safe laser pumped by a Q-switched $Nd:GdVO_4$ laser. *Applied Physics B* 94 (3): 483–487.

189 Chang, H., Huang, S., Chen, Y.-F. et al. (2009). Efficient high-peak-power AlGaInAs eye-safe wavelength disk laser with optical in-well pumping. *Optics Express* 17 (14): 11409–11414.

190 Chen, Y.-F., Lee, Y., Huang, S. et al. (2012). AlGaInAs multiple-quantum-well 1.2-μm semiconductor laser in-well pumped by an Yb-doped pulsed fiber amplifier. *Applied Physics B* 106 (1): 57–62.

191 Chen, Y.-F., Su, K., Chen, W. et al. (2012). High-peak-power optically pumped AlGaInAs eye-safe laser at 500-kHz repetition rate with an intracavity diamond heat spreader. *Applied Physics B* 108 (2): 319–323.

192 Wen, C.-P., Tuan, P.-H., Liang, H.-C. et al. (2015). Compact high-peak-power end-pumped AlGaInAs eye-safe laser with a heat-spreader diamond coated as a cavity mirror. *IEEE Journal of Selected Topics in Quantum Electronics* 21 (1): 148–152.

193 Wen, C., Tuan, P., Liang, H. et al. (2015). High-peak-power optically-pumped AlGaInAs eye-safe laser with a silicon wafer as an output coupler: comparison between the stack cavity and the separate cavity. *Optics Express* 23 (24): 30749–30754.

3

Single-frequency and High Power Operation of 2–3 Micron VECSEL

Marcel Rattunde, Peter Holl, and Joachim Wagner

Department Optoelectronics, Fraunhofer Institute for Applied Solid State Physics IAF, Freiburg, Germany

3.1 Introduction

By exploiting the group-III antimonide material system, the emission wavelength of vertical external-cavity surface-emitting lasers (VECSELs) can be extended further to the mid-infrared (MIR) region so that high-brightness laser sources in the 2–3 µm wavelength range can be built. Although GaAs and InP and related Al and/or In containing alloys are considered to be more mature semiconductor materials, there were major improvements with GaSb-based VECSELs in recent years. Progress has been made both in terms of efficiency and output power of the semiconductor gain chip itself as well as in terms of demonstration of several setups for low-noise, tunable, single-frequency laser emission.

There are numerous applications that can benefit from these improvements in 2–3 µm VECSEL performance. Because of the strong absorption lines of a variety of relevant gases in this wavelength range, high resolution spectroscopic gas detection for industrial process monitoring or environmental control can be realized using such lasers [1, 2]. The ability to realize low-noise, tunable 2–3 µm laser sources with very narrow emission linewidths in the 100 kHz regime [3] makes these lasers ideal tools for quantum optic experiments [4], heterodyne-based light detection and ranging (LIDAR), or the use as seed lasers for high-power solid-state lasers in the 2–3 µm regime, such as holmium- or thulium-based laser systems [5]. Besides, GaSb-based VECSELs have proven to be able to deliver high output powers (up to 20 W in continuous-wave (CW) operation in a lab experiment [6]), enabling applications like material processing (e.g. processing of transparent plastic material), medical surgery (laser scalpel), or infrared countermeasures. Further on, due to the good beam quality, 2.X µm VECSEL can be used as pump source for solid-state laser material like Cr:ZnSe [7] or holmium [8, 9] or provide pump energy to nonlinear optical materials like periodically poled GaAs or zinc germanium phosphide (ZGP) in order to realize tunable coherent sources at even longer emission wavelength (4–10 µm range). The latter two materials benefit especially from the long-wavelength (>2 µm) pumping a GaSb-based VECSEL can provide, as their optical losses greatly decrease with pump wavelength [10].

Vertical External Cavity Surface Emitting Lasers: VECSEL Technology and Applications, First Edition.
Edited by Michael Jetter and Peter Michler.
© 2022 WILEY-VCH GmbH. Published 2022 by WILEY-VCH GmbH.

This chapter provides a basic overview of MIR VECSELs based on the (AlGaIn)(AsSb) materials system and introduces recent key developments concerning conversion efficiency and output power as well as single-frequency operation (SFO) and tunability of such laser sources.

In the following section, a general overview of semiconductor-based lasers for the >2 µm MIR wavelength range is given as an introduction. Then, some key aspects of the (AlGaIn)(AsSb) material system are presented in Section 3.3, followed by design principles for GaSb-based VECSELs and achievable laser efficiencies in Section 3.4. Section 3.5 reviews the different VECSEL-chip mounting technologies and output power levels that can be reached for the different designs and mounting options. Section 3.6 finally details different ways to achieve low-noise SFO with different key requirements.

3.2 Semiconductor Lasers for the MIR Range

Beside the GaSb-based VECSELs, there exist now a plurality of semiconductor-based laser sources for the MIR wavelength range, each with its own strength and weaknesses and specific performance limitations. The key parameters of these different sources are discussed in the following, revealing also the unique possibilities of the GaSb-based VECSEL sources.

The most advanced and commercially available technology are the electrically driven GaSb-based type-I quantum well (QW) semiconductor diode lasers [11, 12]. They span the wavelength range from 1.9 to above 3.7 µm emission wavelength [13, 14] with a distinct drop in performance toward longer emission wavelengths. This behavior can be seen in the internal laser parameters [15] but also on the maximum achievable output power in CW-operation at room temperature, as plotted in Figure 3.1 (full black squares): for broad area lasers with a typical width of the emitting facet of 100–150 µm (which implies a poor beam quality), output power values close to 2 W are achievable around 2 µm. Toward longer emission wavelengths, these values drop to 100 mW at 3.3 µm. Below 1.9 µm, InP-based diode lasers show a better performance, and above approximately 3.4 µm, the InP-based quantum cascade lasers (QCL, stars in Figure 3.1) are the best performing semiconductor laser variant with the highest available output powers [16, 17].

An alternative approach also based on cascaded sections but with a type-II QW interband active region are the interband cascade lasers (ICLs) [18]. Due to improvements such as the "W" structure in the active region [19] and fine-tuning of the electron and hole injection [20], very low threshold current devices were achieved (much lower than the QCL counterparts) with moderate output powers (blue downward triangles in Figure 3.1). And recently, a cascading scheme was also successfully implemented for type-I QW diode lasers: Using a GaSb/InAs tunnel junction, the cascaded type-I QW GaSb-based diode laser was introduced in [21], with impressive performance (see green upward triangles in Figure 3.1). Most noticeably they could improve the performance compared to the "standard" GaSb type-I diode lasers in the longer wavelength side with e.g. close to 1 W output power at around 3 µm emission wavelength [22]. This result was obtained from 100 μm wide single emitters, i.e. a diode laser variant with a very poor beam quality. In order to achieve a beam quality parameter M^2 close to the Gaussian limit of 1, the width of the electrically pumped emitter has to be drastically reduced to the order of 4–10 μm, reducing the available output power to the couple of 10 mW regime around 3 μm [22] and to ~70 mW at 2 μm [12].

Compared to these electrically driven MIR lasers described above, the work on GaSb-based type-I VECSELs started much later with the first realization published in 2004 [23]. For high power operation, an intracavity heatspreader (ICH) out of diamond or SiC is almost exclusively used [24], resulting in a maximum of 17 W CW output power at room temperature at 2.0 μm [25] and still around 0.8 W at 2.8 μm [9] (see red circles in Figure 3.1). This means that the GaSb-based VECSELs are the most powerful semiconductor-based single emitters in this wavelength range. Moreover, as the beam quality of the VECSELs is far superior to that of the broad area edge emitters, they are by far the lasers with the highest brightness.

So far, there has been no report of GaSb-based type-II VECSELs in order to expand the emission wavelength of these devices further to the long-wavelength side. But the VECSEL concept has also been realized using the II-VI material system with PbTe QWs and EuPbTe barriers. In this way, VECSEL in the 3–5.3 μm range have been realized with mW-range output power [26, 27] (see magenta crossed circles in Figure 3.1).

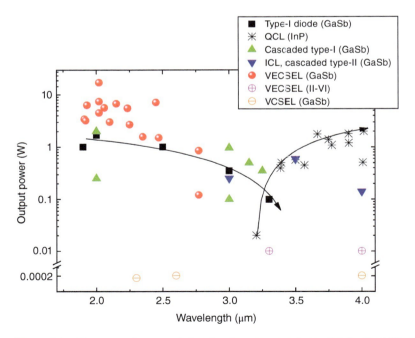

Figure 3.1 Overview of various semiconductor-based laser sources for the MIR regime around 2–4 μm emission wavelength: CW output power at room temperature versus emission wavelength (Refs. [9–31]).

Finally, yet importantly, it is worth to mention that also the electrically pumped version of the VECSEL, the VCSEL, has been realized in the GaSb-material system exploiting type-I [28, 29] and type-II [30, 31] QW active regions, resulting in sub-mW output powers in the 2.3–4 µm wavelength range (see orange open circles in Figure 3.1).

Apart from the QCL and the II-VI VECSEL, all other laser sources are based on the GaSb-material system, whose basic features will be presented in the next chapter with the focus on type-I QW active regions.

3.3 III-Sb Material System

The III-V compound semiconductor material system (AlGaIn)(AsSb) constitutes an ideal basis for the realization of semiconductor lasers at and above 2 µm emission wavelength. With only a few exceptions, GaSb is almost exclusively used as substrate material and starting point for the laser heterostructure.

Due to the higher bandgap compared to GaSb, quaternary $Al_xGa_{1-x}As_ySb_{1-y}$ is well suited for barrier, waveguide, window, or pump absorbing layers. Because of the large width of these layers, they are almost exclusively grown lattice matched to the GaSb-substrate, which is achieved by adding a small amount of As to the AlGaSb to form $Al_xGa_{1-x}As_ySb_{1-y}$ with $y = 0.08x$ [32]. The direct band gap for AlGaAsSb at 300 K, lattice matched to GaSb, is given by $E_g(\Gamma) = 2.297x + 0.727(1-x) - 0.48x(1-x)$ eV [32]. The refractive index for these layers can be found in [33]. For distributed Bragg reflectors (DBRs), $AlAs_{0.08}Sb_{0.92}$/GaSb layer pairs are the most efficient way to achieve the necessary high refractive index contrast and thus reflectivity.

For the active layer, $Ga_{1-x}In_xAs_ySb_{1-y}$ is used that has a direct bandgap for all alloy compositions and is latticed matched to GaSb if $y = 0.913x$ [32], i.e. the As-content is close to the In-content. For the quaternary material, the interpolation scheme introduced in [34] together with the material data from [32] have been proven to deliver material data that reproduce the experimental values quite well [35].

A detailed microscopic simulation revealed that almost ideal type-I QWs can be achieved in this material system [36]: compressively strained ternary $Ga_{0.8}In_{0.2}Sb$ QW embedded between $(Al_{0.30}Ga_{0.70})(As_{0.02}Sb_{0.98})$ barriers show a large spacing of the electronic subbands in the QW and almost equal electron and hole effective masses. This constitutes the ideal conditions for high optical gain, even superior to the well-known GaInAs/GaAs or GaInAs/AlGaAs QWs. These $Ga_{0.8}In_{0.2}Sb$ QWs yield an emission wavelength around 2.0 µm and in agreement to the finding above, very low lasing thresholds have been reported in the literature for diode lasers (50 A/cm² [37]) or VECSELs (0.9 kW cm^{-2} [38]) around 2 µm.

In order to increase the emission wavelength beyond 2 µm, the In-content in the GaInSb QWs can be increased further. But as this increases also the strain to values beyond the critical layer thickness for pseudomorphic growth, this strand is limited to emission wavelengths below approximately 2.2 µm. Beyond that, As has to be added that further reduced the bandgap and also the strain of the quaternary GaInAsSb QWs. Unfortunately, adding As also reduces the hole confinement as can be seen in Figure 3.2a. Here, the relative band alignment for

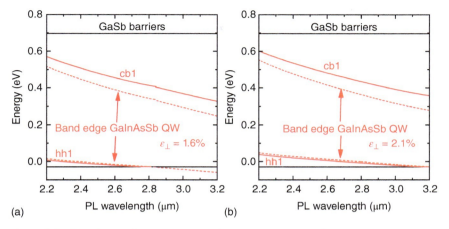

Figure 3.2 Relative position of the conduction and heavy hole band and the first electron and hole subbands (cb1, hh1) for a 10 nm wide GaInAsSb QWs embedded in GaSb barrier layers for a compressive strain of (a) $\varepsilon_\perp = 1.6\%$ and (b) $\varepsilon_\perp = 2.1\%$.

10 nm wide GaInAsSb QWs with a compressive strain of $\varepsilon_\perp = 1.6\%$ is plotted versus the simulated photoluminescence (PL) emission wavelength. Also shown are the band edges of the GaSb barriers. With increasing emission wavelength, the valence band offset ΔE_V is reduced and reaches 0 meV at approximately 2.8 μm. Depending on the carrier density inside the QW at and above threshold, a low band offset can lead to increased carrier leakage and successive recombination or absorption in the surrounding layers, limiting the laser performance. The hole confinement can be increased by using more strained QW material, as can be seen in Figure 3.2b. Here, the same parameters are plotted for a strain of $\varepsilon_\perp = 2.1\%$, leading to a valence band offset of 30 meV at 2.8 μm.

Another option to increase a low valence band offset in GaSb-based type-I QWs is the increase of the Al-content in the barrier layer, as illustrated in Figure 3.3a. The upper limit to that, at least for optical pumped devices such as VECSELs, is the

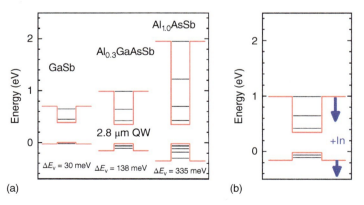

Figure 3.3 (a) Band edge profile of 2.8 μm GaInAsSb QWs embedded between different barrier material. (b) Schematic of the change in the relative band edge position for In incorporation into AlGaAsSb barrier.

photon energy of the pump source when the barriers should act as pump-absorbing layers (an alternative to that is the QW pumping as discussed in Section 3.4.2). From Figure 3.3a one can also clearly see that in all cases, the electron confinement is much higher than the hole confinement for these 2.8 μm QW, no matter what the Al-content is. In order to balance the offset and thus actually increase the valence band offset, Indium can be added to the barrier layers as can be seen schematically in Figure 3.3b. These quinternary (AlGaIn)(AsSb) barriers were successfully used in long-wavelength GaSb diode lasers [14] and electrically pumped VCSELs [39], but not in optically pumped VECSEL so far.

3.4 2–3 μm VECSEL Design

3.4.1 Standard Barrier Pumped Structures

The most widely used general structural layout for GaSb-based VECSEL in the 2–3 μm wavelength range is a barrier pumped structure. Around 5–15 type-I QW are used, which gain is maximized by using the concept of resonant periodic gain [40], i.e. by placing the QW in the antinodes of the standing wave of the microcavity, formed by the DBR and the top interface of the semiconductor gain chip toward the surroundings. The barriers and pump-absorbing spacer layers are normally made out of the same material, which is either GaSb or AlGaAsSb with an Al-content up to 35% [23] (see Figure 3.4a). With this material selection, readily available and powerful 808–980 nm diode lasers can be used for optical barrier pumping. The DBR is normally made out of quaterwave GaSb/AlAs$_{0.08}$Sb$_{0.92}$ layer pairs. In order to achieve the necessary high reflectivity of >99.8%, around 20 pairs are used (18.5–24.5 reported in the literature [15, 41, 44]). This number is lower than the typical number of DBR layer-pairs used in GaAs-based 1 μm VECSEL, which is due

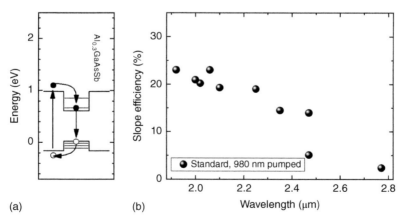

Figure 3.4 (a) Standard VECSEL design for a barrier-pumped active region. With an Al-content around 30% in the barrier layers, 980 nm pump diodes can be used. (b) Resulting slope efficiency in CW operation at room temperature vs. emission wavelength for various VECSEL structures with ICH from different groups [6, 41–43].

to the higher refractive index contrast that can be achieved in the Sb-based material system. The total thickness of the DBR section is, however, much greater as each quarterwave pair in the 2–3 µm range is much thicker than those designed for 1 µm. Further on, the thermal conductivity of the GaSb-based alloys is significantly lower than that of the GaAs-based counterparts. Therefore, heat-extraction through the DBR is limited in GaSb-based VECSEL. The use of an intracavity heatspreader (ICH) is thus the standard mounting option to maximize heat extraction in order to achieve high output powers in CW. Single-crystalline diamond or silicon carbide (SiC) are the most prominent materials for these ICHs, as they are transparent for the pump and laser wavelength and offer excellent thermal conductivity. All following data shown in this section are from VECSEL chips bonded to an ICH, unless otherwise stated (other and more advanced mounting options are discussed in Section 3.5).

For this standard design, Figure 3.4b summarizes the achievable slope efficiency of the VECSEL devices versus the emission wavelength. Output couplers with a transmission of 1–4% were used in these cases, optimized for the individual VECSEL structure. Similar to the behavior found for GaSb-based type-I diode lasers (see e.g. [15]), the efficiency degrades with emission wavelength. At around 2 µm emission wavelength, the lasers reach up to 23% slope efficiency [38], dropping to 14% around 2.5 µm [42] and 2% at 2.8 µm [45]. Similar to that, the maximum achievable output power in CW operation at room temperature drops from 4.5 W to 1.5 W and 0.12 W at 2.0 µm, 2.5 µm, and 2.8 µm, respectively. The threshold pump power density for these devices is quite low with reported values of ∼1 kW cm^{-2} [38], with the exception of the long-wavelength 2.8 µm VECSEL where the threshold increases significantly to 3 kW cm^{-2} [45]. As these VECSELs were pumped with 980 nm pump diodes, the quantum efficiency η_Q, i.e. the ratio between laser and pump photon energy $\eta_Q = h\nu_{laser}/h\nu_{pump} = \lambda_{pump}/\lambda_{laser}$, is rather small (i.e. the quantum deficit $(1 - \eta_Q)$ is high). The value for the quantum efficiency η_Q for the 980 nm pumped VECSEL is 50% at 2.0 µm, dropping to 35% at 2.8 µm. This means, that even for a perfect VECSEL without any carrier or optical losses, the achievable power slope efficiency is limited to the values given above. As this severely limits the achievable efficiency of these devices, two ways to increase the quantum efficiency are described in the next paragraphs.

3.4.2 In-well Pumping

If barrier pumping leads to a large difference between pump and laser wavelength, in-well pumping [46] can be a convenient way to reduce that difference and thus to increase the efficiency of GaS-based VECSEL [47]. As illustrated in Figure 3.5, this approach allows not only to increase the quantum efficiency η_Q but also to select the barrier material independent of the pump wavelength, so that the band offsets can be increased. In this way, in-well pumped 2.3 µm VECSEL have been realized [48] with 100% Al-content in the $Al_{1.0}AsSb$ barrier layers. The slope efficiency was successfully increased in that way to 32% at room temperature using a 1.96 µm Tm-fiber laser as the pump source ($\eta_Q = 87\%$ for this configuration).

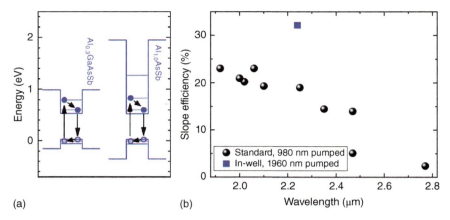

Figure 3.5 (a) In-well pumped VECSEL design for an increased quantum efficiency. The barrier height can be chosen independent of the pump wavelength. (b) Resulting slope efficiency for a 1960 nm in-well pumped structure (blue square), compared to the standard barrier pumped VECSELs [6, 41–43, 48].

Despite this positive result, this approach has some severe drawbacks: first, the low pump absorption in the QWs requires either a complex pump recycling optics [49] or a complex VECSEL structure with a double resonance design (i.e. internal pump recycling) [48]. If the latter concept is used, the resonance for the pump light strongly depends on the angle of incidence of the pump beam and the pump wavelength. Therefore, a pump source and a pump optic with a low numerical aperture (<0.07) and low spectral width (<10 nm) is required. Tm-fiber lasers can easily meet these criteria, but not 1.9–2.1 μm GaSb-based diode lasers or diode laser stacks, which would be the practical pump source for the in-well pumped VECSEL. On the other hand, the realization of the first concept (use of pump recycling optics) results in a bulky and complex laser setup.

3.4.3 Low Quantum Deficit Barrier Pumping

The two drawbacks mentioned above can be avoided with barrier pumping, exploiting the large absorption in the spacer region between the QW in the VECSEL active region. In order to increase the quantum efficiency for this pump configuration, the barrier bandgap has to be reduced. The binary GaSb can be used for this purpose, which bandgap (0.72 eV) coincidences well with readily available and powerful 1470 nm pump diodes (0.84 eV photon energy). This leads to a significant increase of the quantum efficiency (e.g. $\eta_Q = 73\%$ at 2.0 μm emission wavelength) but also to a reduction in the band offsets between the GaInAsSb QWs and the GaSb-barriers (see Figure 3.6a). Nevertheless, the slope efficiency was successfully increased for several of these "low-quantum deficit" (LQD) structures, as can be seen in Figure 3.6b, where the slope is again plotted against the emission wavelength and compared to the other two GaSb VECSEL active region concepts described above [6, 9, 42]. From 2.0 to 2.5 μm, a very high slope efficiency was reached with a value of around 30%,

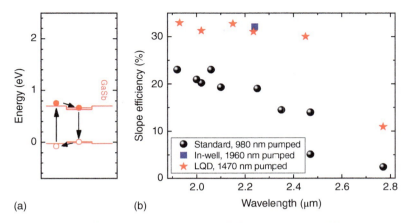

Figure 3.6 (a) LQD barrier pumping, using GaSb-barriers and 1470 nm pump diodes. (b) Resulting slope efficiency vs. emission wavelength for various LQD samples (red stars) in comparison to the results discussed above [6, 41–43, 48].

dropping to 11% at 2.8 μm. These values (red stars in Figure 3.6b) are well above the slope efficiencies of standard barrier pumped structures (black dots) and even similar to the slope of in-well pumped VECSELS (blue squares). Comparing the threshold pump power density of a 2.0 μm LQD structure and a standard 980 nm pumped VECSEL revealed almost unchanged values at room temperature. Only at elevated heatsink temperature >60 °C, the threshold power density increased more rapidly for the LQD structure, which is a hint of increased thermionic emission from the QWs to the barrier layers due to the reduced band offset.

The high achievable slope efficiency of the LQD-VECSEL samples translates in a much higher output power, as can be seen in Figure 3.7a. Here, the power characteristics of a standard 980 nm pumped (black line) and LQD 1470 nm pumped

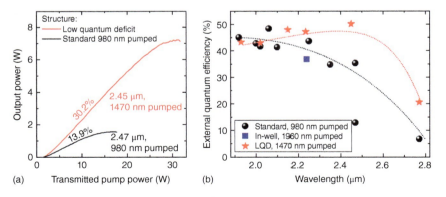

Figure 3.7 (a) Output power characteristics of a 2.5 μm LQD-VECSEL compared to a standard barrier pumped VECSEL, both with SiC intracavity heatspreader. (b) EQE of the LQD-VECSEL versus emission wavelength. The EQE is the slope efficiency of the laser characteristics divided by the quantum efficiency η_Q. This value describes the fraction of emitted photon per pump photon.

(red line) VECSEL structure, both emitting at 2.5 µm, are compared. Both samples exhibit a SiC intracavity heatspreader and are operated in CW mode at 20 °C heatsink temperature. The maximum output power is 1.6 W for the standard structure and above 7 W for the LQD structure, pumped at 1470 nm [42].

One reason for the degrading slope efficiency with emission wavelength, as seen in Figure 3.6b, is of course the decrease of the quantum efficiency with the laser wavelength for a fixed pump wavelength. A more direct comparison of the "pump-photon to laser-photon" conversion efficiency can be made when dividing the slope by the quantum efficiency, which results in the external quantum efficiency (EQE), plotted in Figure 3.7b. For the LQD-VECSEL structures (red stars) pumped at 1470 nm, this graph reveals a quite unusual behavior for GaSb-based devices: an initial increase of the EQE from 42% around 2.0 µm to 50% at the longer laser wavelength of 2.5 µm. This behavior can be attributed to an increase of the valence band offset ΔE_V with wavelength for these structures (42 meV for the 2.0 µm and 72 meV for the 2.5 µm structure, see [42]) due to an increase in QW strain. Together with the excellent performance of the 2.5 µm LQD-VECSEL, this leads to the conclusion that a valence band offset around $\Delta E_V = 70$ meV is sufficient for the hole confinement in GaSb-based devices. And even with a value of $\Delta E_V = 42$ meV as for the 2.0 µm LQD structure, excellent laser performance can be achieved, given that the carrier density at and above threshold is not too high. Structures with a high barrier offset are less affected by high QW carrier concentration, which are e.g. needed to overcome the losses for high output coupling. An example for that is the in-well pumped structure with $Al_{1.0}AsSb$ barriers [48] that can be operated efficiently at high outcoupling rates of 10%.

The impressive efficiencies of the 1470 nm pumped LQD-VECSEL structure are a pronounced improvement compared to the designs reported above. The high slope efficiencies above 30% (see Figure 3.6) and concomitant external quantum efficiencies in the 45–50% range (see Figure 3.7b) for 1.9–2.5 µm VECSELs lead to a reduced thermal load and thus higher possible output powers for a given mounting technology (see following section). However, even with this improvement, the efficiencies of these GaSb-based VECSELs is still inferior compared to state-of-the-art GaAs-based VECSEL around 1 µm emission wavelength [50] with a slope efficiency of 62% and an EQE of 78%. This is in fact the major difference between VECSELs from the two material systems and the main reason for the inferior performance of the GaSb-based 2.X µm devices compared to the 1 µm GaAs-based counterparts.

3.5 Mounting Technologies

GaSb-based VECSELs are generally thermally limited, as there is a large amount of waste energy to be dissipated, which heats up the chip and impair laser parameters. Therefore, efficient heat extraction from the VECSEL chip is crucial to enable high power operation. Two heat removal approaches are commonly used, the thin device and the intracavity heatspreader (ICH) approach [24]. They use different pathways to remove the heat from the active region (Figure 3.8): The thin device approach uses a

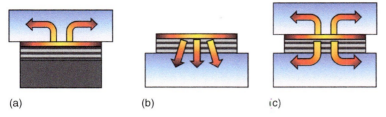

Figure 3.8 Three heat removal approaches and their dominant heat removal pathways: (a) intracavity heatspreader (ICH). (b) Thin device. (c) Double-sided heatspreader (DSH).

heatspreader on the backside of the chip, whereas the ICH is placed in direct contact to the front side of the VECSEL chip. The ICH is therefore in close proximity to the active region, where the majority of waste heat is generated and the heat removal to the front circumvents the DBR and potential remaining substrate. The thin device approach on the other hand removes the heat through the DBR to the backside of the chip, and special care has to be taken to minimize the thermal resistance of this pathway, which is especially challenging for GaSb-based DBR structures (see Section 3.5.2).

A rather new approach is the double-sided heatspreader (DSH), which is a combination of a front- and a backside heatspreader together with a thinned or removed substrate [6].

A crucial parameter for any heat management technique for VECSELs is its power scalability upon increasing the pump spot diameter. An ideally power scalable device has a constant temperature increase relative to the heatsink, as long as the pump power density in the pumped area is kept constant. This means, that even though larger pump spot diameters deposit more total power in the active region of the chip, the maximum temperature does not change.

In order to compare the different approaches described above, Figure 3.9 shows the results of a thermal simulation for a standard 980 nm pumped 2.0 μm GaSb-based VECSEL structure [6]. Plotted is the temperature increase relative to the heatsink temperature in dependence of the pump spot diameter for a constant pump power density of 20 kW cm^{-2}. This simulation reveals that the ICH (black line) has advantages over the thin device (red line) especially for small pump spot diameters. This is an effect of the three-dimensional (i.e. vertical and lateral) heat transfer in the front side heatspreader, which leads to a bottleneck in the lateral heat removal path for larger pump spot diameters. The thin device approach on the other hand largely depends on one-dimensional heat transfer to the backside of the chip and thus exhibits an almost ideal flat curve. The temperature rise is four times higher compared to the ICH approach for small pump spot diameters around 100 μm. On the other hand, for larger pump spots >700 μm, the thin device should be advantageous, even for these GaSb-based VECSELs. For this calculation, a complete removal of the underlying GaSb substrate was assumed. If a 30 μm thick layer of substrate material remains (dotted red line in Figure 3.9), the temperature increases slightly, but this "almost" thin device approach is still superior to the ICH for pump diameters above 1400 μm.

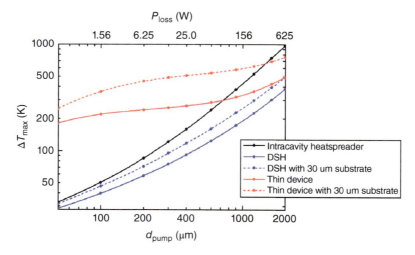

Figure 3.9 Simulated temperature increase relative to the heatsink temperature for different heat management approaches in dependence of the pump spot diameter, simulated for constant pump power density. A standard, 980 nm pumped 2.0 µm GaSb-VECSEL structure was used for the simulation.

The DSH approach (blue line in Figure 3.9) combines both heat removal pathways and therefore outperforms both approaches independent of the pump spot diameter. As above, the straight blue line was calculated assuming complete substrate removal and the dotted blue line assuming 30 µm of remaining substrate material. Also in the latter case, the DSH approach should outperform both other approaches and lead to a lower temperature rise inside the active region.

The realization and results for these three heat removal strands for GaSb-based VECSEL are discussed in the following sections.

3.5.1 Intracavity Heatspreader

As the DBR is a thermal bottleneck for GaSb-based VECSEL (see Section 3.4.1), the ICH approach is the most commonly used heat extraction method for these devices. In order to ensure optimum thermal contact to the chip, the heatspreader is bonded to the as-grown front surface of the VECSEL chip by liquid capillarity bonding [51]. The material used for the ICH has to fulfill several requirements: It has to be transparent for both the laser and pump wavelength, it has to be highly thermally conductive, and also it has to be available with an excellent surface quality and flatness in order to make liquid capillarity bonding possible. These requirements are fulfilled by diamond and SiC. The thermal conductivity of single-crystal diamond is much higher, but both the surface quality of commercially available, epi-ready SiC wafers as well as the availability of SiC are far superior. Also for emission wavelengths longer than 2.6 µm, only SiC can be used, as the absorption coefficient of diamond increases for higher wavelengths due to two-phonon absorption [52].

Figure 3.10 Emission spectra of an ICH sample for different heatsink temperatures. The used heatspreader was made from SiC and has a thickness of 500 μm.

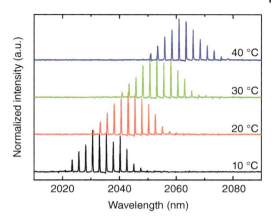

The ICH forms a low-resistivity pathway for the waste heat generated in the active region, but it also acts as an optical etalon and restricts the emission wavelengths to those emission maxima. The spectral position of these maxima is given by the optical thickness of the etalon, which can only be changed by varying the temperature of a given heatspreader sample. This leads to the typical spectrum of a sample with an ICH in high-power, multimode emission shown for different heatsink temperatures in Figure 3.10. The maximum of the envelope of the multitude of individual lasing modes shifts with temperature (∼1.3 nm K^{-1}) due to temperature dependence of the QW bandgap. The maximum of each individual etalon mode is also red-shifted upon increasing temperature due to the chance in the effective refractive index n. But this shift occurs at a much lower rate of ∼0.03 nm K^{-1}.

By using ICHs made from SiC, output powers up to 5.6 W for 980 nm pumped standard structures are possible (black open circles in Figure 3.11). The highest output powers are reached around 2.0 μm emission wavelength, whereas longer emission wavelengths result in decreasing output power. The 1470 nm pumped LQD structures show a different behavior (red filled circles in Figure 3.11) with rather constant

Figure 3.11 Output power vs. emission wavelength of VECSEL chips with ICH made from SiC (circles) and diamond (stars) at room temperature. Standard 980 nm pumped VECSEL structures (open black symbols) and LQD 1470 nm pumped structures (filled red symbols) are shown. All data was measured at room temperature [6, 9, 25, 43, 53].

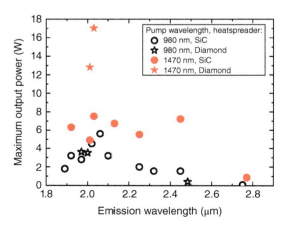

performance in the wavelength range between 2.0 and 2.5 µm, where maximum output powers between 6 and 7.5 W were possible.

Using a single-crystal diamond ICH can greatly reduce the thermal resistance of the setup (red stars in Figure 3.11), resulting in a CW output power as high as 17 W at room temperature for a 1470 nm LQD-VECSEL structure emitting at 2.0 µm [25]. The use of diamond heatspreaders is severely hampered by the limited availability and the resulting cost of high-quality samples with low optical loss, low birefringence, and an atomically flat surface.

3.5.2 Thin Device

The thin device approach relies on heat removal through the backside of the VECSEL chip, so it is beneficial to minimize the thermal resistance of this pathway by completely removing any remaining substrate and using a heatspreader with the highest possible thermal conductivity. As there are no constraints regarding the optical properties of such a backside heatspreader, commonly polycrystalline diamond is used, which is commercially available in large quantities and at reasonable cost and has a very good thermal conductivity.

The thin device setup is the state of the art for GaAs-based VECSELs emitting around 1 µm, where the bottom-up process is used in order to completely remove the substrate [50]. For this, the chip is grown in reversed order, i.e. first an etch stop layer, then the active region, and last the DBR, and then soldered DBR down onto the backside heatspreader. The last step is to remove the substrate from the front by chemical wet-etching [54]. This procedure is very reliable and reproducible when using the mature GaAs material system, where combinations of etch solution and etch stop materials with selectivities over 1 : 100 000 are available [55]. The highest output power of a VECSEL up to date of 106 W was demonstrated with a GaAs-based VECSEL using the thin device approach [50].

In the GaSb material system, the best available combinations of etch stop and etch solution have much lower selectivities around 1 : 100 [56], making the reverse order growth approach hard to realize. Nevertheless, this approach has been demonstrated for GaSb-based VECSELs [55, 57] with an output power of 12 mW out of a 120 µm diameter pumped spot for a 2.3 µm emitting VECSEL, barrier pumped at 980 nm.

An alternative approach to realize an "almost" thin device for GaSb-based VECSELs was introduced recently [58]. Here, the structure is grown in the standard sequence (substrate – DBR – active region), and only mechanical thinning is used in order to remove most of the substrate. As this process cannot be controlled with a sub-µm accuracy, a ~30 µm thick substrate layer is left in order to avoid any damage to the DBR. This thin device is subsequently soldered to a backside, polycrystalline diamond heatspreader. The power characteristic and emission spectra of a standard 980 nm pumped 2.0 µm VECSEL fabricated this way are shown in Figure 3.12. The minimum pump spot diameter, where this laser could be operated was ~200 µm. For smaller diameters, the device could not be operated in CW mode in contrast to the device with the substrate completely removed, described in the last paragraph [55]. This is a clear indication of the thermal impedance induced by the residual ~30 µm

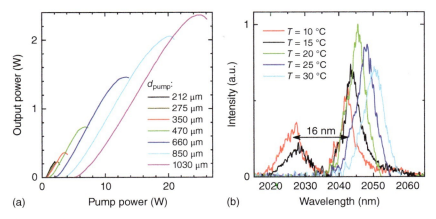

Figure 3.12 (a) Output power characteristics of a thin device for different pump spot diameters at 20 °C heatsink temperature. (b) Emission spectra of the thin device VECSEL at different heatsink temperatures. The visible substrate mode spacing corresponds to a substrate thickness of 31 µm.

thick substrate layer. But for larger pump diameters of >200 µm on the other hand, this device exhibits an almost ideal scalability with almost unchanged threshold pump power density, slope efficiency, and maximum output power density with increasing pump spot diameter. This is in accordance to the flat line in Figure 3.9 for this device (red dotted line) and further indicates a homogeneous removal of the GaSb substrate without any damage to the epitaxial layer sequence. For the maximum pump spot diameter of over 1000 µm, this results in a CW output power of 2.4 W for this "almost" thin device (see Figure 3.12a).

The remaining substrate layer on the backside of the DBR has effects on the emission spectra of the thin device chip in high power, multimode emission. The DBR and the gold metallization on the backside of the chip form a low-finesse cavity with a mode spacing depending on the GaSb substrate thickness. In the specific sample shown in Figure 3.12b, the modes separation correspond to a cavity length of 31 µm. This value is confirmed by thickness measurements of the chip, from which the thickness of the remaining substrate layer was determined to the same value.

3.5.3 Double-sided Heatspreader

The DSH is basically a combination of the two mounting techniques described above. It was realized using a SiC ICH on the front side, a polycrystalline diamond heatspreader on the backside, and a mechanically thinned VECSEL chip with a residual substrate thickness of ~30 µm [6]. Thermal simulations predicted an improvement of the thermal properties compared to the ICH alone even for the above given residual substrate thickness (see Figure 3.9, blue dotted line), with the largest improvement for rather large pump spot diameters.

In Figure 3.13, the power characteristics of a standard VECSEL structure pumped at 980 nm (Figure 3.13b) and an LQD-VECSEL structure pumped at 1470 nm (Figure 3.13a) are shown, comparing two mounting options: the ICH alone and

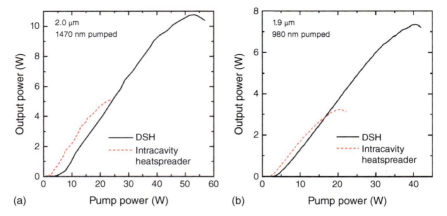

Figure 3.13 Comparison of output power characteristics for two heat management techniques. Each graph compares chips from the same wafer, but one with ICH only (red dotted line) and one with double-sided heatsinking (black solid line). Graph (a) shows data of a VECSEL structure with LQD emitting around 2.0 μm, and graph (b) data for a 980 nm pumped standard structure emitting around 1.9 μm. All data measured at 20 °C heatsink temperature.

the DSH approach. In both cases, the slope is unchanged while the pump power at thermal rollover is almost doubled. This leads in both cases to an increase of the maximum output power by a factor of ∼2 (from 5 to 10.7 W for the LQD 2.0 μm VECSEL and from 3.2 to 7.3 W for the standard 1.9 μm VECSEL). It is noteworthy, that the pump spot diameter for maximum output power increases for the DSH approach. This behavior is expected, as the pump power density at thermal rollover should be comparable and the DSH approach enables and favors larger pump diameters as the temperature reduction (compared to ICH) is highest for large pump spot diameters.

The current implementations of the DSH approach could be further improved by removing also the residual substrate layer on the backside of the VECSEL chip, even though difficult to achieve as detailed above, and by using a front side ICH made from single-crystal diamond instead of SiC.

3.6 Single-frequency Operation (SFO) of 2–3 μm VECSEL

There are different strands to achieve single lateral and longitudinal mode operation of a VECSEL. At the same time, different applications requiring SFO, such as quantum optic experiments, LIDAR, or laser seeding, have quite different requests concerning relevant laser parameters. They have diverse requirements concerning the single-frequency output power, emission linewidth, total or continuous tuning range, tuning speed, and the long-term stability in the relevant environment. Therefore, there is not one optimal configuration for single-frequency VECSELs, but the different routes to SFO available help to cover the various and sometimes conflicting needs of the different applications.

The different optical setups for SFO are closely linked to the underlying mounting technology of the VECSEL chip. As already mentioned in the preceding section, the emission spectra of a chip with intracavity heatspreader (ICH) in a standard linear cavity (Figure 3.10, cavity optimized for high power operation) differs quite significantly to that of an "almost" thin device chip (Figure 3.12b). Therefore, different optimization strategies have to be applied in order to convert the shown spectra into a pure single-frequency emission, and different behavior can be expected concerning the tunability of these laser sources. This chapter is thus organized according to the used gain chip mounting technology (Sections 3.6.2–3.6.5), preceded by some general remarks on narrow-linewidth emission in the following section.

3.6.1 Key Parameters for Single-Frequency Operation

Compared to other laser sources like edge-emitting semiconductor diode lasers or solid-state lasers, the VECSEL combines several features, making this laser concept very advantageous for low-noise SFO: due to the semiconductor QW gain region, the spontaneous lifetime of the upper laser state is rather short with typical values in the ns regime. This is much shorter than e.g. in solid-state lasers, where this value can approach the ms regime. On the other hand, due to the vertical emission and therefore short overlap of the optical mode with the QW gain media, VECSELs have to be operated in a high finesse cavity with just a few percent output coupling. This results in a long photon lifetime or long cavity decay time, which can actually exceed the value of the upper-state lifetime [59]. In this case, the VECSEL exhibits a class A laser dynamics: the optical gain can react instantaneously to variations in the photon density, leading to an exponential relaxation to the steady state and a strong damping and thus reduction of any intensity fluctuations [60]. The opposite case (upper-state lifetime much longer than the photon lifetime, so called class B regime) applies to most solid-state lasers, leading to relaxation oscillations and spiking. And even most edge-emitting semiconductor lasers operate in the class B regime due to the short cavity and rather large outcoupling rate (resulting in a photon lifetime below 1 ns), leading to damped relaxation oscillations in the GHz range.

The linewidth of a single-frequency laser is ultimately limited by the Schawlow-Townes limit [61], given by

$$\Delta \nu_L = \frac{2\pi h \nu (\Delta \nu_{cav})^2 \mu (1 + \alpha^2)}{P}$$

This value is thus proportional to the square of the linewidth of the passive laser cavity $\Delta \nu_{cav}$ and inverse proportional to the output power P. It depends further on the linewidth enhancement factor α (~3) [62] and on the ratio of spontaneous emission to stimulated emission μ (~1.5) [60]. The passive laser cavity linewidth $\Delta \nu_{cav}$ itself is inversely proportional to the cavity length and increases with increasing cavity losses (outcoupling or internal absorption). A long, high-finesse optical cavity with low outcoupling together with a high output power thus decreases this ultimate limit. Using standard values for GaSb-based devices, this limit is already in the

MHz regime for 2 μm Distributed Feedback Diode (DFB) diode lasers but orders of magnitudes lower in the Hz or sub-Hz regime for 2 μm VECSELs with a few centimeter long optical cavity [15]. The actual linewidth of a laser setup will always be above the Schawlow-Townes limit due to technical noise (vibrations, pump power and temperature fluctuations, etc.), which will be discussed in more detail below.

In order to achieve stable SFO from semiconductor edge emitters, a wavelength-sensitive filter is always applied, most typically in the form of a DFB grating or an optical grating in an external cavity setup (the longer cavity in the latter case helps also to reduce the minimum achievable laser linewidth, as discussed in the previous paragraph). For VECSELs, a stable SFO can be also achieved without any intracavity filter [63]. The necessary requirements for the cavity setup and the underlying physics of the nonlinear multimode dynamics of QW VECSELs were discussed in detail in [60] and [64]. It was shown that the VECSEL can be described as an almost ideal homogeneous gain laser where the initial multimode operation (after start of the emission or a large perturbation) collapses to a spectral width below the mode spacing (i.e. single-mode operation) after a characteristic time t_c. This time is proportional to the cube of the cavity length and the square of the gain or filter bandwidth (whichever is smaller) and is in the ms range for a ~2 cm long cavity and typical values for a GaSb-based VECSEL without ICH or any other optical filter [60]. If any process disturbs the laser dynamics before t_c has elapsed, the laser will remain multimode in CW operation. If the timescale of any subsequent perturbation is longer than t_c, the laser will reach SFO. Figure 3.14 summarized the most important noise sources and their typical time scales [65]. For the GaSb-based VECSEL, this translates into the following: the cavity length should be no longer than a few centimeters in order to keep t_c at or below 1 ms, so that the majority of the technical noise sources do not affect the laser mode dynamics. For larger cavity lengths, the gain bandwidth has to be reduced, i.e. intracavity filters have to be inserted in order to decrease t_c. This is actually the most widely used case, with cavity lengths in the ~20 cm range and a birefringent filter (BRF) or etalon (or combination of both) used as wavelength selective elements. A lower limit for the cavity length, where stable SFO can be achieved without intracavity filter, is given by the switch to class B laser dynamics for shorter photon lifetimes in shorter cavities. For GaSb-based VECSEL, this switch occurs roughly below a few millimeter cavity length [60]. Theoretically, the laser is still single mode in CW operation but will amplify any fluctuations and is much

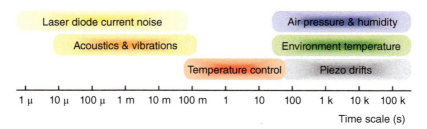

Figure 3.14 Noise sources and their typical time scales adversely affecting the laser linewidth [65].

more sensitive to optical feedback. This will render stable device operation difficult to achieve [60]. Nevertheless, with very small cavities in the ~50 μm range, stable SFO was reported [66]. This is due to the fact that in these small cavities only one (or a few largely spaced) longitudinal modes are supported.

Experimentally determined laser linewidths can be directly compared only if the sampling time used in the respective experiments was the same. The sampling time determines the lower frequency limit of the noise sources, which contributes to the measured linewidth. In other words, the chosen sampling time categorizes all noise sources (see Figure 3.14) into two categories: the one on a shorter time scale will affect the linewidth and the other one with longer characteristic times will be classified as "drift" effects. In datasheets and scientific publications, typical integration times are in the 5–100 μs range. Measuring the linewidth at different sampling times cannot only provide interesting insights into the predominant noise sources. Moreover, it can reveal the linewidth at that time scale which is relevant for the specific application (see Section 3.6.2 below).

A prerequisite for single-frequency emission is a single transverse mode operation in TEM_{00}. Higher order lateral modes are not only affecting the beam quality but also exhibiting slightly higher resonance frequencies due to the different Gouy phase shift [67, 68]. These higher order lateral modes can be suppressed by using the pump spot as a gain aperture: the emergence of higher order modes is suppressed if the pump spot size undercuts the fundamental mode spot, i.e. if the ratio $d_{pump}/d_{mode} \leq 1$. If the pump spot is not a perfect top-hat but rather Gaussian or super-Gaussian like, then the effective "above-threshold" pumped area will increase with increasing pump power, even if the $1/e^2$ diameter is not changed. This is schematically illustrated in Figure 3.15, indicating that higher order modes can be supported at increased pumping levels if the pump diameter was optimized for single lateral mode operation only at or slightly above threshold.

Another important effect that can alter the pump to fundamental mode size ratio is thermal lensing [69, 70]. Internal heating inside the gain chip leads to a refractive index profile determined by the pump intensity distribution. This thermal lens can be approximated to first order by a spherical one with a dioptic power, i.e. the inverse of the focal length, that increases with increasing pump power. Therefore, the VECSEL gain chip has to be approximated by a curved mirror, with the focal length decreasing from infinity at low pump powers to values around 100 mm and below for high CW pump intensities [70]. This in turn reduces the on-chip beam diameter of the fundamental mode, leading also to higher order lateral modes at increasing pump intensities. Because of both effects, it is important to optimize the pump spot diameter at the operating point of the laser in order to keep higher order lateral modes below threshold.

3.6.2 SFO with Intracavity Heatspreader

As explained in Section 3.5.1, an ICH is the most straightforward and mostly used way to achieve a good thermal management in GaSb-based VECSELs. Therefore, the majority of experiments with GaSb-based VECSELs to achieve single-frequency

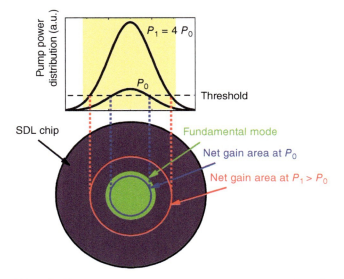

Figure 3.15 If the pump spot is not a perfect top-hat profile, the effective area pumped above threshold (blue and red circles) increases with increasing pump power even though the $1/e^2$ pump diameter (yellow area, upper diagram, same for both curves) remains constant. In this way, regions outside the fundamental mode diameter (green area) are pumped at higher power levels, which can provoke higher order modes.

emission was done using gain chips bonded to ICHs. Besides the possibilities to achieve high output powers, the ICH also acts as an optical element that affects the lasing spectrum: if both surfaces are plane parallel, it acts as an etalon with lasing restricted to well-defined etalon modes (see Figure 3.10).

3.6.2.1 Laser Cavity Setup

SFO can be achieved using a heatspreader-bonded VECSEL in a simple linear cavity without other optical elements: using a 2.3 μm standard-design VECSEL chip, a diamond heatspreader, and a 50 mm radius of curvature (ROC) mirror in a 4.9 cm long linear cavity, SFO was reported in [71]. The transition from multimode to single-mode operation, while carefully aligning the cavity and the pump to cavity mode overlap, was associated with a noticeable reduction in intensity noise due to the suppression of mode competition. The SFO was stable for several minutes. A 980 nm diode laser module, coupled to a multimode fiber, was used as a pump source focused to a pump spot diameter of 200 μm. Although a multimode pump setup was reported to introduce intensity fluctuations on quite short timescales (~100 μs) [64] hindering SFO, the etalon effect of the heatspreader seems to narrow the effective gain bandwidth sufficiently to achieve SFO.

A more flexible and indeed the most widely used cavity setup for a tunable and SFO is shown in Figure 3.16. This V-shaped cavity setup with two external mirrors offers the advantage of an almost collimated beam in the second arm bounded by a flat outcoupling mirror (see in Figure 3.16b), reducing reflection losses for a wavelength selective filter. A BRF out of Quartz placed at Brewster's angle is mostly used

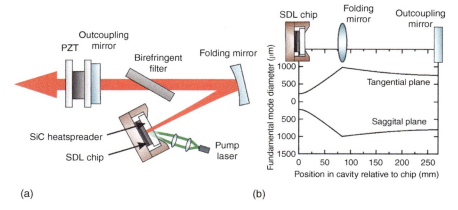

Figure 3.16 Typical V-shaped cavity setup for a tunable VECSEL (a) and the corresponding fundamental mode diameter vs. distance from the VECSEL chip (b). A Quartz BRF at Brewster's angle is placed in the almost collimated part of the setup. With the outcoupling mirror mounted on a PZT, fine-tuning can be achieved by varying the cavity length.

with a typical width in the range of ~3 mm [71, 72]. The outcoupling mirror can be mounted on a piezo-electric transducer (PZT) in order to fine-tune the cavity length and thus the emission wavelength. Further on, the PZT can be controlled by a feedback loop in order to stabilize the cavity, which is subject to a section below.

The two optical filters (heatspreader etalon and BRF) included in this setup enable SFO also in this rather long optical cavity. A typical emission spectra, measured with a Fourier transform spectrometer (FTIR) is shown in Figure 3.17a, revealing a side mode suppression ratio (SMSR) of 30 dB [73]. With the typical resolution limit of an FTIR in the few GHz range, the cavity modes (around 500 MHz mode spacing for a 27 cm long cavity) or higher order lateral modes cannot be resolved. For that purpose, a scanning Fabry-Perot interferometer (FPI) is used. Given that the free spectral range (FSR) of the FPI is wider than the cavity mode spacing, SFO can be confirmed if no side bands are visible (see Figure 3.17b). The inset of this figure shows a detailed measurement of one FPI resonance, showing a full width at half maximum (FWHM) around 2.3 MHz, which is the resolution limit of this confocal FPI. For resolving the true linewidth of the laser, different techniques must be employed that will be discussed below.

3.6.2.2 Wavelength Tuning

By rotating the BRF, the emission wavelength can be tuned over the whole gain spectrum of the VECSEL chip. Due to the heatspreader, the emission wavelength is not scanning the densely spaced cavity modes (~500 MHz) while tuning, but the much wider spaced heatspreader etalon modes (~160 GHz, 2.2 nm) as can be seen in Figure 3.18. The emission is still single frequency, i.e. only on one cavity mode at the maximum of the etalon transmission. The maximum tuning range depends mostly on the gain bandwidth of the used VECSEL structure and the total optical losses of the cavity (internal and outcoupling losses). In [71], a total tuning range of 70 nm (4 THz) was achieved for a standard design 2.3 μm VECSEL. At 2.0 μm central

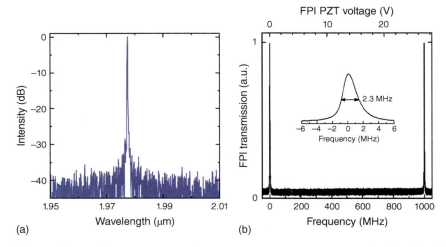

Figure 3.17 (a) FTIR measurement of the emission spectra, revealing an SMSR of 30 dB. (b) Spectrum of a scanning FPI confirming SFO as no side bands are visible with the FSR of 1 GHz. The inset shows a linewidth of 2.3 MHz, which is FPI-resolution limited.

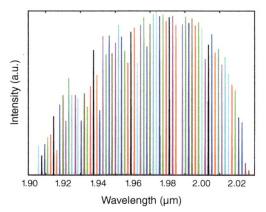

Figure 3.18 Coarse wavelength tuning by rotating the BRF. A 2.0 μm standard 980 nm pumped VECSEL with a 300 μm thick SiC ICH and a 3 mm thick Quartz BRF was used in this case. Each color represents a single-frequency spectra at a different rotation angle of the BRF.

emission wavelength, 120 nm (9 THz) of total tuning was achieved [3], also for a standard 980 nm pumped device. Using a special broad-bandwidth VECSEL design that incorporates QWs with the same composition but different widths and thus different wavelengths of their gain peaks, an increased tuning range of 156 nm (11.6 THz) was achieved around 2.0 μm [41].

In general, the maximum tuning range of a VECSEL is limited not only by the gain bandwidth of the QWs (which is the case for diode lasers) but also by the longitudinal modal gain. This value is also affected by the longitudinal confinement factor for the resonant periodic gain and the microcavity resonance (both effects don't exist in edge-emitting diode lasers). The maximum tuning range is therefore expected to be slightly smaller than that for diode lasers, placed in an external cavity. For GaSb-based 2.3 μm edge emitters, 177 nm (10 THz) of tuning were achieved

[74] using an optical grating as feedback element, which is higher than the values reported above for VECSELs (for identical QWs).

For a mode-hop free fine-tuning, the optical length, i.e. the product of mechanical length and average refractive index, must be changed. This can be accomplished in three ways: by varying the resonator length, a continuous tuning on one cavity mode can be achieved. With the resonator geometry shown in Figure 3.16a, this results in a 0.6 GHz µm^{-1} wavelength shift [3]. By changing the heatspreader temperature, the refractive index inside the VECSEL chip and that of the ICH is changed, leading to a tuning in the range of 3.3 GHz K^{-1} for GaSb-based VECSEL with a SiC heatspreader. The bandgap of the QW changes of course too, but this effect is negligible on a few GHz wavelength scale and does not affect the cavity mode. Increasing the pump power also increases the refractive index inside the active medium as its temperature rises, leading to an increase in emission wavelength. On the other hand, at increasing pump powers the carrier density in the barrier layers increases, which reduces the effective refractive index and thus the emission wavelength. Experimentally, it was found that the latter effect dominates at small pump power changes, leading to a fine-tuning of −830 MHz mW^{-1} [3].

Using one of these three effects individually will result in a limited mode-hop fine-tuning range in the ∼GHz range. E.g. when varying the resonator length alone, the cavity mode shifts away from the fixed heatspreader etalon mode. If this shift is large enough, the adjacent cavity mode experiences considerably less optical loss, resulting in a mode-hop to this cavity mode. This problem could be solved by tuning the etalon mode too in a synchronized way. However, as the heatspreader etalon is directly mounted onto the chip, the only possible way to do so in this configuration is to change the chip temperature. This method has several drawbacks: temperature tuning is per se slow, and the maximum temperature change is limited for practical reasons and will have a large impact on the output power (e.g. at high temperatures). A control of the etalon effect itself without affecting the VECSEL chip, e.g. by tilting the etalon or varying only its temperature, would result in a much more practical way to synchronize the different effects and achieve a large mode-hop free tuning. To enable this possibility, however, the etalon has to be decoupled from the VECSEL gain chip and a layout without ICH has to be used (see Section 3.6.5). In fact, the above outlined limitations for wide mode-hop free tuning is the most severe drawback of using a VECSEL chip with ICH in a single-frequency laser setup. Further on, setting the emission wavelength of the laser to an arbitrary value in particular in between the heatspreader etalon modes shown in Figure 3.18 can be a complex task. It may include adjusting the temperature of the VECSEL plus heatspreader stack (in order to shift the refractive index of the etalon) or to change the pump and mode spot position on the VECSEL chip to realize a change of the etalon mode by very small local thickness variations of the heatspreader [75].

3.6.2.3 Emission Linewidth

The determination of the true laser linewidth in the sub-MHz regime requires sophisticated measurement techniques. As stated above, a typical confocal FPI with a few centimeter cavity length has a resolution in the GHz range and cannot resolve the linewidth of a stable single-frequency VECSEL setup. Only high-finesse plano-FPIs with long cavity length (40 cm) can offer a resolution in the ~20 kHz range [63], but they are not commercially available and hard to align properly. When actively stabilized VECSELs are used, the linewidth can be estimated from the error signal of the control loop [71, 76]. However, this is only an indirect way and impossible to apply for a free-running setup. A direct measurement technique for very narrow linewidths is the heterodyne detection scheme. At wavelengths where optical fibers with negligible transmission losses are available, the self-delayed heterodyne technique can be applied [77]. The prerequisite for this setup is a long fiber-based delay line that must be longer than the coherence length. For a 10 kHz linewidth, this translates in more than 10 km of optical length. Even though the length can be reduced with more sophisticated setups [78], this technique fails in the MIR due to the lack of low-loss fibers. Therefore, a true heterodyne beat-note experiment is the most direct way to measure a small laser linewidth, which requires two almost identical lasers [79]. A schematic of the measurement configuration setup is depicted in Figure 3.19a. In order to measure the beat note, the difference in emission frequency of both VECSELs must be smaller than the bandwidth of the photodiode. High-speed extended InGaAs PIN diodes are readily available with a 60 MHz bandwidth, which translates to a maximum detuning of the emission frequency of <0.0008 nm at 2.0 μm.

An example of the beat-note frequency spectra of two 2.0 μm free-running VECSELs is shown in Figure 3.19b. In order to reduce technical noise, the VECSEL resonator was placed inside a sealed module housing machined from a solid aluminum block [72]. Both modules were operated at 100 mW output power running at 2019.1582 nm and 2019.1587 nm emission wavelength, respectively. The measured beat-note spectrum recorded with 100 μs integration time is centered

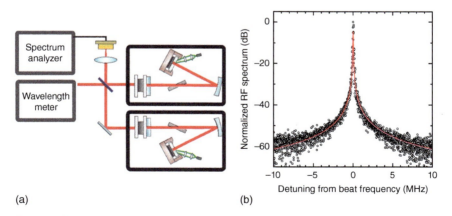

Figure 3.19 (a) Measurement setup for a heterodyne beat note. Two identical VECSEL setups are used with their laser frequencies detuned by some MHz, as surveyed by a wavelength meter. Both beams are then superimposed on a fast photodiode, whose signal is fed into a sampling oscilloscope or spectrum analyzer. (b) Measured RF spectrum (black dots) of a beat note of two free-running 2.0 μm VECSEL laser modules at 100 mW output power with a 100 μs integration time. The red line is a Voigt fit to the experimental data.

around the 35 MHz difference frequency and can be well described by a Voigt profile (red line in Figure 3.19b). Technical noise is associated with a Gaussian lineshape, while broadening due to phase fluctuations (spontaneous emission) is described by a Lorentzian function [80]. Since the Voight profile is a convolution of both, a curve fit using this function simultaneously yields the linewidth contribution from both. Under the assumption that the noise sources and level are similar for both lasers, the Lorentzian (Gaussian) linewidth Δv_L (Δv_G) of the individual laser is calculated by dividing the corresponding fitting parameter by a factor of 2 ($2^{0.5}$). The resulting Voigt linewidth Δv_V can then be calculated [81] using $\Delta v_V \approx 0.535 \Delta v_L + (0.217(\Delta v_L)^2 + (\Delta v_G)^2)^{0.5}$. In this way, a Gaussian linewidth of 40 kHz, a Lorentzian linewidth of 8 kHz, and a total Voigt linewidth of 45 kHz were deduced for the individual free-running VECSEL.

By optimizing the laser cavity and using a larger fundamental mode diameter on the VECSEL chip of 520 µm, the output power for SFO at 2.0 µm could be increased to 960 mW [73]. The beat-note measurement still revealed an extreme narrow linewidth of 60 kHz of one individual VECSEL at 100 µs integration time. Again the laser module was free running, only the temperature of the whole setup was stabilized to room temperature. At the same time, the SFO of this laser module is not limited to several minutes: it shows an excellent long-term stability as can be seen in Figure 3.20. There the central emission wavelength, measured with a 20 MHz resolution wavemeter over a period of over 18 hours, is plotted. The 2.0 µm VECSEL remains in single-mode operation over the whole time span. Further on, no mode-hop to an adjacent cavity mode (500 MHz mode spacing) is visible and the total drift over 18 hours of operation is as small as 320 MHz.

The ability to achieve ~1 W output power at <100 kHz emission linewidth at 2.0 µm with a rather simple and still very stable setup makes these VECSELs a quite unique laser source. DFB laser exhibits much larger linewidths in the MHz regime, and external cavity diode laser do not reach this power level without further amplification stages at this wavelength range.

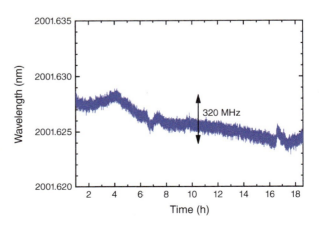

Figure 3.20 Long-term stability of a free-running VECSEL module at 2.0 µm: the emission wavelength, measured with a 20 MHz resolution wavemeter, is plotted against the operation time. Within the observed 18 hours, the VECSEL remains in SFO, no-mode-hop to an adjacent cavity mode was recorded, and the total wavelength drift is only 320 MHz during the 18 hours.

3.6.2.4 Active Stabilization and Influence of Sampling Time

As described above, a linewidth of <100 kHz (100 μs sampling time) of a free-running VECSEL is sufficient for most applications but still much larger than the Schawlow-Townes limit due to technical noise. In order to reduce the linewidth further, active stabilization techniques can be used that lock the laser wavelength to a reference frequency. The simplest and most flexible way is to use a passive cavity (e.g. an FPI) as reference.

When the cavity length of the FPI is adjusted so that the laser emission wavelength is at the edge of a transition maximum, any change in laser wavelength will translate into an intensity change of the light transmitted through the FPI. In this way, this so called side-of-fringe method can produce an error signal that can drive a feedback loop, e.g. adjust the cavity length via the PZT onto which one of the cavity mirrors is mounted (see Figure 3.21a for the schematic setup). Although an estimation of the laser linewidth through an analysis of error signal is possible [71, 76], a direct way to determine the linewidth is again the heterodyne beat-note measurement (see Figure 3.19a). Active stabilization has to be applied to both laser modules used [72] in this experiment. Linewidths measured this way for different sampling times are collected in Figure 3.22 for a 2.0 μm VECSEL emitting 350 mW of output power [3]. The black squares and lines indicate linewidths of a free-running VECSEL module, only temperature stabilized to 20 °C. There is an overall increase in linewidth with sampling time from 20 kHz (10 μs sampling) to 220 kHz (100 ms sampling), which is most pronounced between 100 μs and 1 ms. When the side-of-fringe locking mechanism is turned on (blue triangles and lines in Figure 3.22), the linewidth becomes much narrower, especially at longer integration times. Nevertheless, there is still a pronounced increase between 100 μs and 1 ms.

One limiting factor in this setup is the bandwidth of the PZT that controls the laser wavelength via the cavity length: the PZT bandwidth is typically in the kHz regime; therefore, any noise source with higher frequencies than this cutoff frequency cannot be compensated. A way to increase the bandwidth would be to include the pump laser drive current into the feedback mechanism as another mean to fine-tune the emission wavelength. However, a fundamental problem with the side-of-fringe locking scheme is that it cannot distinguish between frequency or intensity noise, as both would lead to a change of the light intensity transmitted through the reference cavity. A variation in pump power translates into a change of the VECSEL output power, which in turn is interpreted by the side-of-fringe error signal as a false change in wavelength. Hence, the potential benefit of that faster pump diode laser control stage needs a stabilization technique that is immune to intensity fluctuation. This can be achieved using the Pound-Drever-Hall (PDH) technique.

The experimental setup for the PDH stabilization technique [82] that was used for 2.0 μm VECSEL [3] is shown in Figure 3.21b. The PDH error signal is based on a frequency modulation of the laser emission via an electro-optic modulator, producing sidebands in the laser frequency spectrum. The reflected light from the FPI reference cavity was detected by a photodiode, and this signal was mixed with the local oscillator used to modulate the laser light. This multiplication yields the PDH error signal, given that the phase between the local oscillator and the photodiode

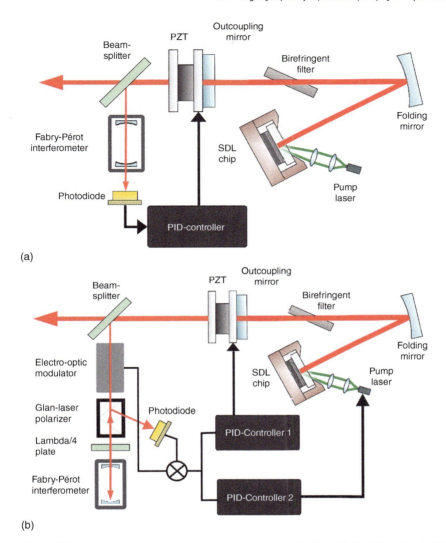

Figure 3.21 Setups for active wavelength stabilization: (a) Side-of-fringe setup where the laser is locked to an edge of a transition maximum of a reference cavity (e.g. an FPI). The proportional–integral–derivative (PID) feedback loop is adjusting the cavity length. (b) PDH setup. The laser is frequency modulated FM modulated, and the signal reflected from the reference cavity is mixed with the local frequency modulation FM oscillator signal yielding the PDH error signal. As the PDH error signal is immune against intensity noise, two feedback loops can be used in order to control the cavity length and the pump power.

is properly adjusted (see [83] for further details). The advantages of the PDH setup over the side-of-fringe locking scheme is that existing intensity noise of the laser will not lead to false wavelength correction (and thus potentially an actual increase of the laser linewidth). As discussed above, for the same reason also the pump diode current can be integrated into the control loop (see Figure 3.21b), which greatly

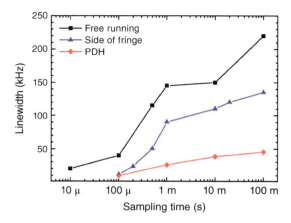

Figure 3.22 VECSEL linewidth obtained by beat-note spectra vs. the integration time used in that measurement. Data are shown for free-running (black), side-of-fringe-stabilized (blue), and PDH-stabilized (red) operation. All lasers used in these experiments were operated at 350 mW output power.

enhances the bandwidth of the feedback to the MHz regime. A further advantage is the steeper slope at the set point for a given finesse of the reference cavity (approximately 2.7 times higher [83]). All this leads to a superior suppression of technical noise within the laser cavity, as can be seen from the experimental data displayed in red in Figure 3.22. They show the laser linewidth, measured with the heterodyne technique, as a function of the integration time for a 2.0 μm VECSEL with active PDH stabilization and the same reference cavity as used before. The laser linewidth is reduced to <50 kHz for all sampling times up to 100 ms. Further on, the sharp increase of the linewidth between 100 μs and 1 ms is greatly reduced.

A detailed analysis revealed a mechanical resonance of the module housing around 1.4 kHz as the origin of this increase [3]. In order to suppress the frequency noise stemming from a narrow-bandwidth resonance, specific challenges for derivative and high frequency gain arise, which can be met effectively only with a two-stage feedback loop [84]. This is another reason for the superior performance of the PDH control scheme, using the cavity length and the diode current as independent means to stabilize the laser wavelength.

For a 100 mW VECSEL at 2.0 μm, a 9 kHz linewidth (100 μs integration) was reported in [72] using a side-of-fringe stabilization scheme. With PDH stabilization, the linewidth of a 960 mW VECSEL at the same wavelength was reduced to 20 kHz (30 kHz) at 100 μs (1 ms) integration time [3]. With a more careful design of the laser module housing and by using more sophisticated and more stable reference cavities (see e.g. [85]), a further reduction of the linewidth in the Hz regime for e.g. metrology purposes seems to be possible for these VECSELs.

3.6.2.5 Conclusion

Using a VECSEL chip with ICH as the gain element in a V-shaped single-frequency cavity setup ~1 W of output power at a narrow linewidth of <100 kHz can be readily achieved at 2.0 μm using a 980 nm pumped standard VECSEL design and a SiC heatspreader. So far, no gain chip with more advanced thermal management (e.g. DSH mounting, see Section 3.5.3) or higher efficiency (1470 nm LQD design, see Section 3.4.3) has been operated in a single-frequency setup. Both of these improvements applied successfully to high-power multimode VECSELs could eventually lead also to further improvements in achievable narrow-linewidth output power at or around 2 μm.

SFO VECSEL modules can be built with an excellent long-term stability also in free-running operation making them an ideal laser source for seeding, quantum physics, or metrology. The major drawback, that arises from the used mounting technology (which is that the heatspreader etalon is directly attached to the VECSEL chip), is the limited tuning performance: The mode-hop free tuning range is very narrow and the rather fixed mode pattern imposed by the heatspreader makes it difficult to address all wavelengths within the gain regime. Possible ways to improve on this will be discussed below. This includes the reduction of the etalon effect (Section 3.6.3), avoiding the coupled subcavities by using a homogeneously filled external cavity without air space (Section 3.6.4) or by removing the heatspreader (Section 3.6.5) inside the laser cavity

3.6.3 SFO with Wedged Heatspreader

As discussed above, a plano-parallel ICH severely limits the tuning performance of a single-frequency VECSEL setup. In order to suppress these etalon modes, a wedged and antireflection (AR)-coated heatspreader can be used. Typical wedge angles are in the 2–3° range (an AR coating alone would not be sufficient to suppress the etalon effect) [86]. However, this approach has two drawbacks: due to the wedge, the upper surface of the heatspreader is not perpendicular to the laser emission axes (see Figure 3.23b), leading to reflection losses at the surface and "walk-off" losses inside the heatspreader that are still relevant even for good AR coatings. Further on, the longitudinal enhancement factor for the microcavity resonance is reduced, leading to a higher threshold and a reduced maximum tuning range.

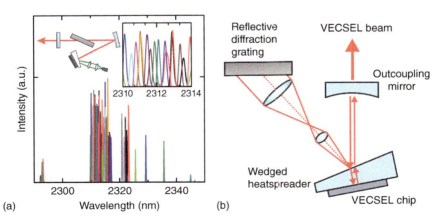

Figure 3.23 (a) Emission spectra of a VECSEL with wedged and AR-coated heatspreader inside a V-shaped cavity tuned by rotating the BRF. (b) Schematic drawing of a VECSEL setup, where the reflection of the upper surface of the wedged heatspreader is used to self-seed the laser.

Closely spaced lasing spectra of a 2.3 μm VECSEL chip with a wedged and AR-coated SiC heatspreader are shown in Figure 3.23a [87]. The chip was incorporated in a standard V-shaped cavity with a BRF that was used to tune the emission wavelength. Due to the successful suppression of the heatspreader etalon effect, the lasing wavelength could be set to any arbitrary value within the overall tuning range without any jumps to adjacent etalon modes (compare Figure 3.23a with Figure 3.18). The lasing wavelength was, of course, still locked to the densely spaced cavity modes, not resolved in this measurement. But the total tuning range was limited to 57 nm, and the maximum output power (170 mW) was only half of that which was achieved with a plane-parallel heatspreader and the same gain chip material.

Self-seeding, i.e. a spectrally filtered feedback from outside of the laser cavity, is another means to tune the emission wavelength that was successfully demonstrated for GaSb-based VECSELs. In [88], a 2.0 μm VECSEL was self-seeded by means of a diffraction grating placed behind an high-reflection (HR) mirror (99% reflectivity), that couples a fraction of the spectrally filtered light back into the resonator. A 35 nm tuning range could be achieved using a standard (plane-parallel) diamond heatspreader. Another setup using a GaAs-based VECSEL and fiber Bragg gratings was reported in [89].

A wedged heatspreader will always introduce reflections at its upper surface that are not directly coupled back into the laser cavity. A way to use this (otherwise lost) portion of the laser radiation to self-seed the linear VECSEL cavity is shown in Figure 3.23b. The back reflected light is spectrally filtered by a diffraction grating and used to tune the emission wavelength. Using a 2.0 μm VECSEL chip and a SiC wedged heatspreader, it was possible to tune the emission to any arbitrary cavity mode. The total tuning range was 30 nm at 400 mW output power. But this setup showed also severe fluctuations of the output power on a ms timescale. Varying the intensity of the self-seeding signal coupled back into the cavity did not yield a regime [90] where self-seeding and a stable power output could be achieved simultaneously.

Both approaches using a wedged heatspreader on top of a GaSb-based gain chip proved successful in removing the tuning limitations imposed by the etalon modes associated with a plano-parallel heatspreader. But with none of them stable operation with a large total tuning range (>50 nm) and a high output power (>200 mW) could be achieved simultaneously. And in fact no measurement proving true SFO (e.g. an FPI spectrum) has been reported so far.

3.6.4 SFO with Microcavity VECSELs

One possible way to decrease the complexity of the laser cavity discussed so far is to omit the air space and realize a setup where the cavity length is reduced significantly and the complete cavity is filled with semiconductor or heatspreader material (Figure 3.24a). The cavity mirrors are then formed by the DBR inside the VECSEL structure and an HR coating, deposited either on top of the heatspreader (see Figure 3.24) or on top of the VECSEL structure itself. By means of pump absorption and internal heating, a thermal lens is established that renders this otherwise plano-parallel cavity stable. This microcavity (μC) can be described to first order as a

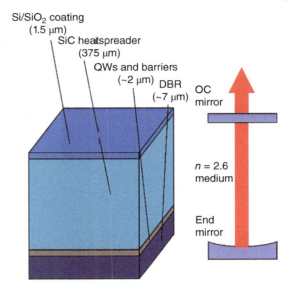

Figure 3.24 Schematic of the μC-VECSEL setup. An HR coating deposited on the top surface of the heatspreader acts as outcoupling mirror. Due to thermal lens effects, the effective cavity can be approximated by a curved end mirror and a planar outcoupling mirror.

linear cavity with a plane mirror and one effectively spherically curved mirror, whose focal length is reduced with increasing pump power. Therefore, the fundamental mode size diameter is also reduced when the pump power is increased, leading to multiple lateral mode operation (see Section 3.6.1).

This monolithic μC-VECSEL approach was realized successfully using a 0.85 μm VECSEL and a SiC heatspreader [91], a 1.3 μm VECSEL with diamond heatspreader [92] and an 0.98 μm VECSEL with a sapphire heatspreader [93]. Using a GaSb-based 2.3 μm VECSEL structure, a μC device was reported in [63] where no ICH was used and the HR coating was deposited directly onto the semiconductor chip. With that approach, up to 3 mW of output power were obtained in SFO using a pump spot diameter of 36 μm. Measuring the FPI trace revealed a linewidth of 5 MHz at 0.5 mW output power. The higher linewidth compared to the standard VECSEL setup (see Section 3.6.2) was expected as the cavity length is greatly reduced.

The influence of the thermal lens on the lateral mode profile (and thus beam quality) for a 2.0 μm GaSb-based μC-VECSEL with SiC heatspreader was studied in [70] and [94]. Two different 2.0 μm VECSEL structures were compared in the μC geometry: a standard 980 nm pumped design and an LQD 1470 nm pumped design. The latter exhibited a higher quantum efficiency, thus higher slope and reduced internal heating (see Section 3.4.3). The output power characteristics for these two devices with different pump spot diameters can be seen in Figure 3.25a, while Figure 3.25b shows the corresponding beam quality parameter M^2 as a measure for the number and intensity of higher order lateral modes. For the standard VECSEL design (black curves), the M^2 value increases rapidly with increasing pump power, revealing a strong increase of the thermal lens. Due to that, the fundamental mode size diameter is strongly reduced with increasing pump power, leading to highly multilateral mode emission and thus a large M^2, which is quite unusual for the VECSEL concept otherwise. The LQD structure (red line in Figure 3.25) shows a rather different

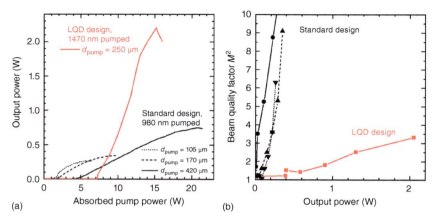

Figure 3.25 (a) Output power of the μC-VECSEL vs. absorbed pump power at room temperature for different pump spot diameters and two different VECSEL active region designs: a standard 980 nm pumped design (black) and an LQD 1470 nm pumped design (red). (b) Measured values of the corresponding beam quality parameter M^2. At higher output powers, this value increases drastically, indicating high lateral mode operation. This effect is much more pronounced for the standard 980 nm VECSEL design where the internal heating is much stronger, compared to the 1470 nm pumped device.

behavior: as the internal heating is reduced due to the higher conversion efficiency of the VECSEL device, the increase of thermal lensing with pump power is less pronounced. This results on the one hand in an increased threshold pump power (as more power is needed to render the cavity stable). But on the other hand the output power reaches very high values of over 2 W in CW operation and the increase of M^2 is only moderate, reaching values around $M^2 = 3.5$ at 2 W output power. Single lateral mode (TEM_{00}) emission was observed up to 105 mW for the standard device (105 μm pump spot diameter) and up to 700 mW for the LQD device (250 μm pump spot diameter).

For single-frequency emission, i.e. the selection of one longitudinal mode, the output power had to be further reduced to 30 mW in case of the standard design and 90 mW for the LQD sample. By changing the submount temperature, a coarse tuning range of 110 nm could be achieved for a 120 °C temperature interval (see Figure 3.26a). Mode-hop free fine-tuning could be observed in a smaller temperature range of $\Delta T = 6$ K, leading to a 30 GHz change in emission wavelength (approximately 20% of the μC FSR of 2.1 nm; see inset in Figure 3.26a). The single-frequency emission of the μC-VECSEL was verified using a scanning FPI with 2.5 MHz resolution limit. The measured linewidth of 7 MHz (Figure 3.26b, 100 μs time interval) is comparable to the results achieved in [63] for a 2.3 μm μC-VECSEL, confirming that the linewidth of these short cavity devices is limited, as expected, to the MHz range.

3.6.5 SFO without Intracavity Heatspreader

As described in Section 3.6.2, the poor single-frequency tuning performance is actually the major drawback of using a VECSEL with ICH bonded directly onto the

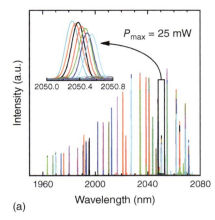 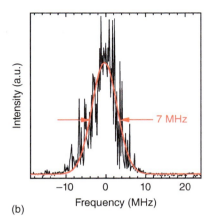

Figure 3.26 (a) Wavelength tuning with a temperature change of 120 °C. The inset shows the mode-hop free tuning range. (b) Measured FPI transmission of the µC-VECSEL revealing a linewidth of 7 MHz.

laser chip. Omitting the heatspreader leads to a better control of the laser modes (i.e. the heatspreader etalon modes do not dominate the spectral behavior) and can lead to a much wider mode-hop free tuning range. The major challenge, especially for GaSb-based devices, is to achieve efficient heat removal from the pumped gain region and thus high output powers (see Section 3.5.2).

Using a 2.3 µm VECSEL without any special means to improve heat extraction, i.e. without ICH and with the ~500 µm thick GaSb-substrate still underneath the DBR, SFO with a maximum output power of 5 mW was reported in [15, 63] and [95]. Due to the high thermal resistance of the DBR plus substrate layer stack, the pump spot diameter was limited to small diameters in the 10–50 µm regime. A short linear cavity was used in all cases (1.4–15 mm cavity length) that enable SFO without additional spectral filtering employed (see Section 3.6.1). The advantage of a short cavity is the larger possible tuning range when changing the cavity length: Standard PZTs have a maximum travel range around ~10 µm and the resulting wavelength change is more pronounced for a short cavity with large FSR ($\lambda/2$ length change is needed to change the frequency by one FSR). In Figure 3.27a, the schematic setup for a tunable, single-frequency 2.3 µm VECSEL (without heatspreader) is shown, using a short 1.4 mm long cavity (108 GHz FSR) with a 2 mm ROC outcoupling mirror [95]. The VECSEL is diode pumped under Brewster's angle using a single mode 830 nm diode laser that produced a pump spot diameter of approximately 23 µm diameter, resulting in an output power of 2.3 mW at 5 °C. Tuning was possible by changing the cavity length (PZT with 8 µm travel range) and by changing the heatsink temperature (by 14 °C). In both cases, a regular pattern of mode-hops occurred after one cavity FSR of 108 GHz, respectively. This pattern is shown in Figure 3.27b (black square for the position of a mode-hop) for the two tuning parameters PZT voltage and heatsink temperature. Using a slightly curved diagonal trajectory in the above-defined parameter space (line in Figure 3.27b), a mode-hop free scan larger than the cavity FSR can be achieved. The synchronous change of both parameters in this way leads to

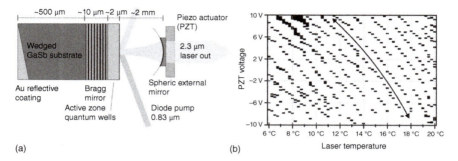

Figure 3.27 (a) Short cavity setup for single-frequency 2.3 μm VECSEL and a large 500 GHz mode-hop free tuning range. (b) Position of mode-hops (squares) in relation to the tuning parameters PZT voltage and heatsink temperature [95].

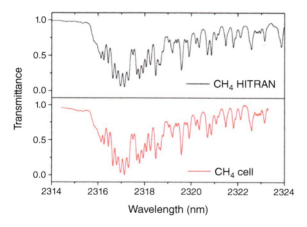

Figure 3.28 Measured CH_4 absorption spectra with the laser setup shown in Figure 3.27 [95].

a 500 GHz (8.8 nm) mode-hop free tuning that was applied to measure a methane absorption in the 2314–2324 nm wavelength range. The resulting absorption spectra are in perfect agreement with high-resolution transmission molecular absorption database (HITRAN) absorption data (Figure 3.28). It has to be noted that the regular mode-hop pattern (Figure 3.27b) could only be achieved when the backside of the GaSb-substrate was wedge polished (angle of 2°, see Figure 3.27a) in order to suppress cavity loss modulations due to the substrate etalon [95].

In order to increase the output power for ICH-free SFO of GaSb-based VECSELs, the heat management has to be improved. As discussed in detail already in Section 3.5.2, this can be at least in part achieved by using an adapted thin device approach. By using a reverse growth order, followed by a complete chemical removal of the substrate, the output power of these thin GaSb-based VECSEL chips could be increased by a factor of 2.5 compared to the results above, resulting in a maximum CW power of 12 mW [55, 57] at 15 °C. The pump diameter was increased to 120 μm for this 2.3 μm device.

For the "almost" thin device introduced in Section 3.5.2 (i.e. exploiting a standard growth order and mechanical thinning to achieve a ~30 μm layer of residual

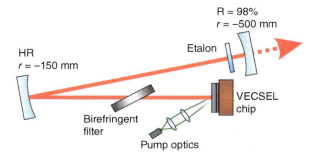

Figure 3.29 Schematic setup of the long V-shaped cavity used for wavelength tuning and SFO of the "almost" thin VECSEL device, whose basic characterization is shown in Figure 3.12. Note that the main difference to Figure 3.16 (typical SFO cavity setup for a VECSEL with ICH) is the position of the etalon away from the VECSEL chip. This enables an adjustment of the tilt angle or etalon temperature indivicually, without affecting the gain chip.

substrate), single-frequency performance was obtained by using the cavity shown in Figure 3.29. In order to achieve a large fundamental mode diameter of >500 μm, where this chip has proven to support high output powers (see Figure 3.12a), a large cavity with 47 cm length was chosen. SFO was established in this long cavity by narrowing the effective gain by using a 3 mm thick Quartz birefringent plate and a 500 μm thick SiC etalon filter. The relatively broad width of the BRF function alone was not sufficient to discriminate between the densely spaced cavity modes (320 MHz FSR) and achieve SFO. An alternative to the additional etalon would be to use a stack of BRF with different thicknesses. The maximum output power in SFO was 500 mW at 2.0 μm and thus over an order of magnitude more than with the previously reported experiments for a VECSEL without ICH. The measured linewidth was limited by the resolution of the scanning FPI used. Although no beat-note experiment was carried out, the real laser linewidth is expected to be also in the ~100–400 kHz range (see Section 3.6.2).

Note that the setup described here is very similar to a typical SFO cavity setup for a VECSEL with ICH (see Figure 3.16). The small but essential difference is the position of the etalon away from the VECSEL chip. This enables an adjustment of the tilt angle and/or temperature of the etalon individually, without affecting the gain chip. In this way, the tuning behavior is simplified and an arbitrary emission wavelength within the gain spectrum is easily addressable.

Coarse tuning of the emission wavelength in this cavity design can be achieved by rotating the BRF. In Figure 3.30, the single-frequency emission wavelength, measured with a 20 MHz resolution wavemeter, is plotted vs. this rotation angle. Regular mode-hops with a 120 GHz spacing can be seen, that correspond to the FSR of the etalon filter. In the detailed section in Figure 3.30b, the position of the cavity modes are indicated by the gray lines (320 MHz spacing). This plot shows that the laser stays at exactly one cavity mode (at the transmission maximum of the corresponding etalon mode) before it jumps to the next etalon mode. Addressing the different cavity modes is achieved by tilting the etalon filter. This is shown in Figure 3.31 where

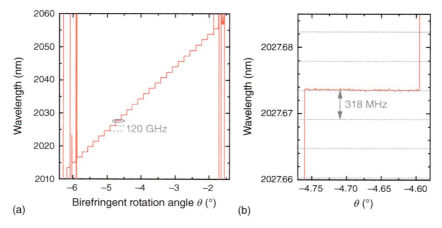

Figure 3.30 (a) Emission wavelength measured with a 20 MHz resolution wavemeter vs. rotation angle for the BRF. (b) Detailed view with the cavity modes displayed as gray lines.

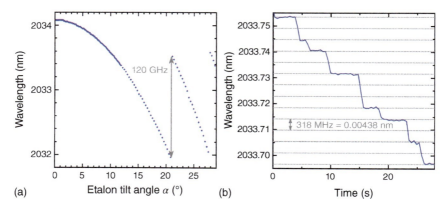

Figure 3.31 (a) Emission wavelength vs. tilt angle of the etalon. (b) Detailed view with the cavity modes displayed as gray lines.

the measured laser wavelength is plotted versus the tilt angle. From the detailed section (Figure 3.31b), it is evident that the laser is not always jumping to an adjacent cavity mode while tilting the etalon, sometimes the mode-hops occur with a spacing corresponding to two or three times the cavity mode separation. This irregular pattern could originate from the plane-parallel 30 μm thick residual substrate (see the discussion about the need for a wedged substrate above) or from mechanical noise as the rotation was done by hand. By tilting the etalon, 157 GHz of total tuning was possible before the laser jumped back to an adjacent etalon mode (see Figure 3.31a). Fine-tuning in between the cavity modes (320 MHz spacing) would be possible in this setup by changing the cavity length (which was, however, not demonstrated). For a large mode-hop free tuning range, the length change (via PZT) and tilt of the etalon would need to be synchronized. However, due to the long cavity, a 500 GHz continuous tuning range would translate into a length change of 1600 μm, which is

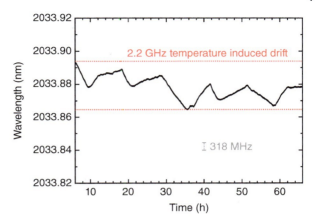

Figure 3.32 Emission wavelength in a 60 hour time span for the free running single-frequency setup using a thin device 2.0 μm VECSEL chip. No mode-hop (320 MHz cavity mode spacing) was observable during the whole operation time. The baseplate was not temperature stabilized leading to the observable wavelength drift.

difficult to achieve with regular PZTs. Finally, Figure 3.32 displays the long time stability of this cavity setup. Again, a very stable operation without cavity mode-hops was achieved over a time span of 60 hours. The temperature-induced wavelength fluctuation (2.2 GHz) is larger than corresponding results shown above for another V-shaped cavity (see Figure 3.20). This is because in the latter experiment, the base plate of the laser module was also temperature stabilized, which was not the case for the setup described here.

3.7 Conclusion

The GaSb-based type-I QW VECSEL devices covering the 1.9–2.8 μm wavelength range have reached a considerable level of maturity. Fully exploiting the design flexibility offered by the (AlGaIn)(AsSb) materials system, the highly successful VECSEL concept could be extended to the above given wavelength range with an impressive performance level.

The "low-quantum deficit (LQD)" VECSEL structures, using 1470 nm barrier pumping, have emerged as the most efficient GaSb-based gain chips offering high slope efficiencies around 30% up to an emission wavelength of 2.5 μm. This is considerably higher than the slope of the "standard" 980 nm barrier pumped VECSEL structures (16–22%) for that wavelength range. Despite the low carrier confinement in the LQC structures, the threshold pump power density at room temperature is almost unchanged, leading to higher achievable output powers in the 5–8 W range in CW operation using a SiC intracavity heatspreader (ICH). With a diamond heatspreader in combination with a LQD structure, 17 W CW output power at room temperature could be demonstrated for an emission wavelength of 2.0 μm. On the long-wavelength side, 0.8 W was demonstrated at 2.8 μm emission wavelength with an LQD structure and a SiC heatspreader.

Although the ICH is the most widely used technology for efficient heat removal for GaSb-based VECSEL, two other techniques were also successfully demonstrated: the thin device approach with the substrate completely or partially removed. With the latter approach (~30 μm residual substrate left), over 2 W of CW output power was achieved for very large pumped areas (>1000 μm diameter), while the first approach (substrate completely removed) produced VECSEL chips that are more efficient for smaller pump diameters around 100 μm. The "double-sided heatspreader (DSH)" technology combines the ICH with the thin device approach. With this technique, higher pump powers and thus an improvement of a factor of ~2 for the maximum output power (compared to the ICH alone variant) was demonstrated for a plurality of GaSb-based VECSEL structures.

The VECSEL concept combines several features, which are advantageous for low-noise SFO. Using GaSb-based VECSEL chips with different mounting techniques, a variety of setups for SFO have been demonstrated, all with different strength and weaknesses. Using a VECSEL chip with ICH in a V-shaped cavity setup, high power (~1 W) narrow linewidth (~100 kHz at 100 μs integration time) single-frequency emission was demonstrated around 2.0 μm for a free-running laser module, i.e. without any active stabilization applied. Excellent long-term stability could be achieved (>18 hours with no mode-hops in free-running operation and only a 320 MHz temperature-induced drift), making this setup a stable laser source for seeding, quantum physics, or metrology. Using active stabilization employing a reference cavity (side of fringe or PDH locking), a reduction of the linewidth to the ~10 kHz regime (100 μs integration time) was demonstrated. The drawback of this setup employing planar ICHs is the limited tuning performance: the mode-hop free tuning range is limited to a few GHz or below and the fixed mode pattern of the heatspreader etalon makes it difficult to address wavelengths in between these etalon modes.

By depositing an HR coating directly on the top surface of the VECSEL chip or the ICH, μC-VECSELs are formed with the complete cavity filled with solid-state (heatspreader or semiconductor) material. For these small and simple devices, SFO was demonstrated with up to 90 mW of output power at 2.0 μm emission wavelength, using an LQD-VECSEL structure. Temperature tuning lead to a mode-hop free tuning range of 30 GHz while the linewidth was increased to 7 MHz due to the short laser cavity.

By omitting the ICH, laser cavities allowing a smooth continuous tuning performance and a precise control of the laser modes can be build. Using a short (1.4 mm long) laser cavity and a 2.3 μm VECSEL chip with a wedged substrate, a very large mode-hop free tuning range of 500 GHz was demonstrated by changing the cavity length and the VECSEL chip temperature in a synchronized way. However, due to the poor thermal management, the output power was limited to a few mW, which could be improved to 12 mW for a thin device chip with the substrate completely removed. Using a 2.0 μm "almost" thin device VECSEL chip (~30 μm residual substrate left) in a large optical cavity, the maximum single-frequency output power could be increased to 500 mW and a precise control of the laser modes over the whole gain regime was demonstrated. However, the mode-hop free tuning range was again reduced due to practical limitations (maximum travel range of the PZTs used for varying the cavity length).

From the overview presented here, it is evident that there is not a single optimal configuration for single-frequency VECSELs. Instead, we have a plurality of different single-frequency VECSEL variants at hand, with their respective strengths and weaknesses, from which one can chose the variant delivering the best compromise in performance to meet the, sometimes even conflicting, needs of a specific application.

By using a combination of the more advanced VECSEL gain structure with higher efficiencies (LQD) and/or more advanced thermal management methods (DSH) could eventually lead to an improvement of the single-frequency performance in the future for most of the above discussed single-frequency VECSEL variants.

References

1 Garnache, A., Liu, A., Cerutti, L., and Campargue, A. (2005). Intracavity laser absorption spectroscopy with a vertical external cavity surface emitting laser at 2.3 μm: application to water and carbon dioxide. *Chemical Physics Letters* 416: 22–27.
2 Juranyi, Z., Burtscher, H., Loepfe, M. et al. (2015). Dual-wavelength light-scattering technique for selective detection of volcanic ash particles in the presence of water droplets. *Atmospheric Measurement Techniques* 8 (12): 5213–5222.
3 Kaspar, S., Rattunde, M., Töpper, T. et al. (2013). Linewidth narrowing and power scaling of single-frequency 2.X μm GaSb-based semiconductor disk lasers. *IEEE Journal of Quantum Electronics* 49 (3): 314–324.
4 Kambs, B., Kettler, J., Bock, M. et al. (2016). Low-noise quantum frequency down-conversion of indistinguishable photons. *Optics Express* 24 (19): 22250.
5 Kucirek, P., Meissner, A., Nyga, S. et al. (2017). A single-frequency Ho:YLF pulsed laser with frequency stability better than 500 kHz. *Proceedings of SPIE* 10082: 100821K.
6 Holl, P., Rattunde, M., Adler, S. et al. (2015). Recent advances in power scaling of GaSb-based semiconductor disk lasers. *IEEE Journal of Selected Topics in Quantum Electronics* 21 (6): 1501012.
7 Hempler, N., Hopkins, J.-M., Rösener, B. et al. (2009). Semiconductor disk laser pumped Cr^{2+}:ZnSe lasers. *Optics Express* 17 (20): 18136.
8 Scholle, K., Lamrini, S., Adler, S. et al. (2016). SDL in-band pumped Q-switched 2.1 μm Ho:YAG laser. *Proceeding CLEO (Conference on Lasers and Elektro-Optics), San Jose, USA, 2016*.
9 Holl, P., Rattunde, M., Adler, S. et al. (2017). GaSb-based VECSEL for high-power applications and Ho-pumping. *SPIE LASE 2017, San Francisco, CA, USA. Proceedings of SPIE* 10087: 1008705.
10 Hildebrand, A., Kieleck, C., Tyazhev, A. et al. (2014). Laser damage of the nonlinear crystals $CdSiP_2$ and $ZnGeP_2$ studied with nanosecond pulses at 1064 and 2090 nm. *SPIE Optical Engineering* 53 (12): 122511–122511-6.

11 Kim, J.G., Shterengas, L., Martineli, R.U. et al. (2002). Room-temperature 2.5 µm InGaAsSb/AlGaAsSb diode lasers emitting 1 W continuous wave. *Applied Physics Letters* 81 (17): 3146.

12 Rattunde, M., Schmitz, J., Kaufel, G. et al. (2006). GaSb-based 2.X µm quantum well diode lasers with low beam divergence and high output power. *Applied Physics Letters* 88: 081115–081115-3.

13 Vizbaras, K., Vizbaras, A., Andrejew, A. et al. (2017). Room-temperature type-I GaSb-based lasers in the 3.0–3.7 µm wavelength range. *Proceedings of SPIE* 8277: 82771B.

14 Vizbaras, K. and Amann, M.-C. (2012). Room-temperature 3.73 µm GaSb-based type-I quantum-well lasers with quinternary barriers. *Semiconductor Science and Technology* 27: 032001.

15 Garnache, A., Ouvrard, A., Cerutti, L. et al. (2006). 2-2.7 µm single frequency tunable Sb-based lasers operating in CW at RT: microcavity and external-cavity VCSELs, DFB. *SPIE Semiconductor Lasers and Laser Dynamics II, Strasbourg, France. Proceedings of SPIE* 6184: 61840N.

16 Figueiredo, P., Suttinger, M., Rowel, G. et al. (2017). Progress in high-power continuous-wave quantum cascade lasers. *Applied Optics* 56 (31): H15.

17 Razeghi, M., Zhou, W., Slivken, S. et al. (2017). Recent progress of quantum cascade laser research for 3 to 12 µm at the Center for Quantum Devices. *Applied Optics* 56 (31): H30.

18 Lin, C.-H., Yang, R.Q., Zhng, D. et al. (1977). Type-II interband quantum cascade laser at 3.8 µm. *Electronics Letters* 33 (7): 598–599.

19 Vurgaftman, I., Meyer, J.R., and Ram-Mohan, L.R. (1998). Mid-IR vertical-cavity surface-emitting lasers. *IEEE Journal of Quantum Electronics* 34 (1): 147–156.

20 Vurgaftman, I., Bewley, W., Canedy, C. et al. (2011). Rebalancing of internally generated carriers for mid-infrared interband cascade lasers with very low power consumption. *Nature Communications* 2: 585.

21 Shterengas, L., Liang, R., Kipshidze, G. et al. (2013). Type-I quantum well cascade diode lasers emitting near 3 µm. *Applied Physics Letters* 103: 121108–121108-3.

22 Shterengas, L., Kipshidze, G., Hosoda, T. et al. (2017). Cascade pumping of 1.9–3.3 µm type-I quantum well GaSb-based diode lasers. *IEEE Journal of Selected Topics in Quantum Electronics* 23 (6): 1500708.

23 Cerutti, L., Garnache, A., Ouvrard, A., and Genty, F. (2004). High temperature continuous wave operation of Sb-based vertical external cavity surface emitting laser near 2.3 µm. *Journal of Crystal Growth* 268: 128–134.

24 Kemp, A.J., Valentine, G.J., Hopkins, J.M. et al. (2005). Thermal management in vertical-external-cavity surface-emitting lasers: finite-element analysis of a heatspreader approach. *IEEE Journal of Quantum Electronics* 41 (2): 148–155.

25 Holl, P., Rattunde, M., Adler, S. et al. (2016). GaSb-based 2.0 µm SDL with 17 W output power at 20 °C. *Electronics Letters* 52 (21): 1794.

26 Fill, M., Khiar, M., Felder, F., and Zogg, H. (2011). PbSe quantum well mid-infrared vertical external cavity surface emitting laser on Si-substrates. *Journal of Applied Physics AIP* 109: 093101–093101-6.

References

27 Ishida, A., Sugiyama, Y., Isaji, Y. et al. (2011). 2 W high efficiency PbS mid-infrared surface emitting laser. *Applied Physics Letters AIP* 99: 121109–121109-3.

28 Arafin, S., Bachmann, A., Vizbaras, K. et al. (2011). Comprehensive analysis of electrically pumped GaSb-based VCSELs. *Optics Express* 19 (18): 17267.

29 Cerutti, L., Ducanchez, A., Grech, P. et al. (2008). Room-temperature, monolithic, electrically-pumped type-I quantum-well Sb-based VCSELs emitting at 2.3 μm. *Electronics Letters* 44 (3): 203–205.

30 Veerabathran, G.K., Sprengel, S., Andrejew, A., and Amann, M.-C. (2018). Electrically pumped VCSELs using type-II quantum wells for the mid-infrared. *SPIE OPTP 2018, San Francisco, CA, USA. Proceedings SPIE* 10536: 1053602.

31 Bewley, W.W., Felix, C.L., Vurgaftman, I. et al. (1998). Continuous-wave mid-infrared VCSEL's. *IEEE Photonics Technology Letters* 10 (5): 660–662.

32 Vurgaftman, I., Meyer, J., and Ram-Mohan, L. (2001). Band parameters for III-V compound semiconductors and their alloys. *Applied Physics Reviews* 89: 5815.

33 Alibert, C., Skouri, M., Joullie, A. et al. (1991). Refractive-indexes of AlSb and GaSb-lattice-matched $Al_xGa_{1-x}As_ySb_{1-y}$ in the transparent wavelength region. *Journal of Applied Physics* 69 (5): 3208–3211.

34 Shim, K., Rabitz, H., and Dutta, P. (2000). Band gap and lattice constant of $Ga_xIn_{1-x}As_ySb_{1-y}$. *Journal of Applied Physics* 88: 7157–7161.

35 Rattunde, M., Schmitz, J., Mermelstein, C., and Wagner, J. (2006). III-Sb-based type-I QW diode lasers. In: *Mid-infrared Semiconductor Optoelectronics*, Springer Series in Optical Science (ed. A. Krier). London: Springer.

36 Bückers, C., Thränhardt, A., Koch, S.W. et al. (2008). Microscopic calculation and measurement of the laser gain in a (GaIn)Sb quantum well structure. *Applied Physics Letters* 92: 071107–071107-3.

37 Turner, C.W., Choi, H.K., and Manfra, M.J. (1998). Ultralow-threshold ($50 A/cm^2$) strained single-quantum-well GaInAsSb/AlGaAsSb lasers emitting at 2.05 μm. *Applied Physics Letters* 72 (8): 876.

38 Töpper, T., Rattunde, M., Kaspar, S. et al. (2012). High-power 2.0 μm semiconductor disk laser – influence of lateral lasing. *Applied Physics Letters AIP* 100: 192107–192107-3.

39 Andrejew, A., Sprengel, S., and Amann, M.-C. (2016). GaSb-based vertical-cavity surface-emitting lasers with an emission wavelength at 3 μm. *Optics Letters* 41 (12): 2799–2802.

40 Corzine, S.W., Geels, R.S., Scott, J.W. et al. (1989). Design of Fabry-Perot surface-emitting lasers with a periodic gain structure. *IEEE Journal of Quantum Electronics* 25 (6): 1513–1524.

41 Paajaste, J., Suomalainen, S., Koskinen, R. et al. (2009). High-power and broadly tunable GaSb-based optically pumped VECSELs emitting near 2 μm. *Journal of Crystal Growth* 311 (7): 1917–1919.

42 Holl, P., Rattunde, M., Adler, S. et al. (2016). Optimization of 2.5 μm VECSEL: influence of the QW active region. *SPIE LASE 2016, San Francisco, CA, USA. Proceedings of SPIE* 9734: 97340S.

43 Paajaste, J., Koskinen, R., Nikkinen, J. et al. (2011). Power scalable 2.5 μm (AlGaIn)(AsSb) semiconductor disk laser grown by molecular beam epitaxy. *Journal of Crystal Growth* 323 (1): 454–456.

44 Rösener, B., Schulz, N., Rattunde, M. et al. (2008). High-power, high-brightness operation of a 2.25 μm (AlGaIn)(AsSb)-based barrier-pumped vertical-external-cavity surface-emitting laser. *IEEE Photonics Technology Letters* 20 (7): 502–504.

45 Rösener, B., Rattunde, M., Moser, R. et al. (2011). Continuous-wave room-temperature operation of a 2.8 μm GaSb-based semiconductor disk laser. *Optics Letters* 36 (3): 319–321.

46 Beyertt, S.-S., Zorn, M., Kübler, T. et al. (2005). Optical in-well pumping of a semiconductor disk laser with high optical efficiency. *IEEE Journal of Quantum Electronics* 41 (12): 1439–1449.

47 Schulz, N., Rattunde, M., Ritzenthaler, C. et al. (2007). Resonant optical in-well pumping of an (AlGaIn)(AsSb)-based vertical-external-cavity surface-emitting laser emitting at 2.35 μm. *Applied Physics Letters* 91: 091113–091113-3.

48 Schulz, N., Rösener, B., Moser, R. et al. (2008). An improved active region concept for highly efficient GaSb-based optically in-well pumped vertical-external-cavity surface-emitting lasers. *Applied Physics Letters* 93: 181113–181113-3.

49 Mateo, C., Brauch, U., Kahle, H. et al. (2016). 2.5 W continuous wave output at 665 nm from a multipass and quantum-well-pumped AlGaInP vertical-external-cavity surface-emitting laser. *Optics Letters* 41 (6): 1245.

50 Heinen, B., Wang, T.L., Sparenberg, M. et al. (2012). 106 W continuous-wave output power from vertical-external-cavity surface-emitting laser. *Electronics Letters* 48 (9): 516.

51 Liau, Z.L. (2000). Semiconductor wafer bonding via liquid capillarity. *Applied Physics Letters* 77 (5): 651–653.

52 Mildren, R.P. (2013). Intrinsic optical properties of diamond. In: *Optical Engineering of Diamond*, 1–34. Wiley-VCH GmbH & Co. KGaA.

53 Hopkins, J.M., Hempler, N., Rösener, B. et al. (2008). High-power (AlGaIn)(AsSb) semiconductor disk laser at 2.0 μm. *Optics Letters* 33 (2): 201–203.

54 Kuznetsov, M., Hakimi, F., Sprague, R., and Mooradian, A. (1999). Design and characteristics of high-power (>0.5-W CW) diode-pumped vertical-external-cavity surface-emitting semiconductor lasers with circular TEM_{00} beams. *IEEE Journal of Selected Topics in Quantum Electronics* 5 (3): 561–573.

55 Devautour, M., Michon, A., Beaudoin, G. et al. (2013). Thermal management for high-power single-frequency tunable diode-pumped VECSEL emitting in the near- and mid-IR. *IEEE Journal of Selected Topics in Quantum Electronics* 19 (4): 1701108.

56 Rehm, R., Walther, M., Schmitz, J. et al. (2012). Substrate removal of dual-colour InAs/GaSb superlattice focal plane arrays. *Physica Status Solidi C* 9 (2): 318.

57 Perez, J.-P., Laurain, A., Cerutti, L. et al. (2010). Technologies for thermal management of mid-IR Sb-based surface emitting lasers. *Semiconductor Science and Technology* 25: 045021. (6pp).

58 Adler, S., Holl, P., Lindner, C. et al. (2017). Continuous-tunable single-frequency 2 μm GaSb-based thin device semiconductor disk laser. *Proceedings CLEO Europe, München, Germany* (25–29 June 2017).

59 Myara, M., Sellahi, M., Laurain, A. et al. (2013). Noise properties of NIR and MIR VECSELS. *Proceedings of SPIE LASE, Vol. 8606, 86060Q-1-13, San Francisco, CA, USA* (18 February 2013).

60 Garnache, A., Ouvrard, A., and Romanini, D. (2007). Single-frequency operation of external-cavity VCSELs: non-linear multimode temporal dynamics and quantum limit. *Optics Express* 15 (15): 9403–9417.

61 Schawlow, A.L. and Townes, C.H. (1958). Infrared and optical masers. *Physical Review* 112 (6): 1940–1949.

62 Shterengas, L., Belenky, G.L., Gourevitch, A. et al. (2002). Measurements of a α-factor in 2–2.5 μm type-IIn(Al)GaAsSb/GaSb high power diode lasers. *Applied Physics Letters* 81: 4517–4519.

63 Ouvrard, A., Garnache, A., Cerutti, L. et al. (2005). Single-frequency tunable Sb-based VCSELs emitting at 2.3 μm. *IEEE Photonics Technology Letters* 17 (19): 2020–2022.

64 Jacquemet, M., Domenech, M., Lucas-Leclin, G. et al. (2007). Single-frequency cw vertical external cavity surface emitting semiconductor laser at 1003 nm and 501 nm by intracavity frequency doubling. *Applied Physics B* 86: 503–510.

65 Neuhaus, R. (2013). Application Note "Diode Laser Locking and Linewidth Narrowing", https://www.toptica.com

66 Khiar, A., Rahim, M., Fill, M. et al. (2010). Continuously tunable monomode mid-infrared vertical external cavity surface emitting laser on Si. *Applied Physics Letters* 97: 151104–151104-3.

67 Siegman, A.E. (1986). *Lasers*. Mill Valley, CA: University Science.

68 Feng, S. and Winful, H.G. (2001). Physical origin of the Gouy phase shift. *Optics Letters* 26 (8): 485–487.

69 Kemp, A.J., Hopkins, J.-M., Maclean, A.J. et al. (2008). Thermal management in 2.3-μm semiconductor disks lasers: a finite element analysis. *IEEE Journal of Quantum Electronics* 44 (2): 125–135.

70 Kaspar, S., Rattunde, M., Schilling, C. et al. (2013). Micro-cavity 2-μm GaSb-based semiconductor disk laser using high-reflective SiC heatspreader. *Applied Physics Letters* 103: 041117–041117-4.

71 Hopkins, J.-M., Maclean, A.J., Riis, E. et al. (2007). Tunable, single-frequency, diode-pumped 2.3 μm VECSEL. *Optics Express* 15 (13): 8215.

72 Rösener, B., Kaspar, S., Rattunde, M. et al. (2011). 2 μm semiconductor disk laser with a heterodyne linewidth below 10 kHz. *Optics Letters* 36 (18): 3587–3589.

73 Kaspar, S., Rattunde, M., Töpper, T. et al. (2012). Semiconductor disk laser at 2.05 μm wavelength with <100 kHz linewidth at 1 W output power. *Applied Physics Letters* 100: 031109–031109-3.

74 Geerlings, E., Rattunde, M., Schmitz, J. et al. (2006). Widely tunable GaSb-based external cavity diode laser emitting around 2.3 μm. *IEEE Photonics Technology Letters* 18 (18): 1913–1915.

75 Paboeuf, D. and Hastie, J.E. (2016). Tunable narrow linewidth AlGaInP semiconductor disk laser for Sr atom cooling applications. *Applied Optics* 55 (19): 4980.

76 Kaspar, S., Rösener, B., Rattunde, M. et al. (2011). Sub-MHz-linewidth 200-mW actively stabilized 2.3-μm semiconductor disk laser. *IEEE Photonics Technology Letters* 23 (20): 1538–1540.

77 Okoshi, T., Kikuchi, K., and Nakayama, A. (1980). Novel method for high resolution measurement of laser output spectrum. *Electronics Letters* 16 (16): 630–631.

78 Tsuchida, H. (1990). Simple technique for improving the resolution of the delayed self-heterodyne method. *Optics Letters* 15 (11): 640–642.

79 Freed, C. (1968). Design and short-term stability of single-frequency CO_2 lasers. *IEEE Journal of Quantum Electronics* 4 (6): 404.

80 Stephan, G.M., Tam, T.T., Blin, S. et al. (2005). Laser line shape and spectral density of frequency noise. *Physical Review A* 71: 043809.

81 Olivero, J.J. and Longbothum, R.L. (1977). Empirical fits to the Voigt line width: a brief review. *Journal of Quantitative Spectroscopy and Radiative Transfer* 17 (2): 233.

82 Drever, R.W.P., Hall, J.L., Kowalski, F.V. et al. (1983). Laser phase and frequency stabilization using an optical resonator. *American Journal of Physics B* 31 (2): 97.

83 Black, E.D. (2000). An introduction to Pound-Drever-Hall laser frequency stabilization. *American Journal of Physics* 69 (1): 79.

84 Hall, J.L., Taubman, M.S., and Ye, J. (2001). Laser stabilization. In: *Handbook of Optics: Fiber Optics and Nonlinear Optics*, vol. 4 (eds. M. Bass, J.M. Enoch, E.W.V. Stryland and W.L. Wolfe), 27.1. New York: McGraw-Hill.

85 Stoehr, H., Mensing, F., Helmcke, J., and Sterr, U. (2006). Diode laser with 1 Hz linewidth. *Optics Letters* 31 (6): 736.

86 Maclean, A.J., Kemp, A.J., Calvez, S. et al. (2008). Continuous tuning and efficient intracavity second-harmonic generation in a semiconductor disk laser with an intracavity diamond heatspreader. *IEEE Journal of Quantum Electronics* 44 (3): 216–225.

87 Rösener, B., Kaspar, S., Rattunde, M. et al. (2010). Continuous tuning and narrow-linewidth operation of GaSb-based semiconductor disk lasers. *Poster and Proceedings of MIOMD X (Mid Infrared Optoelectronics: Materials and Devices), Shanghai, China* (5–9 September 2010).

88 Härkönen, A., Guina, M., Rößner, K. et al. (2007). Tunable self-seeded semiconductor disk operating at 2 μm. *Electronics Letters* 43 (8): 457–458.

89 Pereira, D., Rautiainen, J., Härkönen, A., and Okhotnikov, O. (2007). Spectral and spatial mode control in self-seeded semiconductor disk laser using optical feedback from fiber Bragg grating. *Proceeding of Quantum Electronics and Laser Science Conference*, Baltimore, USA (6–11 May 2007), pp. 1–2, JWA120.

90 Lenstra, D., Verbeek, B.H., and Den Boef, A.J. (1985). Coherence collapse in single-mode semiconductor lasers due to optical feedback. *IEEE Journal of Quantum Electronics* QE-21 (6): 674–679.

91 Hastie, J.E., Hopkins, J.-M., Jeon, C.W. et al. (2003). Microchip vertical external cavity surface emitting lasers. *Electronics Letters* 39 (18): 507–508.
92 Smith, S.A., Hopkins, J.-M., Hastie, J.E. et al. (2004). Diamond-microchip GaInNAs vertical external-cavity surface-emitting laser operating CW at 1315 nm. *Electronics Letters* 40 (15): 935–936.
93 Kemp, A.J., MacLean, A.J., Hastie, J.E. et al. (2006). Thermal lensing, thermal management and transverse mode control in microchip VECSELs. *Applied Physics B: Lasers and Optics* 83: 189–194.
94 Kaspar, S., Rattunde, M., Holl, P. et al. (2014). 2-μm high brilliance micro-cavity VECSEL with >2W output power. *SPIE LASE, San Francisco, CA, USA. Proceedings of SPIE* 8966: 89660T.
95 Triki, M., Cermak, P., Cerutti, L. et al. (2008). Extended continuous tuning of a single-frequency diode-pumped vertical-external-cavity surface-emitting laser at 2.3 μm. *IEEE Photonics Technology Letters* 20 (23): 1947–1949.

4

Highly Coherent Single-Frequency Tunable VeCSELs: Concept, Technology, and Physical Study

Mikhael Myara

Université Montpellier, IES-CNRS UMR 5214, Montpellier Cedex 05, France

4.1 Introduction: Lasers for Applications

Applications working with continuous light wave need specific properties concerning the light that is emitted by the source. For many of these applications, the electrical field has to be precisely controlled (coherent) regarding its localization in space, its spectrum (or energy state), and its polarization state. This is what makes more efficient the interaction between light and matter or between light and various optical devices. For example, injecting light at an optical resonance, with large power density and with the good polarization state results in higher efficiency or sensitivity in lots of situations (atomic-clocks [1], detection of gravitational waves [2], trace gas specrtroscopy [3] etc.). In other situations, this control is especially critical on at least one of these three physical properties of the field:

- For long-range applications, a good coherence of the spatial distribution of the light is mandatory, and it's the same for injecting efficiently the laser light in a single-mode optical fiber or in a Fabry–Perot cavity or to increase the amount of encodable information for a given surface in data-storage applications.
- Time coherence/spectral resolution is necessary for a good signal-to-noise ratio in telecommunication systems [4], for gas spectroscopy [3], or for most interferometry-oriented applications such as telemetry (OFDR [5]).
- In this last case, the polarization state stability is crucial too.

Moreover, most applications need to exist "out-of-the lab" in real conditions and thus need to be embedded in a compact and integrated package, with good power efficiency. All the properties of the light should allow easy injection in the targeted optical system, with the fewer adaptations possible, in order to minimize the amount of external elements: this is important because a complex setup usually leads to reduced performances. For example, the output beam of an edge-emitting laser diode is generally so astigmatic and divergent that it requires complex wide-angle reshaping optics to be used in acceptable conditions in free space or for injection in an optical fiber, to avoid significant optical aberrations and/or optical

Vertical External Cavity Surface Emitting Lasers: VECSEL Technology and Applications, First Edition.
Edited by Michael Jetter and Peter Michler.
© 2022 WILEY-VCH GmbH. Published 2022 by WILEY-VCH GmbH.

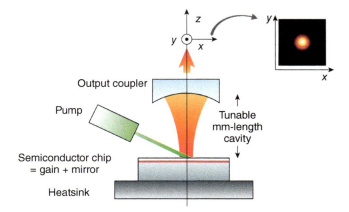

Figure 4.1 Basic design of a VeCSEL. The longitudinal (z) axis and the transverse (x, y) plane are displayed.

losses; another telling example is the one of some technologies of single-frequency fiber lasers that require external optical amplification to reach a suitable level of optical power, which is at the expense of the intensity noise [6]. In short, we can say that it is usually more efficient for a given property to come from the laser source itself rather than to be obtained outside, because it takes advantage of the amplification by the laser.

All these properties of the light source are critical for most applications, but to fit the most advanced requirement, the laser source should also exhibit some extra functionalities. An important example is laser spectroscopy. For this purpose, the laser source has also to demonstrate an added feature: the *continuous tunability*. By moving the laser wavelength λ to accurately each wavelength of interest in a given range, the continuous tunability enables the probing of various materials' spectral properties without missing any variation, for example, in an absorption spectrum involving multiple gas species.

At first glance, the aforementioned features define the specifications of "the ideal laser" that we want to describe here.[1] In this chapter, we propose a design and discuss our implementation of this "ideal" laser source by means of the VeCSEL technology (see Figure 4.1), a laser that redefines the state of the art, by outclassing the other laser technologies for this purpose.

A VeCSEL is a laser based on a linear, paraxial cavity. A specificity of this kind of laser topology is that its wavefunction can be divided in two distinct parts. The first one concerns the *longitudinal* axis (z) of the cavity that defines the temporal properties of the laser. The second one concerns the *transverse* plane (x, y) that defines the

---

1 This definition of the "ideal laser" concerns a stereotyped vision of the light coherence. We describe here the light emitted by a class of very demanding lasers, as it suits lots of applications. Because the coherence is mathematically defined in a more general frame [7, 8], other kinds of sources can be said to be coherent following other criteria. However, these sources are not the purpose of this chapter and will be discussed, for some of them, in part II about modelocking, or, for CW lasers, in the literature in [18, 19, 46, 57–60].

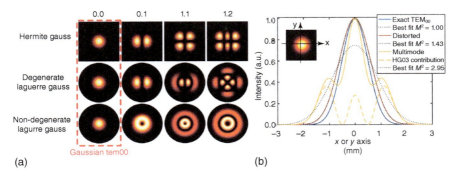

Figure 4.2 (a) Transverse intensity distribution of some possible transverse modes of Fabry–Perot cavities. (b) Exact TEM$_{00}$ compared with distorted and multimode beam, with associated M^2 values.

spatial profile of the light. The VeCSEL is composed of a semiconductor chip, an output coupler, and in most cases an optical pump (electrical pumping is often possible) [9]. The semiconductor chip embeds both the gain medium and a multilayer High-Reflectivity (HR) backside mirror. The mm–cm long cavity can be adjusted for the continuous-wavelength tunability purposes, over wide spectral ranges (close to THz demonstrated [10]). In the following, we focus on the design parameters of a VeCSEL that demonstrates tunability, compactness, and high coherence.

4.2 The "Ideal" Laser

The purpose of what we call here "the ideal laser" in continuous-wave regime is to generate a high-power light (from some mW to multiwatt operation), which is strongly confined on three physical observable quantities:

- *Space*: the generated light has to be focusable in the smallest possible spot, as predicted.[2] by the diffraction. This property is called "spatial coherence" and leads to "brightness" in the case of high-power lasers. The spatial coherence also induces the possibility to collimate the beam over a very long distance. Most spatially coherent lasers emit solely the *transverse* TEM$_{00}$ mode from the Laguerre–Gauss or Hermite–Gauss basis [11], (see Figure 4.2a). An exact TEM$_{00}$ emission is said to be "diffraction limited." Any mismatching observed between the emitted light profile and the exact TEM$_{00}$ mode is usually estimated by means of the propagation factor $M^2 \geq 1$. $\sqrt{M^2}$ can be understood like the unexpected radius increase of the beam [12], compared with the diffraction-limited (theoretical) one, defined to be w_0 in near field and θ_c in far field.[3] Both are related

---

2 In extreme focusing conditions, its radius tends to be $2\lambda/\pi^2 \approx 0.2\lambda$ where λ is the wavelength.
3 Both w_0 and θ_0 are evaluated at $1/\exp(2)$ from the maximum on the intensity profile. However, choosing an appropriate indicator to define the size of a beam is not something easy. We encourage the reader to examine the very interesting work of Siegman [12] for free-space beams, and the works of Petermann [13, 14] for confined/guided light.

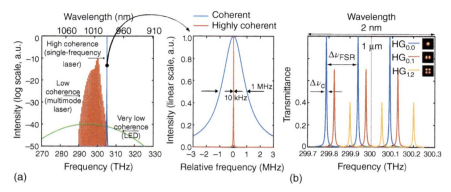

Figure 4.3 (a) Various degrees of time coherence, depending on multimode operation and also on the ability of the cavity to filter the spontaneous emission, thus on its Finesse $\mathcal{F}$. Please note the range ratio ($\approx 10^7$) between the two frequency ranges. (b) Longitudinal resonances of HG_{00} (TEM_{00}) for a passive Fabry–Perot cavity, as well as the resonances of two higher-order transverse modes. We also depict two fundamental parameters of the cavity transmittance spectrum: its periodicity $\Delta\nu_{FSR}$ and its FWHM linewidth $\Delta\nu_c$.

by the invariant parameter from the Gaussian beam theory $w_0\theta_0 = \lambda/\pi M^2$ [11]. However, it makes more sense, for better quantification or physical analysis, to perform a more advanced investigation, by observing the unexpected phase variations in the transverse plane, as well as performing a Zernike decomposition of the wavefront [15].

- *Time/spectrum*: the spectrum of the light source must be as narrow as possible. This property is called "monochromatic wave" or "long coherence time" and means that the electric field follows a perfect, pure, and unmodulated sine wave oscillation at the optical frequency ν_0,[4] that was born a long time ago and that will die in a very far future. Whatever the physical origin of the spectral width (multimode operation and/or temporal phase distortion, see Figure 4.3a and b), a classical way to quantify this time coherence is the Full-Width-Half-Maximum (FWHM) of the laser spectrum, which is, from high to low coherence, in the kHz–GHz range for most single-frequency free-running lasers.[5] However, because the possible physical reasons for broadening are many, studying the coherence of a laser by means of the spectrum FWHM is not always relevant. For better analysis and more accuracy, one may look at the time-domain fluctuations of the laser field instant frequency, often quantified its the Allan Variance, but more usually through its spectral density [17]. That's what we do in the following.

- *Polarization*: the polarization state of the electrical field must be unique in space and time. In most cases, it will be linear, following a specific direction, but it can also be circular, left or right-handed, or elliptical. This property is called "polarization coherence."

---

4 This frequency is in the 100–1000 THz range for usual ultraviolet to infrared wavelengths.
5 It can go close to Hz level for lasers actively locked on a frequency reference [16].

Aiming at this kind of coherence, the laser design will have to *induce the selection of a single transverse, longitudinal, and polarization mode* of the cavity (see Figure 4.3b). Of course, this ability for the laser strongly depends on its technology and design; however, *reaching a single-mode operation is only the first step*: it is necessary but not sufficient for high coherence, and it needs to be discussed in the following. These two topics are the purpose of the two next sections.

4.3 Toward Single-Mode Operation

In most cases, each mode of a passive cavity is identified by a specific optical resonance frequency (see Figure 4.3b). These resonances can however be displayed in a more convenient way (see Figure 4.4a-c), by using the dispersion relation that exists between the wave vector k and the optical frequency ν in free space:

$$\nu = c/\lambda = c/2\pi \times k \tag{4.1}$$

On the longitudinal axis, these resonance frequencies are well-known (-a) as each longitudinal mode resonates at a frequency located by one of the tooth of a frequency comb, periodicity of which is given by the Free Spectral Range, defined as follows:

$$\Delta \nu_{\text{FSR}} = c/(2nD) \tag{4.2}$$

where c is the speed of light in vacuum, n the optical index of the medium in the cavity, and D the cavity length.

Each mode of the transverse plane supports a frequency comb with exactly the same periodicity $\Delta \nu_{\text{FSR}}$ on the longitudinal axis. However, the frequency comb associated to a given transverse mode is shifted by an offset that depends on the base of modes supported by the cavity. For a VeCSEL, it is usually the Hermite–Gauss or

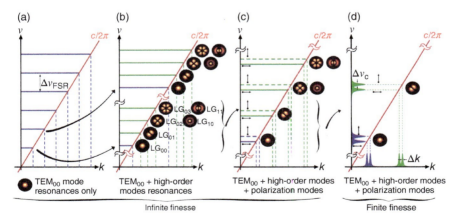

Figure 4.4 Relation of dispersion (Eq. 4.1) for optical modes. (a) Longitudinal modes for TEM_{00}. (b) *Zoom of (a)*: Transverse modes added. (c) *Zoom of (b)*: Polarization modes added. (d) *Zoom of (c)*: Impact of a finite Finesse on the wave vector k.

Laguerre–Gauss base. In the last case, the absolute resonance frequencies are [11]:

$$\nu_{q,p,l} = \Delta\nu_{\text{FSR}} \left(q + (1 + 2p + |l|)\varphi/\pi \right) \tag{4.3}$$

where q, p, and l are the modes' quantum numbers, and $\varphi = \cos^{-1}\left(\sqrt{1 - D/R_c}\right)$ is known as the Gouy phase shift (R_c is the radius of curvature of the mirror). For that reason, the resonance frequencies for a given transverse mode can be equal or different to the ones of other transverse modes, depending on their quantum numbers.

For the polarization modes (-c), the cavity anisotropy (birefringence) lifts the frequency degeneracy of the two orthogonal polarization states, splitting the Fabry–Perot mode into two distinct polarization modes at different frequencies.

We have now finished describing the passive cavity: we now power the laser on and aim at selecting a single longitudinal, transverse, and polarization mode, chosen among the ones supported by the cavity. This selection is possible because the modes undergo dissimilarities concerning their own gain (called *modal gain*) or their own losses, and there are numerous physical reasons for that. First of all, as described above, most modes are not located at the same resonance frequency, and then, the simple spectral dependency of the gain (see Figure 4.5a) creates gain differences between the modes. But other reasons exist; for example, if some gain dichroism occurs, a polarization state will be more amplified than another one. For the transverse modes, the spatial overlap between the pump and each mode (see Figure 4.5c) leads to more or less efficient pumping of each individual mode, leading to more or less gain.

Considering this discrimination that exists in the gain and the losses for the cavity modes, the first effect that makes mode selection happen is simply the threshold condition: all the modes that do not experience a gain stronger than their losses are not eligible for lasing. Lots of laser designs are based on this principle, and their single-frequency operation usually relies on the introduction of filters inside the laser cavity.

The other modes that fulfill the threshold condition will experience "mode competition." The mode competition stems from the fact that all the modes of an optical cavity share the available population inversion (or gain) and thus compete for it. The rise of a mode will generally reduce the available gain for other modes [18]. To illustrate this idea, Hermann Haken used the biological natural selection analogy in his book "Laser light dynamics" [20] to explain this process: we visualize here the different modes as different kinds of animals, and the inversion of population as food, which is continuously fed into the system. The kind of animal that has a better access to the food sees its own population growing more quickly and can thus eat more food. The other kinds cannot compete in eating and eventually perish. This is what is illustrated in Figure 4.5a and b for the simple case of two longitudinal modes.[6] In this example, we assume a very small gain (food) difference between two modes: 0.2%. During the very beginning of the laser start-up in time, the two populations grow and are quite indistinguishable, because the gain difference is very weak. However it exists, and once a critical population size is reached, this difference in gain (or food)

---

6 A good starting point for such a modeling is [21].

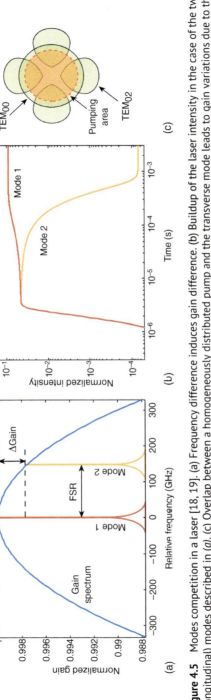

Figure 4.5 Modes competition in a laser [18, 19]. (a) Frequency difference induces gain difference. (b) Buildup of the laser intensity in the case of the two (longitudinal) modes described in (a). (c) Overlap between a homogeneously distributed pump and the transverse mode leads to gain variations due to the pump-to-mode overlap. Depicted in the transverse plane here, but in real conditions this is to be understood in volume.

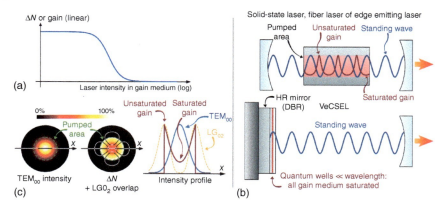

Figure 4.6 Spatial Hole Burning in lasers. (a) Gain saturation as a function of the laser intracavity intensity. (b) Longitudinal Spatial Hole Burning. (c) Transverse Spatial Hole Burning for a homogeneous pumping over the TEM$_{00}$ waist.

has significant effects, leading to the domination of one of the two species, sacrificing the other. That is due to the *saturation of the gain medium by the laser power (see*[7] *Figure 4.6a)*: because the rate at which the food regenerates is limited (i.e. the pump power[8] is finite), the population that experiences the strongest gain can grow enough to eat all the food at quite the same speed at which the food regenerates, so that other populations vanish. This line of thinking leads to a very important conclusion: two different kinds can coexist only if they live from separated resources of food,[9] in other words *two optical modes can only coexist if they consume separated population inversions* ΔN. However, we have to mention that, in the foregoing, we only considered the "photon population" nature of the optical modes, and thus we forgot that optical modes are waves too. Due to their wave nature, two modes can be indistinguishable if the spectral distance between them is so short that $\Delta \nu_{q,p,l} < 2\pi \Delta \nu_c$, where $\Delta \nu_c$ is the cold-cavity mode width, see Figure 4.6. If this condition is respected, then the laser can lock the two considered modes together, leading to a multiple frequency operation without requiring separated population inversion for each. In the following, we will never consider this so specific case and will always work with modes in terms of photons populations.

In a laser, various reasons can lead to the existence of multiple independent population inversions and thus to the unwanted multimode operation:

- First of all, in some gain media, all the atoms do not provide gain in accurately the same spectral area, which is similar to say that the global gain bandwidth

---

7 *The gain saturation usually responds to a nonlinear law close to $1/(1 + S/S_0)$, with S being the laser intensity in the gain medium (not the pump intensity) and S_0 the saturation intensity. This is the law depicted in Figure 4.6a.*

8 The optical power can be understood like a rate of photons per second, and thus, the pump power, like a rate of excitation per second. This is discussed in Section 4.11.1 and illustrated by Eq. 4.15.

9 Or if they experience rigorously identical gain and losses, which is never obtained in practical cases.

is the combination of many independent population inversions, each addressing a given spectral region: those materials are said to experience "inhomogeneous spectral broadening" [11]. For that reason, the light emitted at a given wavelength can not saturate the other population inversions, leading to multimode operation: this what is called "Spectral Hole Burning." In semiconductor materials, quantum wells exhibit homogeneous gain properties due to the homogeneity of the atomic cristal, while quantum dots technology lead typically to a nonuniform atomic organization, thus to inhomogeneous gain broadening. Because VeCSELs are based on quantum wells, they do not experience this kind of inhomogeneity.
- The inhomogeneity can also come from the fact that two uncoupled spin states exist in the gain medium: each of them can feed two independent polarization modes at the same time, leading to gain inhomogeneity [22].
- Finally, inhomogeneity can exist because two modes eat in two separate regions of the gain medium. This last case is called "Spatial Hole Burning" (SHB) and is very common in lasers, especially concerning the longitudinal axis. However, we can take advantage of the VeCSEL technology to make it vanish. As a matter of fact, semiconductors allow to work with very thin gain layers such as quantum wells, the size of which is far below λ, because the gain in semiconductors is huge: this obviously leads to spatial homogeneity on the longitudinal axis (see Figure 4.6b).

As a result, a laser launches a more or less pure single-frequency light, depending on the gain medium homogeneity and on (complex) nonlinear effects [23, 24], and this purity has to be quantified.

In the case of longitudinal modes, it is made by computing the ratio between the power obtained in the main lasing mode and the power in the strongest side mode: that is the "Side Mode Suppression Ratio" (SMSR). The fundamental limit, due to the amount of spontaneous emission, is nearly identical for all the cavity modes. For the mode situated at $q \times \Delta v_{FSR}$ from the strongest mode, the SMSR limit in dB is given by [24], in the case of homogeneous broadening of the gain:

$$\text{SMSR}_q \simeq 10 \, \log \left[\frac{P_{out} \, \lambda}{h \, c \times 2\pi \Delta v_c \times \xi} \times \left(\frac{q \, \Delta v_{FSR}}{\Delta v_{gain}} \right)^2 \right] \tag{4.4}$$

where P_{out} is the output power, ξ the spontaneous emission factor, and Δv_{gain} the optical gain bandwidth. In the case of the VeCSEL, for typical internal losses of 1%, $\Delta v_{gain} = 6$ THz and $\xi \approx 1$, this leads to $\text{SMSR}_q > 60$ dB for a cavity length $D > 1$ mm. A important point here is that SMSR_q increases when the cavity mode width Δv_c narrows, which is a manifestation of the filtering ability of a good cavity.

In the transverse plane (see Figure 4.6c), the single-mode operation is easily obtained by pumping the VeCSEL chip with a waist that is a little smaller than the TEM$_{00}$ waist of the cavity: most high-order modes do not experience enough gain to rise, while low-order modes are eliminated by the competition.[10] The cavity waist is imposed by the study of the cavity stability, due to the relation between the cavity parameters and the waist of the Gaussian beam. In the very usual case of a

---

10 A model of this transverse behavior is described in [19, 25].

plano-concave VeCSEL cavity, the relation between the waist on the plane mirror surface and the cavity parameters is given by the Gaussian beam theory [11], as follows:

$$w_0 = \sqrt{\frac{\lambda}{\pi} \sqrt{D(R_c - D)}} \tag{4.5}$$

Still concerning the transverse axis, another advantage of the gain thinness is that the pump can be efficiently absorbed through a very narrow depth of field ($< \lambda$), making quite easy the use of pump lasers whatever their beam quality, because there is no longitudinal pumping, but only surface pumping. The VeCSEL is thus, intrinsically, an efficient converter, from low-coherence pump light to high-coherence laser light.

The linear polarization state is selected thanks to the gain dichroism: in III-V semiconductor nanostructures, the gain anisotropy leads to birefringence as well as gain dichroism, following the axes [110] and [1$\bar{1}$0] for a growth on usual substrate [100] [18]. In InGaAs/GaAs and InGaAsSb/GaSb quantum wells, this leads to a gain ratio close to 10% between both axes. This gain dichroism can be reinforced by a slightly asymetric pumping over the cristal axis. The big axis of the pump must correspond to the [110] axis of the cristal, where the gain is the strongest. Again, the two polarization modes compete, and the one exhibiting the strongest gain will rise. Like for SMSR, the strength of the single-polarization operation can be quantified by evaluating the suppression ratio $S_\perp$ between the main polarization mode and the orthogonal one. The fundamental limit of $S_\perp$ is fixed by the spontaneous emission:

$$S_\perp \simeq 10 \log \left[\frac{P_{out} \, \lambda}{h \, c \times 2\pi \Delta v_c \times \xi} \times \frac{\Delta G_\perp}{\overline{G}} \right] \tag{4.6}$$

where $\Delta G_\perp / \overline{G}$ is the gain dichroism, close to 10%. Again, $S_\perp$ increases with decreasing Δv_c.

We conclude that the spectral dynamics of the mode competition in VeCSEL is far more simple compared with most other lasers, making the single-frequency operation more natural for this kind of device. Of course, multimode operation can occur even with a clean VeCSEL configuration because of other nonlinear phenomena [24]. However, well-designed VeCSELs can reach the single-mode operation at large power without any intracavity filter [26], which is very unusual in the state of the art of lasers. This specific feature is strongly helpful to reach broadband continuous tunability and contributes to the coherence properties, since "securing" single-frequency operation with optical filters does impact also the wanted mode losses, decreasing its Finesse, thus limiting the coherence. In this context, the VeCSEL coherence properties are boosted compared with other technologies: this is the point of the next section.

4.4 Toward High Coherence

We examined how a laser can select a single mode, which is a key point for high-coherence emission. Of course, the stimulated emission process plays a crucial function in the construction of the emitted light. However, to achieve a high-coherence laser, the cold-cavity modes need a special attention because they are the very starting condition with which the stimulated emission plays. But only a few laser designs and technologies give the opportunity to reach nearly ideal cold-cavity modes. The impact of the technology on the cold cavity will be discussed in Section 4.5, after the following description of the parameters that influence the cold-cavity modes.

Concerning the transverse modes, we first have to mention that waveguide-based lasers always exhibit high spatial coherence if the waveguide is single-mode. One needs to know that this kind of design is not meant for high power because the waveguide dimension is a restriction to increase the fundamental mode size. In many practical cases, it also leads to very divergent beams (for example, in the case of laser diodes), which makes their use more complex. In the case of free-space cavities, power scaling and low divergence beams are easily obtained by increasing the cavity waist (see Eq. 4.5). However, obtaining the exact Laguerre–Gauss or Hermite–Gauss basis, thus the exact Gaussian TEM_{00} mode, is only possible with an infinite Finesse[11] $\mathcal{F}$ of the cold (Fabry–Perot) cavity, which is defined as follows:

$$\mathcal{F} = \frac{\Delta \nu_{FSR}}{\Delta \nu_c} = \pi/2 \left(\arcsin \left(\frac{1 - \sqrt{\mathcal{L}}}{2\sqrt[4]{\mathcal{L}}} \right) \right)^{-1} \tag{4.7}$$

where $\mathcal{L} = R_1 R_2 \exp(-2\alpha_i D)$ is the round-trip losses in intensity, with R_1 and R_2 the mirror reflectivities in intensity, α_i the intracavity losses coefficient, and D the cavity length. Of course, infinite Finesse cannot be obtained in practice, for various technical reasons and for fundamental ones because we want to extract a beam from the cavity to use it. Therefore, the beam formed by the cavity can be more or less close to the wanted TEM_{00} mode (see Figure 4.7a) and that will limit the spatial coherence. High Finesse also means long photon lifetime τ_c and numerous round trips $\mathcal{N}$:

$$\tau_c = \frac{1}{2\pi \Delta \nu_c} = \frac{\mathcal{F}}{2\pi \Delta \nu_{FSR}} = \frac{nD\mathcal{F}}{2\pi c} \quad \mathcal{N} \equiv \frac{\tau_c}{\tau_{RT}} = \frac{\mathcal{F}}{2\pi} \tag{4.8}$$

where $\tau_{RT} = 1/\Delta \nu_{FSR}$ is the round-trip time.

The impact of the Finesse $\mathcal{F}$ on the spatial modes can be described by a spreading around the eigenvalues, as depicted by the passage from Figure 4.4c-d. To explain this, we start again from the dispersion relation (Eq. 4.1). Because of the finite-Finesse spreading, the spectral width of the cavity modes $\Delta \nu_c$ is not zero. $\Delta \nu_c$ can be understood like an uncertainty on a given resonant frequency ν, which leads to an uncertainty Δk on the wave vector k, leading itself to a wider angular acceptance for each mode: the result is a lower ability to filter the light distribution in the transverse plane (see Figure 4.7a). The consequence is that the fundamental mode is not exactly the wanted Gaussian TEM_{00} mode, and this is worse if $\mathcal{F}$ is low, which occurs usually if the losses $\mathcal{L}$ are strong, as shown by Eq. (4.8).

In the high-coherence laser design, the intracavity losses α_i are usually the difficult point. Their physical origins are, in most cases, the absorption – for example in the gain medium itself – and some unwanted diffraction of the beam inside the cavity, for example, because of thermal lensing, which can change the stability condition of

---

11 Infinite Finesse thus means zero losses and zero-transmission cavity.

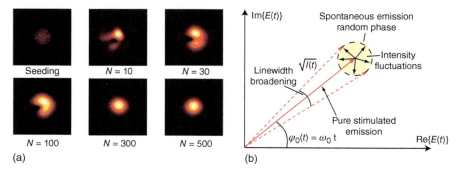

Figure 4.7 (a) Ability of a cavity to filter a noisy seeding as a function of the Finesse (here in terms of number of round trips $\mathcal{N} \equiv \mathcal{F}/2\pi$): observation of the output beam. This was obtained thanks to the Fox–Li method [11, 18, 27], by seeding the cavity with a waist-size disc. Inside the disk, the field was modulated in phase and intensity by low-frequency spatial noise. (b) Phasor illustration of the laser line width [28], in the complex plane.

the cavity (Eq. 4.5). For that reason, the impact of the thermal lens should be taken into consideration during the study of the cavity stability [26]. Moreover, strong thermal lensing can break the regular base of optical modes. In these cases, the spatial distribution of the light is a certain pattern of light, often called improperly "mode" of the cavity. The shape of these "modes" can be similar to a distorted TEM_{00} mode for example. Because they require additional optical gain (compared with regular cavity modes) to keep going, they exhibit more losses than regular modes; however, they can find a better overlap with the laser gain, making them eligible for lasing. The result is a deteriorated coherence emission. This can occur in VeCSELs, but it is easy to avoid because the absolute value of the gain is small, and thus in lots of VeCSEL configurations, these "modes" cannot reach the laser threshold.

Concerning the longitudinal modes, our starting point concerns the most fundamental limit of the time coherence in lasers. Such a limit contributes to the spectral width of single-frequency lasers and is known to originate from the time domain random variation of the phase of the spontaneous emission coupled to the lasing mode [28, 29]. This can be depicted by describing the laser field like the sum of a pure stimulated emission field with a pure spontaneous emission field. The stimulated emission field is perfectly defined concerning its instantaneous phase: because all the photons generated through this process are in phase coincidence, it oscillates following an exact and pure sine wave at the optical pulsation $\omega_0 = 2\pi \nu_0$ and thus exhibits an infinite coherence time. Due to the additional spontaneous field (see[12] Figure 4.7b), the overall laser field has not a perfectly defined phase in time, which leads to a broadening of the laser line for obvious Fourier transform reasons.

---

12 This representation is known for RF oscillators as the "Rice" representation [30]. For a laser: the pure stimulated emission vector describes perfect circles in time with periodicity $2\pi/\omega_0$, corresponding to the optical pulsation. The random spontaneous emission is added to this motion, leading to a global imperfect circular motion, denoting an imperfect sine-wave electrical field, limiting the coherence.

This broadening of the laser line, due to the spontaneous emission, is called the Schawlow–Towmes limit. Its expression is given below [28, 29]:

$$\Delta \nu_{\text{limit}} = \frac{\pi h \nu (\Delta \nu_c^2)}{P_{\text{out}}} \left\{ \xi(1 + \alpha_h^2) \right\} \qquad (4.9)$$

where $\Delta \nu_{\text{limit}}$ is the Full-Width-Half-Maximum of the laser spectrum,[13] h is the Planck constant, ν the average optical frequency, P_{out} the emitted power, and $\Delta \nu_c$ the already described cold-cavity linewidth. The term between accolades contains additionnal linewidth broadening factors: ξ is the spontaneous emission confinement factor, and α_h the well-known Henry factor, the value of which is usually stronger in the case of semiconductor materials [28] ($\alpha_h \approx 4$ in most cases) than with other materials. Depending on the laser design, the Schawlow–Townes limit can be, or not be, the predominant reason for linewidth broadening. The first step on the path to high temporal coherence is thus to minimize the impact of this contribution, which is related to the cold-cavity linewidth $\Delta \nu_c$, (Eq. 4.9), requiring thus a high Finesse design.

To summarize, we want here to point out the fact that *a good way to design a highly coherent laser* is first to *build a high Finesse cavity* to create an efficient filter for the light-modes, for spatial, polarization, and longitudinal reasons, as told by Eqs. (4.4), (4.6) and (4.9). Next, the laser oscillation inside this cavity must be maintained by means of a *homogeneous gain medium*, in order to reach a single-frequency regime without inserting any filter that may reduce the Finesse. For structural reasons, the VeCSEL technology is a better starting point than many others in the landscape of laser technologies for high-coherence emission. It should even be specifically designed and optimized for this purpose. Hereafter, the keys and technologies associated to highly coherent VeCSEL design will be discussed, after a brief state of the art of laser technologies and of VeCSELs.

4.5 The VeCSEL in the State of the Art

As pointed out in the previous sections, the coherence properties are deeply related to τ_c, which is proportional to both the cavity Finesse $\mathcal{F}$ and the cavity length D, as indicated by Eq. (4.7). In Figure 4.8a, we list the most available laser technologies[14] regarding this only criterion. It demonstrates that the VeCSEL is in the best competitors about this. But this is not the only interest of this technology.

In fact, the VeCSEL can be understood like a traditional solid-state laser relying on the modern semiconductor technology for gain medium. It can thus benefit

---

13 Again, choosing an appropriate indicator to define the width of laser spectrum is not something easy. We encourage the reader to examine the complex but fundamental works by Mandel and Wolf [7], or a more pragmatic sum-up by Saleh [8], or illustrated in practical cases in [31].
14 *Used acronyms are the following*: DPSSL: Diode-Pumped Solid-State Laser. VeCSEL: Vertical Cavity Surface Emitting Laser. DFB-FL: Distributed Feedback Fiber Laser. ECDL: External Cavity Diode Laser. DFB-SC: Distributed Feedback Semiconductor Laser. μc-VCSEL: microcavity Vertical Cavity Surface Emitting Laser.

from both worlds: high coherence and high power from solid-state lasers; compactness, continuous tunability, and wavelength flexibility from semiconductor lasers. Considering this, the VeCSEL is a candidate of choice for most applications, because it can be developed to fit most requirements in a single design, as depicted in Figure 4.8b. We want to point out that in the state of the art, most lasers are optimized for one or two criteria; however, to our knowledge, the VeCSEL is the only technology allowing to demonstrate a very high level of performance fitting almost all criteria of interest at the same time (Figure 4.8b).

Inside the world of VeCSELs (Figure 4.8c), a very wide variety of emission wavelengths is offered by the flexibility of the semiconductor technology [21], contrary to fiber or solid-state lasers. Moreover, the single-frequency regime is demonstrated for a wide variety of materials and of emission wavelengths. A specific feature of VeCSELs is that most characteristics (beam quality, polarization, phase noise, etc.) remain the same whatever the emission wavelength, except the emitted power.

4.6 Highly Coherent, Tunable VeCSEL Design

We discuss here how to optimize the design of a VeCSEL for high coherence. As explained before, we have to follow two paths: gain homogeneity and high Finesse cavity. We display in Figure 4.9a the main structure of a VeCSEL (here for emission at 1µm), including the semiconductor chip, the air gap, and the output coupler (dielectric mirror), as well as the standing wave and the pump absorption profile.

About the gain homogeneity, most VeCSELs rely on pump absorption in the barriers[15] as depicted in Figure 4.9c. The excited carriers will be captured with a time constant $\tau_{cap} \approx 10$ ps far quicker than the lifetime of the population inversion ($\tau_e \approx 10$ ns) by the quantum wells thanks to the carrier diffusion. For this kind of pumping scheme and for barriers in GaAs, a length of $7\lambda/2$ is required to absorb more than 85% of the pump power in a single travel. Obviously, for high efficiency, the quantum wells have to be placed at the maxima (or anti-nodes) of the standing wave (Figure 4.9a), which is possible thanks to the integration of the gain medium together with the DBR backs-side mirror on the same chip. The exponential profile of the pump absorption leads to nonconstant carrier density along the active sections, and thus it is necessary to place the quantum wells so that their excitation is uniform. A method consists in changing the amount of quantum wells per antinode, but it is limited (1, 2 or 3 usually[16]) if we want to avoid spectral filtering by the gain due to the spatial overlap between the quantum wells and the standing wave.

About the high Finesse design, we said in the previous sections (Sections 4.3 and 4.4) that a highly coherent laser is a laser that cold cavity exhibits high τ_c values. However, we also want wideband continuous tunability as well as compactness, so

---

15 Other designs perform the pump absorption in the quantum wells.
16 The distance between the quantum wells should not be too small to avoid electronic coupling between them. For example, for 1µm emission, their size is close to 8 nm, and the minimal distance should be close to 20 nm.

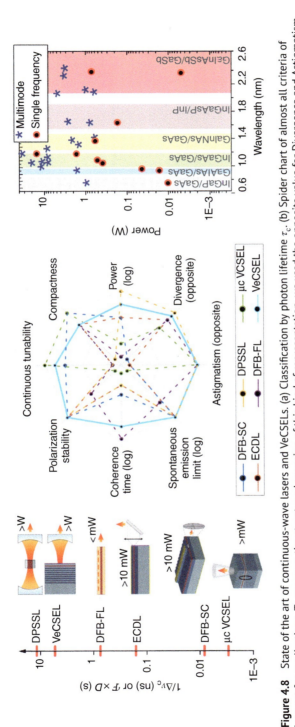

Figure 4.8 State of the art of continuous-wave lasers and VeCSELs. (a) Classification by photon lifetime τ_c. (b) Spider chart of almost all criteria of interest for applications. To preserve the sterotyped meaning of this kind of representation, we used the opposite value for Divergence and Astigmatism and displayed the Power, Coherence Time, and Spontaneous Emission Limit on a log scale. (c) State of the art of the VeCSELs [32], for single-frequency and multimode emission. The material indicated before the slash is the one of the quantum wells, the one after the slash is the one of the substrate.

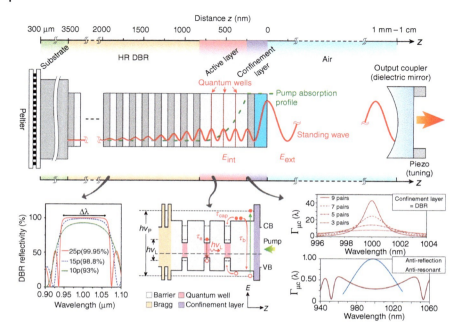

Figure 4.9 (a) Main structure of a VeCSEL. (b) Reflectivity of a DBR. (c) Energy structure of a the semiconductor chip, displaying the conduction band (CB), the valence band (VB), the barriers, the quantum wells, and the characteristic time constants (capture time τ_{cap}, electron lifetime τ_e). (d) Impact of the confinement layer design on Γ in three cases: reflective DBR, antireflection coating, and antiresonant design.

we need to work with short cavity length D. To preserve a low $\Delta \nu_c$ and thus a long photon lifetime τ_c, we have to increase the cavity Finesse $\mathcal{F}$, as indicated by Eq. (4.7). For that purpose, we have to rely on both *highly reflective (HR) mirrors* and *low intracavity losses*.

Concerning the HR mirror design, semiconductor technology benefits from the "Distributed Bragg Reflector" (DBR) structure [21], made of multiple $\lambda/4$ layers performing constructive interferences to enhance the reflection coefficient. As depicted in Figure 4.9b, a reflectivity higher than 99.9% can be achieved by means of some tens of pairs of layers with most semiconductor materials and technologies.

High reflectivity is however not enough, and we also have to minimize the losses inside the cavity. Assuming a stable cavity design, the main origin of the losses in the cavity is the absorption of the laser oscillation inside the active layer barriers. In most usual cases, the standing wave experiences a specific confinement by the micro-cavity formed by the HR-DBR on one side and the semiconductor–air interface on the other side (reflectivity $\approx 30\%$): this can be seen in Figure 4.9a. The strength of this confinement is usually thought in terms of confinement factor $\Gamma_{\mu c}$:

$$\Gamma_{\mu c} = n_{sc} |E_{int}|^2 / |E_{ext}|^2 \qquad (4.10)$$

where n_{sc} is the optical index inside the micro-cavity, $|E_{int}|$ and $|E_{ext}|$ are respectively the magnitude of the electrical field inside the micro-cavity and outside it (in the air). This micro-cavity is not strong enough to enable the laser oscillation; however, it can be designed to enhance the value of the modal gain, or, on the contrary, to decrease the internal losses. For example, the confinement layer can be structured with some pairs of DBR, in order to reinforce reflectivity compared with a simple semiconductor–air interface; this will increase the confinement, thus the modal gain, as depicted in Figure 4.9d. Because it is based on cavity effects, $\Gamma_{\mu c}$ is a function of the wavelength and exhibits resonances. This can be a wanted effect, in order to narrow the modal gain curve and reach a higher SMSR value (because it narrows Δv_{gain}, see Eq. 4.4) or to ensure a single-frequency emission. However, a side effect is that the losses are then larger, due to the absorption of the laser oscillation in the barriers, which widens Δv_c and damages the coherence properties. It also reduces the possibility to obtain wideband continuous tunability.

Fortunately, the confinement layer can also be designed to fit a very different objective: reducing the confinement, by designing it like an antireflection coating or a even an antiresonant structure (see Figure 4.9d). Of course, this leads to a reduction of the gain, but also to reducing the semiconductor chip losses in similar proportions. For that reason, all we have to do is to compensate this reduction of the gain by choosing an output coupler with higher reflectivity. Because the spectral shape of $\Gamma_{\mu c}(\lambda)$ is then much broader, this kind of design also allows to exploit the whole natural gain bandwidth of the quantum wells, without any filtering. This is the way we want to design VeCSELs for high coherence and broadband tunability.

This engineering is possible thanks to the flexibility offered by the semiconductor technology, and this is another advantage of the VeCSEL. Thanks to this technological context, it is also possible to insert on the semiconductor chip various functionalities that enable the generation of unusual light-states, such as two controlled frequency emission, Vortex beams, circular polarization [18, 19, 46, 57–60].

4.7 Limits and Solutions

There are two main limits to this VeCSEL design.

First of all, even if the VeCSEL geometry is candidate for high power emission, this technology is *technically limited concerning the maximum output power*. This limit depends strongly on the emission wavelength, as it is mainly a consequence of the thermal performance of the materials involved. In practice, the output power P_{out} of a VeCSEL working at a temperature $T + \Delta T$ is given by:

$$P_{out}(T + \Delta T) = \eta_d(T + \Delta T) \left(P_p - P_{th}(T) \exp\left(\frac{\Delta T}{T_0}\right) \right) \quad (4.11)$$

where η_d is the differential efficiency of the laser, P_p the pump power, P_{th} the laser threshold, and T_0, the "caracteristic temperature" of the laser. This parameter is phenomenological and hides a lot of complex physics.[17] At a macroscopic level, it takes

---

[17] L. A. Coldren explains in [33] that even very complex theoretical treatments do not allow to predict workable values of T_0. The exp dependency of the threshold (4.11) has no specific physical meaning, it mainly renders the fact that the dependency of the laser threshold is nonlinear with T_0.

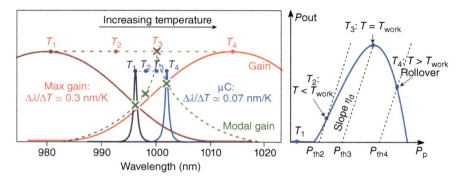

Figure 4.10 Micro-cavity-induced thermal rollover.

into account the evolution of the laser output power as a function of the temperature. If the value of the T_0 is related to microscopic mechanisms [33], its value can be influenced by structural effects too. For example, in the case of the VeCSEL, a usual origin is the thermal shift of the microcavity mode $\Gamma_{\mu c}(\lambda)$, which is not as strong as the one of the gain (see Figure 4.10), leading to temperature-dependent gain variation.

Whatever the physical accurate origin, the pump power almost always warms significantly the semiconductor chip, which in turn leads to a decrease of the modal gain. The most usual impact is the laser threshold P_{th} that can becomes larger and larger all along the $P_{out}(P_p)$ characteristic, limiting the maximum power available. To avoid this, thermal management technology should be used in order to reduce the heating of the structure. In the VeCSEL, the heat source comes from the pumping while, stuck on the backside, a Peltier element maintains a constant temperature $T_{peltier}$. The temperature of the quantum wells T_{QW} can then be evaluated thanks to the thermal resistance $R_{th} = \Delta T / \Delta P_{inc}$ of the structure:

$$T_{QW} = T_{peltier} + R_{th} \times P_{inc} \qquad (4.12)$$

where P_{inc} is the optical power coming from the pump injected into the structure. For a uniform and isotropic medium pumped with a beam of waist w_0, the maximum ΔT_{max} between the Peltier element and the quantum wells is given by [34]:

$$\Delta T_{max} \propto \frac{P_{inc}}{\kappa w_0^2} \left(w_0 + 2d - \sqrt{w_0^2 + 4d^2} \right) \qquad (4.13)$$

where κ is the thermal conductivity of the material and d the thickness of the medium. This gives a good overview of the thermal behavior; however, it is quantitatively not satisfying to develop a real VeCSEL, because the semiconductor chip is not a homogeneous material and also because the pumping geometry plays a role in the heat propagation. The value of the thermal conductivity [35–37], for the semiconductor materials discussed here, is in the range 10– 60 W/m/K, which is not very high. In most cases, the main problems are the thickness and high thermal conductivity of the Bragg section and of the substrate. Fortunately, various thermal management methods are possible, consisting in inserting in the semiconductor structure high conductivity materials (for example, Gold, SiC, and

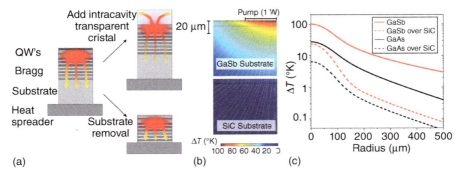

Figure 4.11 Thermal Management. (a) Usual techniques. (b, c) Impact for a pumping power of 1 W and a pump waist of 100 μm.

Diamond exhibit κ in the range 300–2000 W/m/K) with various techniques, as depicted in Figure 4.11a.

All will lead to an increasing of the T_0 of the laser. However, for high coherence design, the intracavity transparent cristal method is not eligible, as it introduces losses and can lead to unstable polarization operation. The substrate removal technique can be used or extended [38] to the reduction of the amount of pairs of DBR in the case of a gold heat spreader ($\kappa = 300$), which thins the amount of semiconductor, thus improves the thermal transfer. In this last case, the gold participates in the global reflectivity of the backside mirror. We checked the efficiency of a host SiC($\kappa = 490$) substrate in Figure 4.11b for a GaSb ($\kappa = 33$) VeCSEL and a 50 μm waist pumping. The result is impressive, reducing ΔT from 100°K down to less than 30°K. As depicted in Figure 4.11c, similar improvements occur on GaAs, with much better absolute performance due to the thermal properties of GaAs($\kappa = 55$).

The second limit concerns the filter-free single-frequency operation. As suggested above, a filter-free operation is attractive as it enables wideband tunability as well as highly coherent operation even with short cavities. However, increasing the cavity length to not so long lengths (above usually 1 cm) and/or working at high optical power reveals longitudinal multimode operation, originating from complex nonlinear spectro-temporal dynamics [24].

4.8 Highly Coherent, Tunable VeCSEL: Main Characteristics

Because the VeCSEL is capable of a very pure light emission, its accurate characterization relies on advanced metrology. In this section, we start with a global overview the VeCSEL features, by exploiting high quality but standard measurement techniques, similar to the one that can be found in any laser optics lab. Next, we will introduce an accurate qualification of the single-frequency emission, as well as more advanced metrology concerning the beam quality. To finish, the time domain coherence is the most complex one and will be described in a dedicated section.

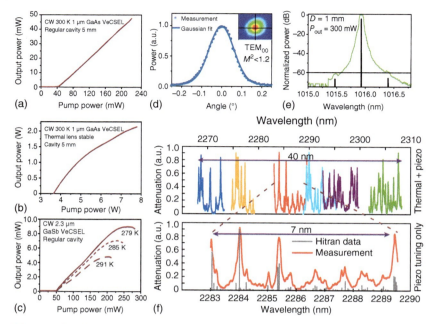

Figure 4.12 Overview of the VeCSEL characteristics. (a–c) Display the output power for various VeCSEL configurations, see main text for comments. (d) Typical beam quality for single-mode VeCSEL. (e) Spectrum of a VeCSEL. The SMSR is close to 60 dB, limited by the spontaneous emission. The displayed width of the main mode is not due to the laser but to the apparatus function of the optical spectrum analyzer, which is a common situation. (f) A demonstration of thermal and piezo tunability in a basic gas analysis setup with a GaSb VeCSEL [39].

Figure 4.12 depicts the main characteristics of various single-frequency VeCSELs. As expected, the power characteristics can change a lot depending on T_0 or on the design. For example, constant η_d above threshold for a low T_0 GaAs structure can be observed (a). A highly variable η_d is observed (b) in the case of a multiwatt highly coherent VeCSEL [26], cavity stability of which is based on thermal lens: because the thermal lens evolves as a function of the pump power, the cavity waist evolves too, inducing threshold variations, caused by the change in the overlap between the pump size and the cavity mode. Finally, for materials exhibiting bad thermal properties such as GaSb structures for 2.3 µm emission (c), the impact of the thermal rollover is strong, because of the Auger effects, limiting the maximal power to less than 10 mW.

Whatever the output power or emission wavelength, high beam quality emission is available with the VeCSEL technology, as observed with a windowless camera (see Figure 4.12d). A stable, single-frequency regime, with SMSR close to the quantum limit, is observable in most cases. Due to high finesse, a cavity as short as 1 mm exhibits SMSR values as high as 60 dB (see Figure 4.12e). This was observed with a 1 m long optical spectrometer by Jobin–Yvon (THR1000). We have to mention that the observable linewidth as well as the tails of the main mode come from the apparatus function of the spectrum analyzer and are not characteristics of the laser. In this figure, the only relevant information is thus the SMSR. Concerning the polarization state, a measurement performed with a commercial polarizer lets us conclude about the rejection of the perpendicular polarization state > 30dB. Again, this result is not a characteristic of the VeCSEL but a limit of the measurement device. For all these quantities, a far better observation can be performed by means of radiofrequency instrumentation, as explained in the following.

Finally, a very interesting and unique feature (to our knowledge) of the VeCSEL is its ability to reach all these features in a filter-free configuration. This enables the broadband continuous tunability over spectral ranges as wide as 7 nm. But beyond that, combining it to thermal tuning opens the exploitation of the full gain bandwidth of the laser, leading to huge tuning bandwidth (close to 40 nm), as demonstrated with a commercial VeCSEL [3, 39] by Innoptics (see Figure 4.12f), design of which relies upon the above described concepts.

4.9 Ultrahigh-Purity Single-mode Operation

As suggested above, because of its high level of performance, the accurate characteristics of the highly coherent VeCSEL cannot be evaluated by means of traditional metrology. For example, observing the SMSR in the case of longer cavities (cm long), or high rejection of the transverse modes as well as the polarization modes, cannot be done by a traditional measurement setup. A much more accurate evaluation can be provided by moving the problem to the radio-frequency domain, because the available instrumentation exhibits huge dynamics as well as very high spectral resolution. As any photodetector is sensitive to the optical intensity (which is $\propto E^2$), radio-frequency instrumentation can be very handy for our purposes with lasers. Nonlinearity of photodetector reveals the beatings between various discrete frequencies into the laser field. Those discrete frequencies come from residual emission in the cavity modes, and this is true for longitudinal modes [40], spatial modes [11, 19], and polarization modes [18, 19].

The principle of the measurement is simple: we observe the beam in a configuration of interest (see Figure 4.13) with a high-frequency photodetector connected to a low-noise amplifier, output of which is observed with a radio-frequency spectrum analyzer. Knowing the detected average optical power (which is simply proportional to the average photocurrent) and measuring the power inside the beating lead to know the relative intensity of the mode, which is identified by its specific beating frequency.

The case of longitudinal mode is straightforward: the beam is injected directly on the high-frequency photodiode. The result is an RF comb with periodicity $\Delta \nu_{FSR}$ displaying all the possible beatings between all the longitudinal modes. The power of this beating contains the information on the SMSR.

The case of transverse mode is a little bit more tricky: because the modes of a resonator form a basis, all the modes are orthogonal and no photodetector could

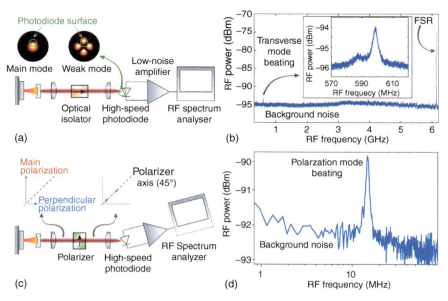

Figure 4.13 (a) Setup for RF measurement of TMSR. (b) Example of result for a ≈2.5 cm long VeCSEL. (c) Setup for polarization rejection measurement. (d) Example of result.

see a beating. To visualize the beating, the detector has to intercept a portion at the intersection between the two modes (see Figure 4.13a) to break the symmetry of the modes basis. Then, the Transverse Mode Suppression Ratio can be evaluated to be:

$$\text{TMSR} = 10 \log \left[2\Gamma_{r,\theta} \left(\frac{i_{\text{ph}}}{i_{\text{beat}}} \right)^2 \right] \tag{4.14}$$

where i_{ph} is the average photocurrent, i_{beat} the effective value of the beating signal and $\Gamma_{r,\theta}$ the theoretical overlap between the two modes under consideration, taking into account all the experimental conditions (beam size, photodetector position and size). In most cases, the VeCSELs developed thanks to our methodology exhibit TMSR > 80 dB. An example of result is given in Figure 4.13b, displaying the TMSR beating as well as the FSR beating.

In the case of the polarization mode, it is necessary to project the two components of the electrical field. For a linear polarization, it is easy to do by putting a polarizer at 45° (see Figure 4.13c) before the photodetector. Because an isolator cannot be used during these measurements,[18] the photodetector should be tilted to avoid optical feedback. Here, the spectral position of the peak gives the birefringence of the cavity, while its power allows to evaluate the rejection of the polarization $S_\perp$, after taking into account the effect of the polarizer. In the case of VeCSELs, this method usually leads to $S_\perp > 70$ dB.

---

18 An isolator plays with the polarization state to perform the isolation.

So, we made the theoretical proof as well as the experimental proof of the exceptional purity of the single-mode operation of the VeCSEL. We revealed its excessively pure photon state on all the three coherence axis thanks to the radio-frequency metrology, which offers much stronger measurement dynamics (by decades compared with usual methods!).

4.10 Spatial Coherence

As demonstrated above, VeCSELs can select a single transverse mode, which is a first important step to reach spatial coherence. However, this pure mode can be polluted by some spatial noise, (see Figure 4.7a). In this section we get a closer look at this *transverse spatial noise*, i.e. the static (time-domain independent) fluctuations superimposed on an ideal wavefront beam: the *time domain phase fluctuations* will be investigated in the next section.

We can consider that a real-world Gaussian TEM_{00} mode is an ideal spherical Gaussian wave, wavefront of which is polluted by some "spatial phase noise" in the transverse dimension of the wave, as illustrated in Figure 4.14. This leads to M^2 propagation factors higher than 1. In the case of VeCSELs, we measured that the wavefront deformation, compared with an ideal spherical Gaussian wave, is less than 1%, leading to excellent M^2 values (typically lower than 1.2). This kind of wavefront deformation typically originates from some residual, nonspheric, thermal lensing, or from the slight irregularities of various surfaces that the beam meets inside the cavity.

4.11 Time Domain Coherence and Noise

An accurate mean for the quantification of the time domain coherence is the study of noise. When talking about laser coherence, two kinds of fluctuations need to be discussed: *magnitude fluctuations* $A(t)$ (or *intensity fluctuation*) and *phase fluctuations* $\varphi(t)$ of the total electrical field $E(t) = (A_0 + A(t))\exp(j(\omega_0 t + \varphi(t)))$ as a function of the time.

The main characteristic of $A(t)$ and $\varphi(t)$ is that they are random functions of the time, i.e. not predictable, so they are called "noise." The word "noise" is often thought in terms of parasitics, which are external to the device. This is a narrow way of thinking: noise has fundamental reasons to be ingrained by the very physics of the device. Lasers are no exception.

4.11.1 Noise in Photonics: Basics

For example, *about* $A(t)$, the mere fact that a photonic source emits photons at random instants[19] leads to *intensity noise*. Actually, the optical power can be described in

---

19 This is the case of most of them, even most lasers. A few counter-example exist in the literature, see a pedagogical example with high-efficiency Light-Emitting Diode [41] or in the case of lasers [42, 43].

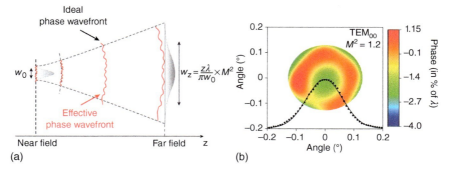

Figure 4.14 Spatial coherence of the VeCSEL beam. (a) Propagation of a nonideal Gaussian beam. (b) Typical transverse phase fluctuation in a VeCSEL.

a pure corpuscular vision like a flux of photons going through a surface, for example, the laser output coupler itself, or the surface of a detector. It is then defined like:

$$P_{out} = h\nu \frac{N}{\tau} \tag{4.15}$$

where N is the number of photon going through the considered surface during the time τ of the experiment. This is a very high rate signal: for a monochromatic emission of 1 mW at $\lambda = 1\,\mu m$, this leads to $N \approx 10^{16}$ for an observation of one second, with an intensity modulation of 100%, because at each instant, a photon is emitted or not, randomly. As we use bandwidth-limited detectors, we filter this high rate signal and observe an average value with moderated fluctuations (see Figure 4.15a). Due to its random nature, it is complex to speak about the time function of a noise because the magnitude in time is totally unpredictable, and we thus prefer to work with a statistically stable indicator such as its variance, or the square root of the variance, called effective value or RMS (Root Mean Squared) value.

Whatever the physical origin of the noise for a given photonic source, it should always be compared with a fundamental, nearly unavoidable, noise reference: *the shot noise*. The shot noise is nearly unavoidable, because optical attenuation,

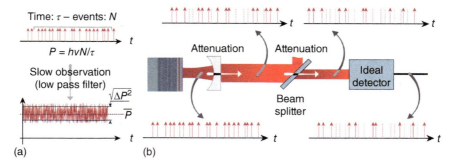

Figure 4.15 (a) Macroscopic intensity noise related to microscopic events. (b) effect of the attenuation on the photon statistics, while extracting light from a cavity or passing through a lossy optical system.

whatever its origin (attenuator, absorption, beam-splitter, diffraction, ...), is nearly inescapable in practical photonic systems. Because attenuation selects randomly the photons that can pass, it alters the statistic of the attenuated beam, compared with the incident one. In the case of lasers, and especially for high finesse cavity lasers, the shot noise is usually unavoidable because the intracavity power experiences a huge attenuation when being coupled to the outside of the cavity [44] (see Figure 4.15b). We must point out that, in most usual cases, the magnitude of the shot noise is low compared with other kinds of noise, and for that reason the existence of the shot noise is rarely a limit[20]: it is more to be understood like a practical reference level of noise. This shot noise is quite easy to quantify. We simply have to evaluate the average value and the variance of N_1, the amount of photons passing through an attenuator of transmittance T [44]:

$$\overline{N_1} = T \times \overline{N} \quad \text{and} \quad \overline{\Delta N_1^2} = \overline{\left(N_1 - \overline{N_1}\right)^2} = \overline{N_1^2} - \overline{N_1}^2 \tag{4.16}$$

Assuming all the attenuation events are independent, we can calculate $\overline{\Delta N_1^2}$ for a single photon and apply a scaling factor by $\overline{N}$ to take into account all the photon selection events: $\overline{\Delta N_1^2} = \overline{N} T(1-T)$. Then, Eq. (4.15) leads to the fluctuation that occurs on an optical power $P_{\text{att}} = P_{\text{out}} \times T$ that experienced some attenuation [44]:

$$\overline{P_{\text{att}}} = h\nu \frac{\overline{N_1}}{\tau} \quad \text{and} \quad \overline{\Delta P_{\text{att}}^2} = h^2 \nu^2 \frac{\overline{\Delta N_1^2}}{\tau^2} = \frac{h\nu}{\tau} \overline{P_{\text{out}}} T(1-T) \tag{4.17}$$

This relation is dependent on the observation time τ, which is not practical. Usually, people prefer to work with the power spectral density $S_P(f)$ of the signal $P(t)$, which is simply related to the variance or its square root P_{RMS} as follows:

$$P_{\text{RMS}} = \sqrt{\overline{\Delta P^2}} = \sqrt{\int_{f_{\text{low}}}^{f_{\text{high}}} S_P(f) df} \tag{4.18}$$

$S_P(f)$ can be understood like the spectral distribution of the noise.[21]

Because observing a signal over a time τ is equivalent to observing it with a spectral bandwidth[22] $\Delta f = (2\tau)^{-1}$, we can deduce from Eqs. (4.17) and (4.18) that the spectral density due to the shot noise is given by $S_{\text{shot}} = 2h\nu \overline{P_{\text{out}}} T(1-T)$. In the case of laser experiments, strong attenuations are usual ($T \to 0$), and thus $T(1-T) \to T$, which leads to:

$$S_{\text{shot}}(f) = 2h\nu \overline{P_{\text{att}}} \tag{4.19}$$

---

20 Except in extreme setup like spin noise spectroscopy [45], or in very long-range doppler detection [46].
21 As depicted above, due to its random nature, the time function of a noise is complex to describe with mathematics. It is exactly the same for the Fourier transform of the noise because it exhibits the same random feature as the time function. We thus use the power spectral density $S_P(f)$, which is a stable indicator of the spectral repartition of the noise.
22 This equivalence is due to the fact that the Fourier transform of a time window τ is a cardinal sine which width is $\Delta f = (2\tau)^{-1}$.

This expression does not depend on the frequency, and this means that the shot noise spectrum is flat, with uniform value for all frequencies: it is a *white noise*. Intuitively, we feel that there is no reason for an attenuation process to have a specific time constant associated. If we take a source with a self noise, spectral density of which is $S_{\text{laser}}(f)$, because the attenuation process is not dependent on the source statistics, the observation of this source through the attenuator should lead to:

$$S_{P_{\text{att}}}(f) \approx S_{\text{laser}}(f) \times T^2 + 2h\nu \overline{P_{\text{out}}} \times T \quad (4.20)$$

Finally, increasing the attenuation ($T \to 0$) leads to observe only shot noise and hides the contribution of $S_{\text{laser}}(f)$.

When detected by a photodiode, a quantum efficiency $\eta_q < 1$ can be understood as if all the photon events were not transfered into electron events. This is similar to the detection with an ideal photodiode (that means $\eta_q = 1$, thus a sensitivity $= hc/q\lambda$ in A/W) placed after an optical attenuator, transmittance of which is equal to the quantum efficiency η_q. And so the simple detection process with a real-life detector can introduce a shot noise contribution too. Because of all the possible attenuation processes described above, it is usually assumed that the detected current contains a part of shot noise, value of which is $S_{\text{shot}}(f) = 2qi_{\text{ph}}$, where i_{ph} is the average detected photocurrent. For that reason, intensity noise is usually evaluated by the "Relative Intensity Noise" (RIN), which does take into account the fact that the shot noise may come from the measurement setup itself:

$$\text{RIN} = \frac{S_{\text{iph}} - 2qi_{\text{ph}}}{i_{\text{ph}}^2} \quad (4.21)$$

where S_{iph} is the spectral density of the photocurrent measured in an experiment similar to the one depicted in Figure 4.15b.

The measurement is usually performed thanks a low-noise-amplified photodiode, output of which will be processed to perform power spectral density spectra averaging. Averaging is necessary because the power spectral density is evaluated during a finite measurement time (the one of the experiment); so, its experimental estimation undergoes a statistical error, which can be reduced M times by accumulating M averages (due to the law of the large numbers). The power spectral density may be estimated by a RF sweeping spectrum analyzer or by measuring the time signal with an acquisition card and some Fast-Fourier-Transform (FFT) processing [47, 48]. The latter is more interesting, when possible, because in most practical cases, the computational efficiency of the FFT algorithm leads to petty computation times.[23] For that reason, the time necessary for the acquisition of a single spectrum is the one necessary for the acquisition of the time domain signal, defined as RBW^{-1} where RBW is the Resolution BandWidth of the spectrum. It is far quicker than with a RF sweeping analyzer, which requires at least $K \times \text{RBW}^{-1}$ with K the amount of points in the spectrum. This necessary time for the acquisition is an important topic because, as explained, accurate noise measurements rely on averaging spectra. An example of

---

23 This is true in lots of practical cases, when the wanted spectral resolution is narrow. In very specific cases, dedicated electronics can be involved to speed up the FFT computation [45].

result of this process is given for S_{iph}, as well as for the corresponding RIN spectrum, in Figure 4.16a.

To sum up, noise is usually quantified through the *power spectral density*, which indicates the spectral repartition of its fluctuations. The attenuation impedes an accurate laser power spectral density measurement because the *shot-noise* is superimposed. For that reason, the intensity noise is usually given by the RIN that takes into account the impact of the shot-noise on the measurement.

4.11.2 Intensity Noise of a VeCSEL

Like in most lasers, the VeCSEL noise has a fundamental cause: the spontaneous emission. However, this contribution is weak due to the high Finesse cavity design, and other technical noise, mainly coming from the pump, usually has a much stronger contribution (see Figure 4.16b and c). The pump noise itself depends on the kind of pump: usually single-mode laser diodes experience mainly $1/f$ noise at low frequency [49] due to the impact of the semiconductor cristal defects on the electronic transport [50]. Multimode pumps can deliver much higher power but may produce extra contributions to noise for optical reasons, coming from the multimode operation itself [51].

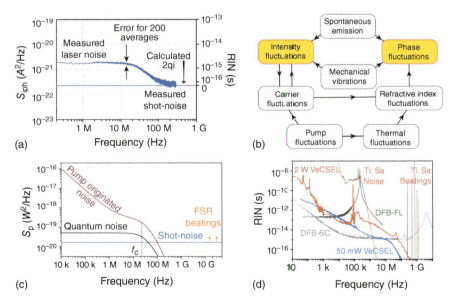

Figure 4.16 (a) Example of noise spectrum, exhibiting both RIN and spectral density. (b) Model of the VeCSEL Noise [18]. (c) Result from the model described in *(b)* in the case of intensity noise. (d) VeCSEL intensity noise compared with other technologies. For the VeCSEL, the cutoff is always below 100 MHz, and the intensity noise inside the VeCSEL bandwidth depends on the kind of pump: single-mode for the 50 mW VeCSEL, multimode in the case of the 2 W VeCSEL. The VeCSEL beatings were outside of the available spectral range (beyond 10 GHz) at the moment of the experiment.

Whatever the kind of pump, the VeCSEL will efficiently filter the pump noise. Actually, even a passive cavity would filter the events quicker than the cavity photon lifetime τ_c, which is related to Δv_c (Eq. 4.8). The ability of the cavity to filter the pump noise is thus linked, again, to $\mathcal{F}$: the higher the Finesse, the lower the noise cutoff frequency, and thus the lower the RMS noise, as written in Eq. (4.18). Lasers dynamic behavior is however a little more complex than the one of a passive cavity: it depends on both the photon lifetime τ_c and the electron lifetime τ_e. Due to the interplay between these two time constants, the laser dynamics may show, in some conditions, resonant amplification of the noise (called "class B" dynamics): this is the case for low τ_c lasers, such as laser diodes, or for low τ_e lasers, such as fiber lasers or DPSSL. This is well illustrated by the transfer function of the pump modulation on the laser power $H(f)$ [9] as a function of the Fourier frequency f:

$$H(f) = \frac{\eta_P/(\eta_P - 1)}{1 - j\frac{2\zeta}{f_0}f - \frac{f^2}{f_0^2}} \quad \text{with} \quad \zeta = \frac{1}{2}\sqrt{\frac{\tau_c}{\tau_e}\frac{\eta_P^2}{\eta_P - 1}} \quad \text{and} \quad f_0 = \frac{1}{2\pi}\sqrt{\frac{\eta_P - 1}{\tau_c\tau_e}}$$

(4.22)

where η_P is the pump rate, defined as $\eta_P = (P_P - P_{tr})/(P_{th} - P_{tr})$ with P_{tr} is the transparency power. ζ is the damping factor, its value is high for high τ_c or small τ_e; for that purpose, the VeCSEL is a unique case as it combines both[24] see Figure 4.16d. This dynamic behavior, leading to a simple short-pass operation for $|H(f)|$, is called "class A" and is not a usual (but usually wanted) feature for a laser. Moreover, often, the cutoff (or resonance) frequency f_0 is low in VeCSELs (below 100 MHz in most cases), being an efficient filter for the pump noise. It grants to the VeCSEL a very original property: its intensity noise is limited by the shot-noise (thus a RIN ≈ 0) over a very wide frequency range, starting at the cutoff frequency and ending at the first FSR beating, above 100 GHz for a mm long cavity. This unique feature makes the VeCSEL very interesting when used as optical carrier for microwave signals, for example, in the context of optical fiber transmissions.

4.11.3 Phase Noise, Frequency Noise, and Linewidth of a VeCSEL

The stability of the laser field frequency through time can be observed thanks to various indicators, which are all related to the phase noise. For RF oscillators, direct measurement of the phase noise $\varphi(t)$ is performed, because it is directly accessible through experiments. In the case of lasers, such a direct measurement is usually not possible. Most measurement systems rely on frequency discriminators and thus the quantity obtained is the "instantaneous frequency" of the laser $v(t)$. Both are equivalent and contain the same information as they are related by the simple relation [52]:

$$v(t) = \frac{1}{2\pi} \times \frac{d\varphi(t)}{dt}$$

(4.23)

---

24 τ_c is high due to high finesse cavity, τ_e is small due to semiconductor gain medium. Fiber or solid-state lasers display τ_e in the ms range, whereas semiconductors display τ_e in the ns range.

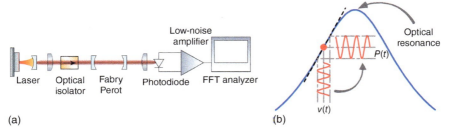

Figure 4.17 (a) Typical frequency-noise setup using a Fabry–Perot interferometer. (b) Principle of side-fringe demodulation.

From both $v(t)$ and $\varphi(t)$, a power spectral density can be calculated. In the case of $v(t)$, the corresponding quantity is called the "Frequency Noise Spectrum," and its physical unit is Hz^2/Hz.

The principle of the measurement is to drive the laser field through an optical device, called "frequency discriminator," in order to convert the frequency variations of the laser $v(t)$ into proportional optical power fluctuations through time. In most cases, this frequency discriminator is simply a device displaying optical resonances features, such as interferometeors like Fabry–Perot or fibered Michelson, but it can also be a gas cell (to exploit)the atomic transition of the gas). This device is used at the side of an optical resonance, in a linear zone, so that laser frequency variations $v(t)$ are converted into proportional optical power fluctuations through time, as depicted in Figure 4.17b. Assuming that the intensity fluctuations coming directly from the laser are known or can be neglected, the observed signal contains, without ambiguity, the frequency noise of the laser, and its spectral density can be measured and computed by the process already used for intensity noise.

Like in other lasers, spontaneous emission is a fundamental limit of the VeCSEL, known as the Shawlow–Townes limit (Eq. 4.9). On a frequency noise spectrum, it contributes as a white noise with magnitude $\Delta v_{\text{limit}}/\pi$. This limit is very low in VeCSELs, and this is why in their case, noise mainly comes more from technical origins (see Figures 4.16b and 4.18a). Two of them are prevalent, both leading to Δv_{FSR} variations[25]: first, the direct cavity length fluctuations due to mechanical vibrations; second, index fluctuations in the semiconductor chip. These index fluctuations come from the pump intensity noise that produces temperature fluctuations, modifying the optical index through time, impacting Δv_{FSR} as follows [9]:

$$\frac{\delta v}{v} = \frac{\delta \Delta v_{\text{FSR}}}{\Delta v_{\text{FSR}}} \approx \frac{2 L_p \Gamma_{\mu c}}{D} \delta n_{\text{sc}} \tag{4.24}$$

where L_p is the penetration length of the laser field inside the semiconductor chip. This technical contribution is thus stronger for high $\Gamma_{\mu c}$ values and so is stronger in the case of low Finesse designs, as discussed in Section 4.6. The index fluctuations

---

25 Δv_{FSR} is the free spectral range of the interferometer, please look at Figure 4.3.

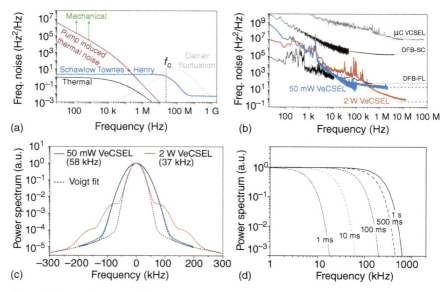

Figure 4.18 (a) Theoretical frequency noise spectrum computed from the model described in Figure 4.16b. (b) State of the art of frequency noise for various laser technologies already discussed. Dashed lines are the calculated Schawlow–Townes–Henry Limits from Eq. (4.9). (c) Laser spectrum computed from frequency noise data for 1 ms of observation time. Please notice that the 2 W laser has a linewidth which is narrower in spite of a much noisier pump. This is due to the fact that the 2 W laser was based on a thermally managed (SiC reported) semiconductor chip. (d) Half laser spectrum shown for various observation times.

δn_{sc} can be related to the pump fluctuations δP_{inc} by:

$$\delta n_{sc}(f) = \left(\frac{\Delta n_{sc}}{\Delta T}\right) R_{th} P_{inc} \delta P_{inc} \Theta(f) \tag{4.25}$$

where $\Theta(f)$ is the thermal transfer function, close to a first-order low-pass filter cutting in the kHz range. Therefore, the index fluctuation is proportional to R_{th}, and a thereby way to reduce the technical source of frequency noise is to perform thermal management. The frequency noise performances of VeCSELs are compared with other technologies in Figure 4.18b.

In the literature, the coherence of lasers is usually quantified by the FWHM linewidth. The laser spectrum is observed at very high resolution, much higher than the one of an optical spectrum analyzer. It can be measured directly, by various beating methods [53, 54] called "heterodyne" methods, or can be computed from the frequency noise. At first glance, the FWHM linewidth seems to be a simple, working indicator: this unique value should represent accurately the time domain coherence of the laser. This is an attractive point of view, compared with the complexity of a full spectrum in the case of the frequency noise. Unfortunately, *this is a very wrong idea*, even if it is a quite common one. As pointed out in [31], the laser spectrum is not a stable information: it depends a lot on the time span on which the spectrum is observed, and thus the FWHM linewidth is observation time-dependent.

The laser spectrum is usually made of two contributions. The Schawlow–Townes–Henry one, which is nearly a white noise, contributes with a Lorentzian shape to the linewidth. Low-frequency contributions, displaying various shapes ($1/f$ like or spikes) on the frequency noise spectrum, lead in most cases to a Gaussian contribution,[26] the width of which increases along the observation time. This is because observing the laser during a longer time implies to be sensitive to events at lower frequencies, which are usually of higher magnitude, as observable here on all the given noise spectra. The overall laser linewidth is therefore very commonly the convolution between a Lorentzian and a Gaussian, leading to a Voigt profile [56].

We estimated the VeCSEL spectrum from frequency noise data. It can be performed by two ways: from the frequency noise spectrum (but it is not always accurate because based on assumptions) or better from the time domain variations of the frequency fluctuation of the laser $\nu(t)$ [31]. According to Eq. (4.24), the technical fluctuations are proportional to $\Delta \nu_{FSR}$ and so depend on the cavity length like D^{-1} (thus D^{-2} in power spectral density). For that reason, the VeCSEL linewidth decreases when the cavity length increases. In the case of the VeCSELs we developed here, with cavities in the mm–cm range, the result is a linewidth of some tens of kHz over observation times as long as 1 ms (see Figure 4.18).

4.12 Conclusion

We stated that the VeCSEL design has all the good features to work like the stereotyped "ideal" continuous wave laser, as depicted in our introduction. For this purpose, the VeCSEL has to be optimized following two principal ideas: gain homogeneity and low losses. However, to reach the best results, thermal management performed outside the cavity may increase power as well as reduce noise.

Following this process, we obtained VeCSELs with remarkable performances in a compact (mm to cm cavity length) design: the single-mode operation is ultrapure, rejecting every unwanted modes (spatial, longitudinal, and polarization) by more than 60 dB in every the configurations developed here. Thanks to single-frequency filter-free operation, we achieved exceptional continuous tunability performance (close to THz) in single-mode regime. Coherence properties are also beyond state-of-the-art expectations: high spatial coherence ($M^2 < 1.2$), low divergence beam (<1°), narrow linewidth (<100 kHz for 1 ms), and shot-noise operation over large span in the RF domain.

These are the reasons why in our opinion, VeCSEL resets the lasers state of the art, as this single design is able to comply with the requirements of a very large diversity of applications.

---

26 We encourage the reader to have a look at the interesting paper of DiDomenico et al. about the β-separation line technique applied to the laser linewidth estimation [55].

Acknowledgements

The authors would like to thank Odile Verdure–Labeille of Ecostrategie, for her kind and valuable help in translating this paper to English.

References

1 Lucas-Leclin, G., Cérez, P., and Dimarcq, N. (1999). Laser-induced noise contribution due to imperfect atomic state preparation in an optically pumped caesium beam resonator. *Journal of Physics B: Atomic, Molecular and Optical Physics* 32 (2): 327.
2 Accadia, T., Acernese, F., Alshourbagy, M. et al. (2012). Virgo: a laser interferometer to detect gravitational waves. *Journal of Instrumentation* 7 (03): P03012.
3 Čermák, P., Chomet, B., Ferrieres, L. et al. (2016). Crds with a vecsel for broad-band high sensitivity spectroscopy in the 2.3 μm window. *Review of Scientific Instruments* 87 (8): 083109. doi: 10.1063/1.4960769.
4 Gimlett, J.L. and Cheung, N.K. (1989). Effects of phase to intensity noise conversion by multiple reflections on gigabit-per-second dfb laser transmission systems. *IEEE Journal of Lightwave Technology* 7 (6): 888–895.
5 Oberson, P., Huttner, B., Guinnard, O. et al. (2000). Optical frequency domain reflectometry with a narrow linewidth fiber laser. *Photonics Technology Letters, IEEE* 12 (7): 867–869.
6 Baney, D.M., Gallion, P., and Tucker, R.S. (2000). Theory and measurement techniques for the noise figure of optical amplifiers. *Optical Fiber Technology* 6: 122–154.
7 Mandel, L. and Wolf, E. (1995). In: *Optical Coherence and Quantum Optics*. Cambridge University Press.
8 Saleh, B. and Teich, M.C. (1991). In: *Fundamentals of Photonics*. Canada: Wiley Interscience.
9 Myara, M., Sellahi, M., Laurain, A. et al. (2013). Noise properties of nir and mir vecsels (invited paper). In: *Proc. SPIE, San Francisco*, vol. 8606, vol. 8606.
10 Triki, M., Cermak, P., Cerutti, L. et al. (2008). Extended continuous tuning of a single-frequency diode-pumped vertical-external-cavity surface-emitting laser at 2.3 μm. *Photonics Technology Letters, IEEE* 20 (23): 1947–1949.
11 Siegman, A.E. (1986). *Lasers*. University Science Books. ISBN 0-935702-11-5.
12 Siegman, A.E., Nemes, G., and Serna, J. (1998). How to (maybe) measure laser beam quality. In: *Diode Pumped Solid State Lasers: Applications and Issues (DLAI)*, vol. 17 (ed. M. Dowley). OSA.
13 Petermann, K. (1983). Constraints for fundamental-mode spot size for broadband dispersion-compensated single-mode fibres. *Electronics Letters* 19 (18): 712–714.
14 Pask, C. (1984). Physical interpretation of Petermann's strange spot size for single-mode fibres. *Electronics Letters* 20 (4): 144–145.
15 Gross, H., Zugge, H., Peschka, M., and Blechinger, F. (2007). In: *Handbook of Optical Systems*, vol. 3. Wiley-VCH.

16 Barwood, G.P., Gill, P., Huang, G., and Klein, H.A. (2007). Observation of a sub-10-Hz linewidth 88sr+2 s1/2-2d5/2 clock transition at 674 nm. *IEEE Transactions on Instrumentation and Measurement* 56 (2): 226–229.

17 (1980) Power spectra estimation. National Semiconductor - Application Note 255. https://www.ti.com/lit/an/snoa719/snoa719.pdf (accessed 6 May 2021).

18 Seghilani, M.S. (2015). Highly Coherent III–V semiconductor laser emitting phase-, amplitude- and polarization-structured light for advanced sensing applications: vortex, SPIN, feedback dynamics. Ph.D. thesis. Université de Montpellier.

19 Sellahi, M. (2014). Laser à semi-conducteur III–V à émission verticale de haute cohérence et de forte puissance: etat vortex, continuum et bifréquence THz. Ph.D. thesis. Université Montpellier II.

20 Haken, H. (1985). In: *Laser Light Dynamics*, vol. 1. Amsterdam: North-Holland.

21 Rosencher, E. and Vinter, B. (2004). Optoelectronics. Cambridge.

22 San Miguel, M., Feng, Q., and Moloney, J.V. (1995). Light-polarization dynamics in surface-emitting semiconductor lasers. *Physical Review A* 52 (2): 1728.

23 Khanin, Y.I. (2006). In: *Fundamentals of Laser Dynamics*. Cambridge: Cambridge Int Sci. Pub. Ltd.

24 Garnache, A., Ouvrard, A., and Romanini, D. (2007). Single-frequency operation of external-cavity vcsels non-linear multimode spectro-temporal dynamics. *Optics Express* 15 (15): 9403–9417.

25 Blin, S., Paquet, R., Myara, M. et al. (2017). Coherent and tunable thz emission driven by an integrated III–V semiconductor laser. *IEEE Journal of Selected Topics in Quantum Electronics* 23 (4). doi: 10.1109/JSTQE.2017.2654060.

26 Laurain, A., Myara, M., Beaudoin, G. et al. (2010). Multiwatt-power highly-coherent compact single-frequency tunable vertical external cavity surface emitting semiconductor laser. *Optics Express* 14 (18): 14631.

27 Fox, A.G. and Li, T. (1961). Resonant modes in a maser interferometer. *Bell Labs Technical Journal* 40 (2): 453–488.

28 Henry, C.H. (1982). Theory of the linewidth of semiconductor lasers. *IEEE Journal of Quantum Electronics* 18 (2): 256–263.

29 Schawlow, A. and Townes, C. (1958). Infrared and optical masers. *Physical Review* 112: 1940–1949.

30 Rice, S.O. (1948). Statistical properties of a sine wave plus random noise. *Bell Labs Technical Journal* 27 (1): 109–157.

31 Bandel, N.V., Myara, M., Sellahi, M. et al. (2016). Time-dependent linewidth: beat-note digital acquisition and numerical analysis. *Optical Express* 24 (24): 27961–27978. doi: 10.1364/OE.24.027961. http://www.opticsexpress.org/abstract.cfm?URI=oe-24-24-27961.

32 Laurain, A. (2010). Sources Laser à semiconducteur à émission verticale de haute cohérence et de forte puissance dans le proche et le moyen infrarouge. Ph.D. thesis. Université Montpellier II.

33 Coldren, L.A. and Corzine, S. (1995) In: *Diode Lasers and Photonic Integrated Circuits*. Wiley.

34 Tropper, A. and Hoogland, S. (2006). Extended cavity surface-emitting semiconductor lasers. *Progress in Quantum Electronics* 30 (1): 1–43.

35 Nakwaski, W. (1988). Thermal conductivity of binary, ternary, and quaternary III-V compounds. *Journal of Applied Physics* 64 (1): 159–166.

36 Borca-Tasciuc, T., Song, D., Meyer, J. et al. (2002). Thermal conductivity of alas 0.07 sb 0.93 and al 0.9 ga 0.1 as 0.07 sb 0.93 alloys and (alas) 1/(alsb) 11 digital-alloy superlattices. *Journal of Applied Physics* 92 (9): 4994–4998.

37 Kemp, A.J., Valentine, G.J., Hopkins, J.M. et al. (2005). Thermal management in vertical-external-cavity surface-emitting lasers: finite-element analysis of a heatspreader approach. *IEEE Journal of Quantum Electronics* 41 (2): 148–155.

38 Gbele, K., Laurain, A., Hader, J. et al. (2016). Design and fabrication of hybrid metal semiconductor mirror for high-power vecsel. *IEEE Photonics Technology Letters* 28 (7): 732–735.

39 Opscan 2300. (2017). Innoptics, Datasheet.

40 Baili, G., Alouini, M., Moronvalle, C. et al. (2006). Broad-bandwidth shot-noise-limited class-a operation of a monomode semiconductor fiber-based ring laser. *Optics Letters* 31 (1): 62–64.

41 Jacubowiez, L., Roch, J.F., Poizat, J.P., and Grangier, P. (1997) Teaching photodetection noise sources in laboratory. *Proceedings of SPIE*, 166–179. doi: 10.1117/12.294378.

42 Marin, F., Bramati, A., Giacobino, E. et al. (1995). Squeezing and intermode correlations in laser diodes. *Physics Review Letters* 75: 4606–4609. doi: 10.1103/PhysRevLett.75.4606.

43 Vey, J.L. and Gallion, P. (1997). Semiclassical model of semiconductor laser noise and amplitude noise squeezing – part I: description and application to Fabry–Perot laser. *IEEE Journal of Quantum Electronics* 33 (11): 2097.

44 Jeremie, F., Vey, J.L., and Gallion, P. (1997). Optical corpuscular theory of semiconductor laser intensity noise and intensity squeezed-light generation. *JOSAB* 14 (2): 250–257.

45 Crooker, S.A., Brandt, J., Sandfort, C. et al. (2010). Spin noise of electrons and hols in self-assembled quantum dots. *Physical Review Letters* 104: 036601.

46 Seghilani, M., Myara, M., Sagnes, I. et al. (2015). Self-mixing in low-noise semiconductor vortex laser: detection of a rotational doppler shift in backscattered light. *Optical Letters* 40 (24): 5778–5781. doi: 10.1364/OL.40.005778. http://ol.osa.org/abstract.cfm?URI=ol-40-24-5778.

47 Press, W.H., Teukolsky, S.A., Vetterling, W.T., and Flannery, B.P. (2007). In: *Numerical Recipes in C – The Art of Scientific Computing – Second Edition*, 3rd edn.. Cambridge University Press.

48 Heinzel, G., Rüdiger, A., and Schilling, R. (2002). Spectrum and spectral density estimation by the discrete fourier transform (dft-fft), including a comprehensive list of window functions and some new flat-top windows. Tech. Rep., Albert Einstein Institut.

49 Fronen, R.J. and Vandamme, L.K.J. (1988). Low-frequency intensity noise semiconductor lasers. *IEEE Journal of Quantum Electronics* 24 (5): 724–736.

50 Hooge, F., Kleinpenning, T.G.M., and Vandamme, L.K.J. (1981). Experimental studies on 1/f noise. *Report on Progress in Physics* 44 (5): 479.

51 Agrawal, G.P. (1988). Mode-partition noise and intensity correlation in a two-mode semiconductor laser. *Physics Review A* 37: 2488–2494. doi: 10.1103/PhysRevA.37.2488.

52 Boashash, B. (1992). Estimating and interpreting the instantaneous frequency of a signal. I. Fundamentals. *Proceedings of the IEEE* 80 (4): 520–538.

53 Signoret, P., Marin, F., Viciani, S., et al. (2001). 3.6-MHz linewidth 1.55 μm monomode vertical-cavity surface-emitting laser. *IEEE Photonics Technology Letters* 13 (4): 269–271.

54 Tourrenc, J.P., Signoret, P., Myara, M. et al. (2005). Low-frequency FM-noise-induced lineshape: a theoretical and experimental approach. *IEEE Journal of Quantum Electronics* 41 (4): 549–553.

55 Di Domenico, G., Schilt, S., and Thomann, P. (2010). Simple approach to the relation between laser frequency noise and laser line shape. *Applied Optics* 49 (25): 4801–4807.

56 Olivero, J. and Longbothum, R. (1977). Empirical fits to the voigt line width: a brief review. *Journal of Quantitative Spectroscopy and Radiative Transfer* 17 (2): 233–236. doi: 10.1016/0022-4073(77)90161-3.

57 Seghilani, M., Myara, M., Sellahi, M. et al. (2016). Vortex Laser based on III-V semiconductor metasurface: direct generation of coherent Laguerre-Gauss modes carrying controlled orbital angular momentum. *Scientific Reports* 6: 38156. https://doi.org/10.1038/srep38156.

58 Seghilani, M.S., Myara, M. Sagnes, I. et al. (2015). Laser generation of coherent photon state carrying Spin-Angular-Momentum with handedness control using electronic spin transfer. In: *CLEO Europe*.

59 Sellahi, M., Myara, M., Beaudoin, G. et al. (2015). Highly coherent modeless broadband semiconductor laser. *Optics Letters* 40 (18): 4301–4304.

60 Blin, S., Paquet, R., Myara, M. et al. (2017) Coherent and tunable THz emission driven by an integrated III-V semiconductor laser. *IEEE Journal of Selected Topics in Quantum Electronics* 23(4).

5

Terahertz Metasurface Quantum Cascade VECSELs

Benjamin S. Williams and Luyao Xu

Department of Electrical and Computer Engineering, University of California Los Angeles, Los Angeles, CA, USA

5.1 Introduction

The challenge of creating lasers that emit with a high power level in a high-quality beam is ubiquitous across the spectrum. Indeed, as is discussed extensively in this book, one of the great advantages of the VECSEL concept is its ability to address this problem for semiconductor lasers. However, the problem of output power and beam quality is particularly acute for terahertz quantum-cascade (QC) lasers. The terahertz range of the electromagnetic spectrum is typically defined as the frequencies between 300 GHz and 10 THz, or wavelengths between 30 µm and 1 mm. The dominant semiconductor laser source in this range is the QC laser – a unipolar intraband laser in which photons are generated as electrons make radiative transitions between quantized "subbands" engineered in stacks of heterostructure quantum wells [1]. The first THz QC lasers were demonstrated in 2001 and shortly thereafter [2–4]; their technological progress has been summarized in various review articles [5–7]. The $GaAs/Al_xGa_{1-x}As$ material system is most commonly used, although other material systems have been demonstrated as well(e.g. InGaAs/AlInAs, InGaAs/GaAsSb, etc.). Various THz QC-lasers currently have been demonstrated between 1.2 and 5.6 THz (and at frequencies as low as 0.6 THz in a strong magnetic field). At this time cryogenic operation is still required for THz QC-lasers; the current temperature record is $T_{max}=200$ K in pulsed mode, and $T_{max}=129$ K in cw mode [8, 9]. While efforts to improve temperature performance of the active material continue, with the continuous improvements in cryogen-free coolers, operation in the 40–90 K range is very feasible for many applications. High output power has been demonstrated in a few devices: current records results stand at over 2 W peak power in pulsed mode, and 230 mW in continuous-wave (cw) mode for devices cooled by liquid helium) [10, 11]. For devices operating above 77 K, milliwatt level powers are more typical. THz QC-lasers have several characteristics which have made it extremely difficult to attain a good beam patterns simultaneously with high output power, particularly in continuous-wave mode above 77 K. This is the challenge that is addressed by the newly developed THz metasurface QC-VECSEL.

Vertical External Cavity Surface Emitting Lasers: VECSEL Technology and Applications, First Edition.
Edited by Michael Jetter and Peter Michler.
© 2022 WILEY-VCH GmbH. Published 2022 by WILEY-VCH GmbH.

5.1.1 Waveguides for THz QC-Lasers

The biggest difference between THz QC-lasers and semiconductor lasers at shorter wavelengths lies in the techniques for waveguiding. Nearly all waveguides in electrically injected visible, near-IR, and mid-IR semiconductor lasers are based upon doped dielectric cladding layers that surround the waveguide core, which provide both modal confinement and a path for current. Because free-carrier losses increase as λ^2, the use of doped cladding layers introduces excessive loss at THz wavelengths. Hence, unique waveguides for THz QC-lasers have been developed so as to minimize the overlap of the mode with any doped cladding layers. These waveguides use metallized layers partially or fully for optical confinement; unlike at shorter wavelengths losses are quite modest for noble metals in the terahertz range. There are two types of waveguides used at present for THz QC-lasers: the surface-plasmon and the metal–metal waveguide [5].

The surface-plasmon (SP) waveguide involves the growth of a thin (0.2–0.8 μm thick) heavily doped layer underneath the 10-μm-thick GaAs/AlGaAs quantum-well active region, but on top of a semi-insulating GaAs substrate [2, 3, 12]. The resulting mode is a compound surface plasmon tighly confined by the top metal contact and loosely bound to the heavily doped lower plasma layer. The mode extends far into the substrate (by tens to hundreds of microns); since the substrate is semi-insulating, the free carrier loss is minimal. The confinement factor to the active region is typically between 10–50%, and in general modes are somewhat loosely confined, which enables one to have relatively wide ridges without supporting multiple lateral modes. The downside is that ridges narrower than ~100 μm tend to squeeze the mode out of the active region and into the substrate. This increases the threshold and effectively puts a floor on the minimum device area (and power dissipation), which in turn limits the maximum achievable cw operating temperature [13].

An alternative to the SP waveguide is the the so-called "metal–metal" (MM) waveguide, also known as a "double-metal" or "double-plasmon" waveguide [14, 15]. In this structure, the waveguide mode is tightly confined between metal cladding placed immediately above and below the 5–10 μm thick epitaxial active region. Such a structure is fabricated using copper-to-copper or gold-to-gold thermocompression wafer bonding, followed by a substrate removal selective wet etch that stops on active region heterostructures [16]. Then, the remaining epitaxial active region can be patterned via photolithography, etched into ridges (or any other geometry), and further metallized via conventional microfabrication processes. Since any doped contact layers are usually quite thin, waveguide losses are dominated by absorption in the metal, and any reabsorption from inside the active region itself (which is often not negligible). The overall result resembles a microstrip transmission line, and indeed transmission-line formalism can be used to analyze the structure [17, 18].

MM waveguides tend to have the best high-temperature performance, mostly as a result of lower overall losses (both absorption and radiative) compared to the SP waveguide. For example, the highest operating temperature in pulsed mode in a metal–metal waveguide is 199.5 K, compared to approximately approximately

120 K in a SP waveguide [8]. Furthermore, the strong modal confinement of MM waveguides allows both the vertical and lateral dimensions to be made smaller than the wavelength. This in turn reduces the total thermal dissipation and required cooling power, which enables cw operation up to 129 K in a metal–metal waveguide, compared to approximately 80 K in a SP waveguide [9, 16]. However, the question arises: how do we get the radiation out of the waveguide? If a MM ridge waveguide is simply cleaved to form a Fabry–Pérot cavity, it performs poorly as a edge emitting laser. The cleaved facet radiates as a subwavelength sized aperture, which exhibits an extremely divergent beam [19]. Even worse, the emitting aperture is poorly impedance matched to free space; as a result, the effective facet reflectivity becomes large $R \sim 0.6$–0.9 (depending on the waveguide dimensions relative to the wavelength). This is much higher than the expected Fresnel value of $R \sim 0.32$ calculated from the GaAs/vacuum index mismatch [13, 17]. For realistic dimensions, the radiation losses are roughly an order of magnitude smaller than absorption losses, which implies the optical coupling efficiency is low (10% or less), and the emitted output powers tend to be milliwatts up to a few tens of mW at best.

A host of different techniques have emerged to improve both the beam pattern and output coupling efficiency – too many to fully survey here. To give one example, by mounting a silicon hyperhemispherical lens on the metal–metal waveguide facet, significant improvement in output power and beam quality can be achieved [20]. As another example, a great deal of effort has been on a variety of distributed feedback (DFB) cavities, where Bragg scattering from periodic structures is used both to provide feedback, and to couple out and redirect the emission into a directive far-field beam. Second-order DFB [21, 22] and photonic crystal cavities [23] use Bragg scattering to redirect in-plane radiation into surface emission. This improves the beam a great deal, since the waveguide surface, not the facet, is now the radiating aperture. However, when one tries to increase the aperture by increasing the width of the waveguide, thermal performance suffers and multiple transverse modes tend to appear. Phase-locking arrays of narrow-ridge second-order DFBs is another option – however, on-chip phase locking is challenging for large numbers of array elements, and grating side lobes appear if the array spacing is larger than the wavelength [24]. End-fire DFB QC-lasers based upon third-order DFBs or antenna DFB cavities are another attractive option; they can achieve very narrow far field beams even when the transverse waveguide cross section is subwavelength, since the beam divergence scales inversely with square root of the cavity length [25–28]. However, once again scaling up the output power is not trivial: if wider waveguides are used the cw performance will degrade, if longer cavities are used the phase matching condition becomes increasingly strict (for third-order DFBs). In summary, many of these DFB approaches are indeed very effective at improving the beam pattern and outcoupling efficiency; however, they show their limitations when one wishes to scale up the output power levels while maintaining a good beam pattern. A secondary issue is that DFB and other Bragg grating approaches are inherently narrowband and largely unsuitable for widely tunable single mode lasers or for broadband frequency combs. This is the landscape that has motivated the development of the THz QC-VECSEL.

5.1.2 Overview of Metasurface QC-VECSEL Concept

While the VECSEL is a natural architecture for achieving high power with a good beam, it is ordinarily not possible to implement this geometry for QC-lasers. This is because since the optical gain occurs via intersubband transitions of electrons within planar quantum wells, which obey a selection rule that only allows interaction with electric field polarized perpendicular to the plane of the wells. This is incompatible with the natural in-plane polarization for surface incident waves in a VECSEL cavity. We address this problem by introducing the concept of an active reflectarray metasurface, which consists an array of metallic microcavity antennas loaded with QC gain material; each antenna efficiently couples in THz radiation, amplifies it via stimulated emission, and reradiates into free space. The amplifying metasurface reflector is then paired with an output coupler to create an external laser cavity. A schematic of this concept is shown in Figure 5.1. Several SEM images of fabricated metasurfaces are shown in Figure 5.2. This approach is fundamentally different from the previous beam shaping techniques for THz QC-lasers – it is the super-mode of the VECSEL cavity, rather than the modes of the individual metallic microcavities on metasurface, that exhibits lasing and shapes the beam to a near-Gaussian profile. The metasurface is deliberately designed to have a low radiative quality factor (Q-factor) so that the microcavities will not self-oscillate in the absence of an external cavity. Furthermore, since each microcavity also supports a highly confined fundamental propagating waveguide mode, absorbing boundary conditions implemented by creating lossy tapers to suppress self-lasing. These tapered regions have a dielectric layer underneath the top metallization, so that no current is injected, and they exhibit loss rather than gain.

Compared to conventional THz QC-lasers, the advantage of the QC-VECSEL can be summarized as: (a) the output power is scalable with the active area on the metasurface as more active microcavities contribute stimulated emission to the VECSEL cavity mode, (b) the beam quality is primarily determined and well shaped by the external cavity and not individual microcavities, (c) compared to monolithic cavities, it is easier to achieve the optimum coupling condition and maximize the output power by choosing the reflectance of the output coupler, and (d) the sparse

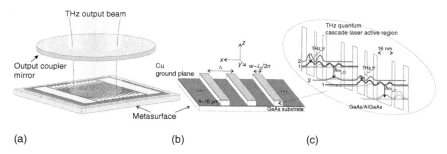

Figure 5.1 (a) Schematic figure of metasurface QC-VECSEL configuration. (b) Close view of amplifying metasurface in its simplest form, which is made up of periodic arrays of microcavity antennas implemented in metal–metal waveguide ridges of width w, height h, and period Λ, each loaded with QC-laser gain material. (c) Band diagram of typical GaAs/AlGaAs THz QC-laser active region.

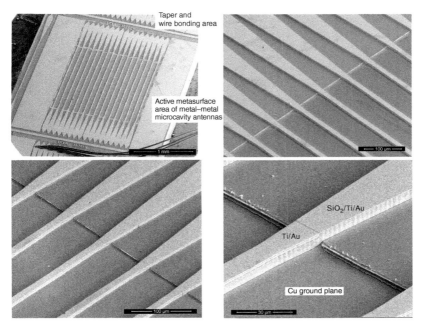

Figure 5.2 Scanning electron microscope images at various magnification of a QC-VECSEL metasurface made up of uniform metal–metal waveguide ridges and tapered lossy terminations.

arrangement of the microcavities reduces the power dissipation density for improved cw performance. While the advantages (a)–(d) are somewhat generic to the VECSEL configuration, a further advantage (e) is somewhat unique, in that the metasurface provides a flexible platform for engineering the phase, amplitude, and polarization response, in both the spatial and spectral domains. That is not to say this comes at no cost – the use of an external cavity adds complexity and complications for stability. (Since the laser must reside in a cryostat, increased size, and weight need not be substantial.) Nonetheless, in many cases, the advantages will collectively outweigh the disadvantages of an external cavity.

The first demonstration of a QC-VECSEL was in April of 2015, in which exhibited lasing at 2.8 THz in a near-Gaussian beam pattern with about significant power output [29]. Since then improvements have been occurring rapidly. Indeed, since the QC-VECSEL is barely two years old at the time of this writing, we expect the field will continue to move quickly. Here, we provide an overview of the QC-VECSEL approach, give a snapshot of some current performance levels, and highlight the opportunity for novel laser engineering made possible by the metasurface approach.

5.2 Metasurface Design

The enabling component for the QC-VECSEL is the active metasurface reflector. It can be considered a THz version of the reflectarray antennas that are common in the mm-wave range [30]; however, here the primary purpose of microcavity antenna elements is to couple the incident THz radiation with the QC laser gain material. Any other functionality, such as control of the reflection phase and polarization, is a "bonus." In the simplest form, the active metasurfaces are composed of a sparse array of identical metal–metal waveguide ridges; a schematic of such a structure is shown in Figure 5.1b. However, while metal–metal waveguides are usually intended to support a confined and guided mode (i.e. the fundamental mode), in this case we wish them to act as surface emitting antennas. This is done by using a very narrow ridge with a width that corresponds approximately to half of the wavelength in the semiconductor ($w \approx \lambda_0/2n$). This produces a resonance at the corresponding frequency – essentially the lowest order standing wave in very short cavity of length w. Alternately, each microcavity can be considered to be a leaky-wave antenna operating at the cutoff frequency of the first higher order lateral waveguide mode, where the group velocity of the mode goes to zero [31, 32]. The structure is best understood as an elongated patch antenna, where the fringing fields at the ridge sidewalls radiate constructively into the far-field in the surface normal direction [17]. When electrically biased, the QC active material provides THz gain described by the bulk material gain coefficient $g(\nu)$. Incident THz radiation (polarized transverse to the ridge axis) is coupled into the standing wave mode within each microcavity, which possesses the necessary vertical E-field polarization needed to satisfy the intersubband polarization selection rule. Interaction with the QC gain material amplifies the THz field via stimulated emission and reradiates it back into free space. The microcavities are spaced with a period Λ which is kept less than the free-space wavelength λ_0 (of intended laser emission). This is necessary to suppress any higher order Bragg diffraction — only zeroth-order (specular) reflection occurs. It is important to note that despite its resemblance to a grating, the primary mechanism of operation for the metasurface is not Bragg scattering. Indeed, periodicity is not strictly necessary. Rather, each microcavity antenna is locally self-resonant on the unit cell level. For $\Lambda \ll \lambda_0$, this resonance frequency is primarily determined by the ridge width w; the choice of period Λ is only a small perturbation.

As an example, a finite element solvers are used to numerically simulate the reflectance of the metasurface – this is done both in passive mode, and when a simulated gain is added to the dielectric via an anisotropic permittivity tensor. Here we present a two-dimensional unit cell, with periodic boundary conditions used to simulate an infinite sized metasurface. Drude model parameters are used to describe the loss of both the metal, and the free-carrier losses inside the semiconductor [33]. Figure 5.3a shows a set of simulated reflectance spectra for a group of metasurfaces with different ridge widths ranging from 11 to 12.5 μm and a fixed periodicity $\Lambda = 70$ μm. Two scenarios are considered: passive with $g = 0$ cm^{-1} and active with $g = 30$ cm^{-1} in the QC-material. As expected, reducing the width w leads to a higher resonance frequency. Figure 5.3b shows the reflectance R_{MS} increase with gain for the metasurface $w = 11.5$ μm and $\Lambda = 70$ μm at its resonance frequency 3.4 THz and two other frequencies above and below the resonance. It is useful to fit these numerical results to the relation

$$R_{MS} = R_1 G = e^{\xi(\nu)(g-g_{tr})}. \tag{5.1}$$

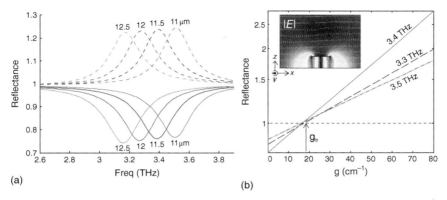

Figure 5.3 (a) Simulated reflectance spectra for four metasurfaces with ridge widths varying from 11–12.5 µm and fixed period $\Lambda = 70$ µm. The solid lines are results for passive metasurface with $g = 0$ cm^{-1}, and the dashed lines are for active metasurface with $g = 30$ cm^{-1}. (b) Simulated reflectance change (plotted in log scale) versus gain coefficient for a metasurface with $w = 11.5$ µm and $\Lambda = 70$ µm at different frequencies on and off resonance. The transparency gain g_{tr} is indicated with an arrow. Source: ©2017 IEEE. Reprinted, with permission, from [34].

where R_1 is the passive metasurface reflectance at frequency ν, G is the intensity gain, g_{tr} is the transparency gain coefficient needed to balance absorption losses (from the metal and semiconductor), and $\xi(\nu)$ is a fitting coefficient that contains information about the metasurface frequency response and quality factor. Gain is more efficiently coupled to incident radiation when the operation frequency is closer to the resonance; this effect is represented by a larger value of $\xi(\nu)$. In these simulations, we assumed a frequency-independent gain for the active medium, so that the metasurface response can be analyzed independently of the choice of active material. In reality, the gain will have its own lineshape; the general design goal is to match the metasurface resonance with the peak gain frequency of the gain medium. This is typically accomplished by first measuring the lasing spectrum of a conventional metal–metal waveguide QC-laser fabricated from the same active material and then designing the metasurface dimensions around the measured values.

We might ask, how "subwavelength" must the period be? Indeed, it is advantageous for thermal reasons to make the metasurface as sparse as possible. One might expect as long as $\Lambda < \lambda_0$, at normal incidence all Bragg diffraction orders are suppressed save the zeroth order (specular reflection). However, even for periods approaching λ_0, coupling occurs between the localized microcavity resonance and a propagating Bloch surface wave which introduces diffraction loss [34]. This effect is more significant than is readily apparent by using simulations with periodic boundary conditions, which approximate the case of an infinite sized metasurface excited by normally incident plane waves. For an infinite-size metasurface reflectance simulation, the Bloch mode appears to be of very high quality factor; however, for a more realistic finite sized metasurface, the opposite is in fact true; as Λ approaches λ_0 there is considerable additional loss present associated with scattering and diffraction. Furthermore, a finite-size Gaussian beam will have additional transverse momentum components which will more readily couple to the Bloch wave. So far, we have developed a design rule of thumb to keep $\Lambda \leq 0.8\lambda_0$ to prevent these effects.

5.3 QC-VECSEL Model

In order to have a framework for understanding QC-VECSEL performance, we review a formalism first presented in Ref. [34]. We consider an idealized QC-VECSEL cavity as illustrated in Figure 5.4, which consists of a a metasurface of area A with passive reflectance R_1 and an output coupler with reflectance R_2 and transmittance $T_2 = 1 - R_2$. The forward and backward circulating intensity is I_+ and I_-. We also include a single-pass transmittance T for the propagation of the cavity length L which includes the effect of diffraction loss, atmospheric and window absorption. The metasurface produces a uniform power gain $G = e^{\xi g}$, so that its active reflectance is equal to $R_1 G$. The bulk gain coefficient of the active QC-laser material at the cavity mode frequency v is $g = g(v)$. We assume for the moment that the intensity is uniform in the transverse direction and that the mode area and the metasurface area are identical; later we will modify our expressions to include a transverse confinement factor which will let this assumption to be relaxed.

The threshold condition is set by requiring the intensity to be unchanged after one round trip:

$$1 = T^2 R_1 R_2 G_{\text{th}} = T^2 R_1 R_2 e^{\xi g_{\text{th}}}. \tag{5.2}$$

Similarly, this sets the threshold gain coefficient g_{th} as

$$g_{\text{th}} = -\frac{\ln(T^2 R_1 R_2)}{\xi}. \tag{5.3}$$

We now introduce a simple rate equation model to describe the QC-laser active material, as shown in Figure 5.4. While this cannot hope to capture the complex

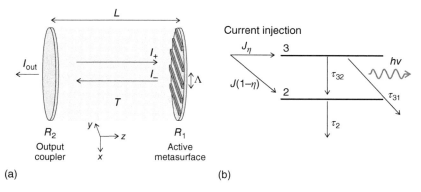

Figure 5.4 (a) Schematic of model VECSEL cavity. (b) Schematic three-level system used to define idealized rate equation model for QC-laser transport. Source: ©2017 IEEE. Reprinted, with permission, from [34].

transport kinetics, it is a standard treatment for intersubband transport similar to that given in Ref. [35]. The upper radiative state 3 is pumped using tunneling current injection at a rate of $J\eta/eL_p$, where J is the current density, η is the injection efficiency, e is the fundamental charge, and L_p is the length of one cascade period. The remaining fraction $(1-\eta)$ of current density is injected into the lower radiative state 2, which is then emptied by some combination of tunneling and electron–phonon scattering. The various nonradiative lifetimes are given by τ_3, τ_{32}, and τ_2. Within the active material, we use a standard expression for the saturated gain coefficient for a homogeneously broadened gain transition:

$$g = \frac{g_0(J - J_{\text{leak}})}{1 + I_0/I_s} = \frac{J\sigma\tau_{\text{eff}}}{eL_p} \frac{1}{1 + I_0/I_s}. \tag{5.4}$$

where $g_0(J)$ is the unsaturated gain coefficient, which we assume is proportional to the pump current density J less an empirical shunt leakage current density J_{leak}. The effective upper state lifetime is given by $\tau_{\text{up}} = \tau_3\left(1 - \tau_2/\tau_{32}\right)$. and the effective lifetime for population inversion is $\tau_{\text{eff}} = \eta\tau_{\text{up}} - (1-\eta)\tau_2$, which accounts for a nonunity injection efficiency η. For the case of $\eta = 1$ we see $\tau_{\text{eff}} = \tau_{\text{up}}$. The circulating intensity within the microcavity ridge is I_0, and the saturation intensity I_s is defined as the intensity sufficient to reduce the population inversion by a factor of two: $I_s = h\nu/\sigma\left(\tau_{\text{up}} + \tau_2\right)$. The stimulated emission cross section at the cavity frequency is $\sigma(\nu)$, such that $g(\nu) = \sigma(\nu)(n_3 - n_2)$, where n_3 and n_2 are the 3D population densities of levels 3 and 2, respectively. Setting the value $g = g_{\text{th}}$, we can obtain an expression for the intensity vs. current density:

$$I_0 = \frac{h\nu}{eL_p} \frac{\tau_{\text{eff}}}{\tau_2 + \tau_{\text{eff}}} \frac{(J - J_{\text{th}})}{g_{\text{th}}}, \tag{5.5}$$

where the threshold current density is:

$$J_{\text{th}} = \frac{n\alpha_{\text{cav}}eL_p}{\sigma\tau_{\text{eff}}\Gamma_1} + J_{\text{leak}}. \tag{5.6}$$

We now must relate the intensity inside the microcavity to the open cavity circulating intensities I_+ and I_-. We assume a form for the cavity mode as a circulating near-plane wave in free space over length L with transverse profile $\psi(x,y)$, and as a standing wave within each QC-laser ridge antenna with intensity I_0. This can be written as

$$\mathbf{E} = \underbrace{\hat{\mathbf{x}}E_+\psi(x,y)\left(e^{ik_0z} + r_1\sqrt{G_{\text{th}}}e^{-ik_0z}\right)}_{\text{open cavity}}$$

$$+ \underbrace{\sum_{i=1}^{N}\hat{\mathbf{z}}E_0\psi(x,y)\sin\left(\frac{\pi}{w}(x - x_i)\right)}_{\text{inside microcavities}}, \tag{5.7}$$

where we assume for the moment that the circulating field is polarized in the x-direction transverse to the N microcavity ridges each centered at x_i. This form neglects the reactive fringing near fields in the air near the ridges. As mentioned above, we assume for the moment that the field is uniform so that $\psi(x,y) = 1$ over the area (i.e. a top-hat beam where we neglect phase front nonuniformities). Conservation of energy requires

$$A(I_- - I_+) = \frac{dU}{dt} = \frac{\omega_0 U_0}{Q_{abs}}, \tag{5.8}$$

where U_0 is the electromagnetic energy stored inside the antenna microcavity, ω_0 is the resonance frequency, and Q_{abs} represents the nonradiative quality factor of the microcavity that accounts for the absorption loss. We write U_0 based upon (5.7):

$$U_0 = AF\frac{1}{4}\varepsilon E_0^2 h, \tag{5.9}$$

where h is the microcavity height determined by the active material thickness such that $h = N_p L_p$, where N_p is the number of cascaded periods. The permittivity of the active region material is ε, and F is the fill factor of the biased antenna area over entire metasurface area ($F = w/\Lambda$ for the metasurface shown in Figure 5.1b). We further write $Q_{abs} = \frac{\omega_0 n}{(g_{tr}-g)c}$, where transparency gain is $g_{tr} = \xi^{-1} \ln R_1^{-1}$, and is obtained from numerical simulations such as described in Figure 5.3. Using the relation $I_- = R_1 G I_+ = c\varepsilon_0 E_-^2/2$ and combining Eqs. (5.8) and (5.9), we can define a field enhancement factor M as

$$M = \frac{|E_0|^2}{|E_-|^2} = \frac{2(1-R_1 G)}{R_1 G(g_{tr}-g)nhF}. \tag{5.10}$$

The output intensity is written as $I_{out} = (1-R_2)TI_-$. With $I_0 = nc\varepsilon_0 E_0^2/4$ and $I_+ = c\varepsilon_0 r^2 E_+^2/2$, we can write the output intensity as

$$I_{out} = \frac{2(1-R_2)T}{nM}I_0. \tag{5.11}$$

Substituting (5.5) and (5.10) in (5.11), using the laser threshold condition $R_1 G_{th} = (R_2 T^2)^{-1}$, and muliplying by the metasurface area A, we obtain the total output power as

$$P_{out} = N_p \frac{h\nu}{e} \underbrace{\frac{\tau_{eff}}{\tau_2 + \tau_{eff}}}_{\eta_i} \underbrace{\frac{T(1-R_2)\ln(R_2 T^2)}{(1-R_2 T^2)\ln(R_1 R_2 T^2)}}_{\eta_{opt}} (I - I_{th}), \tag{5.12}$$

where

$$I_{th} = AF\left[\frac{eL_p}{\sigma\tau_{eff}}\frac{-\ln(R_1 R_2 T^2)}{\xi} + J_{leak}\right]. \tag{5.13}$$

The derivation so far does not account for the effects of modal nonuniformity and spatial hole burning. For example, one effect of modal nonuniformity is to reduce the slope efficiency near threshold, as the injected current is effectively wasted in regions with low modal intensity. Spatial hole burning is particularly acute in THz QC-lasers, due to the long length scale of the standing wave (~10 μm) compared to the lateral diffusion lengths of the inverted carrier population (a few hundred nanometers). This effect results in a nonlinear P–I curve; however, near threshold a linearized expression can be derived, and included through an additional "uniformity efficiency" factor $\eta_u \leq 1$. The slope efficiency near threshold can then be written:

$$\frac{dP}{dI} = N_p \frac{h\nu}{e} \eta_{opt} \eta_i \eta_u. \tag{5.14}$$

Since the electric field is polarized almost entirely in the z-direction within the metal–metal waveguide, the uniformity factor associated with the microcavity mode can be written as

$$\eta_u = \underbrace{\frac{\left(\int_{act} |E_z|^2 dA\right)^2}{A_{act} \int_{act} |E_z|^4 dA}}_{\text{microcavity}} \underbrace{\frac{\left(\int_A |\psi(x,y)|^2 dA\right)^2}{A \int_A |\psi(x,y)|^4 dA}}_{\text{cavity mode shape}}. \tag{5.15}$$

The first factor describes the modal uniformity within each microcavity, and the second factor describes the uniformity of the slowly varying cavity mode profile incident upon the metasurface. While η_u can be solved exactly from numerical results, for a uniform cavity mode and the sinusoidal dependence of the mode within the microcavity described by Eq. (5.7), $\eta_u = 2/3$ (very close to the value of 0.65 extracted from a finite element simulation (inset of Figure 5.3b). If the transverse beam $\psi(x,y)$ within the cavity is not uniform, then η_u will be further reduced. This function is plotted for a Gaussian beam with spot size w_0 on a square metasurface, where only the center circular area of diameter $2a$ is biased. The uniformity is shown in Figure 5.5 and shows that underfilling a metasurface with the beam will cause a reduction in efficiency. Transverse mode confinement factor Γ_t as defined in Eq. (5.19) is also plotted in Figure 5.6, which exhibits a trend opposite to η_u. This suggests a tradeoff between these two factors in the metasurface bias area design.

The effects of modal uniformity and spatial hole burning illustrate one of the largest differences between the metasurface VECSEL and conventional Fabry–Pérot waveguide QC-lasers. In a Fabry–Pérot laser multimode oscillation "washes-out" the overall field nonuniformity and allows for the most efficient use of the available

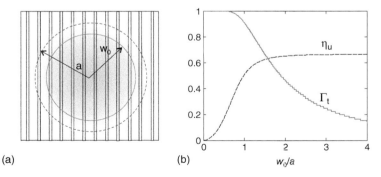

Figure 5.5 (a) Diagram of square metasurface with biased circular area of radius a illuminated by a Gaussian beam with waist w_0. (b) Calculated uniformity factor η_u and transverse confinement factor Γ_t for scenario in (a).

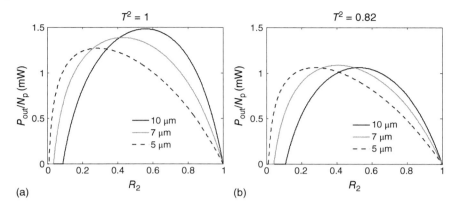

Figure 5.6 Calculated output power per period P_{out}/N_p vs. output coupler reflectance R_2 for $T^2 = 1$ (a) and $T^2 = 0.82$ (b) for three metasurfaces designed with different ridge heights from 5 to 10 μm at the identical resonant frequency of 3.4 THz. A fixed bias area of $A = 0.3$ mm^2 is assumed for all.

gain. Hence, η_u typically does not appear in most expressions for the slope efficiency. However, in a QC-VECSEL cavity with uniform microcavities, all of the various longitudinal modes in the external cavity interact with the active material through the same metasurface resonance and have the same uniformity factor η_u. This may lead to a suppression of multimode operation – more work is needed to fully understand this phenomenon. We also note that this effect may be "engineerable" by designing spectrally and spatially inhomogeneous metasurfaces.

5.3.1 Confinement Factor

Conventional formalism for describing semiconductor and QC-lasers defines the threshold current density in terms of loss coefficients per unit length and modal confinement factors. We can link the above derivation to this more conventional terminology, if we defines the threshold gain g_{th} in terms of a loss coefficient α_{cav} (prorated over the cavity round trip length $2L$) according to:

$$g_{th} = \frac{n}{\Gamma}\alpha_{cav} = -\frac{n}{\Gamma}\frac{\ln\left(T^2 R_1 R_2\right)}{2L}, \quad (5.16)$$

where $\Gamma = \Gamma_l \Gamma_t$ is a modal confinement factor which describes the overlap of the mode with the QC-active material (satisfying the polarization selection rule). It can be defined using a standard expression,

$$\Gamma = \Gamma_l \Gamma_t = \frac{\int_{act} \varepsilon(\mathbf{R})|E_z(\mathbf{R})|^2 dV}{\int_V \varepsilon(\mathbf{R})|E(\mathbf{R})|^2 dV}, \quad (5.17)$$

which can be conceptually separated into a longitudinal confinement factor $\Gamma_l(\nu)$ and a transverse confinement factor Γ_t. The longitudinal confinement factor contains the field enhancement effects of the microcavity resonance $M(\nu)$. For large cavity lengths, and using the field in Eq. (5.7), we can approximate

$$\Gamma_l(\nu) \approx \frac{h n^2 M(\nu) F}{2L(1 + R_1 G_{\text{th}})}. \quad (5.18)$$

This expression has undesirable feature that it depends upon the threshold gain G_{th}. However, in the limit of a high finesse cavity, R_1, R_2, T, and G_{th} are all close to unity, and $\Gamma_l(\nu)$ is directly proportional to the fitted $\xi(\nu)$ parameter (see Eq. (5.1)). When the transverse extent of the mode is smaller than the biased area of the metasurface, the entire mode experiences reflective gain and the transverse confinement factor Γ_t is unity. However, when the transverse extent of the mode is larger than the active "bias area" of the metasurface, the wings (or other portions) of the beam do not experience gain, and we can define

$$\Gamma_t = \frac{\int_{\text{bias area}} |\psi(x,y)|^2 dA}{\int |\psi(x,y)|^2 dA} \quad (5.19)$$

See Figure 5.5, for a plot of the variation of Γ_t and η_u for a fundamental Gaussian beam incident upon a metasurface with a circular bias area.

Note, if we assume a lossless transmittance for the external cavity, i.e. $T = 1$, the optical coupling efficiency in Eq. (5.12) reduces to where $\eta_{\text{opt}} = \alpha_m/\alpha_{\text{cav}}$, where $\alpha_m = \ln R_2^{-1}/2L$ is the prorated output mirror loss coefficient. The threshold current density can then be expressed

$$J_{\text{th}} = \frac{e L_p}{\sigma \tau_{\text{eff}}} \frac{n}{\Gamma} \alpha_{\text{cav}} + J_{\text{leak}} \quad (5.20)$$

Thus, we recover the "classic" formulas for J_{th} and dP/dI for waveguide-based QC lasers. The factor of n in these definition results from the fact that the loss coefficients is prorated over a length $2L$ in vacuum, while the gain coefficient is defined in the semiconductor medium with refractive index n.

5.3.2 Metasurface and Cavity Optimization

Two parameters describing the metasurface characteristics are involved in modeling the QC-VECSEL performance at a given frequency: the transparency gain g_{tr} and the fitting parameter ξ. As described above, the $\xi(\nu)$ is a parameter extracted from the linear curve fit of Eq. (5.1), which physically reflects the modal confinement (or equivalently the field enhancement) with the QC-laser medium. In practice, this is typically limited by the radiative quality factor of the metasurface, and hence $\xi(\nu)$ is peaked at the resonance frequency. The transparency gain g_{tr} not only depends on ξ but also contains information the metasurface reflection loss. Compared with the fundamental guided mode within a metal–metal waveguide, g_{tr} tends to be slightly larger for a metasurface. This is because the metasurface resonance is based upon a higher-order mode within the metal–metal ridge, which has significant fringing fields concentrated around the corners and edges, which slightly increases the metallic loss.

We present now an example case of how metasurface design can affect g_{tr} and ξ, and how QC-VECSEL performance is influenced in turn. Specifically, we consider

Table 5.1 Values of g_{tr} and ξ extracted via simulation for three metasurface designs at 3.4 THz.

Design	g_{tr} (cm^{-1})	ξ (cm)
$h = 10$ μm, $w = 11.5$ μm, $N_p = 163$	16.5	0.0163
$h = 7$ μm, $w = 11.6$ μm, $N_p = 114$	18.9	0.0229
$h = 5$ μm, $w = 11.7$ μm, $N_p = 82$	22.2	0.0327

three nominally identical metasurface designs where the microcavity height h is varied, and how the this affects the choice of the output coupler reflectance R_2 that one needs to maximize the laser output power P_{out} at a fixed injection current. We consider three values of the height $h = 10, 7$, and 5 μm, (see Figure 5.1b), with (nearly) identical widths w so that the resonant frequency of 3.4 THz is kept constant. g_{tr} and ξ are extracted at the resonant frequency and summarized in Table 5.1. While all of the experimental data in this chapter refer to devices with $h = 10$ μm, changing the height is a straightforward way to increase the radiative Q-factor and ξ as the height of the of the radiating sidewall aperture is decreased [17]. However, the trade-off for a thinner active region is an increased transparency gain as h decreases. This is a result of the fact that the loss from the metallic cladding scales as h^{-1}.

Inserting these values into Eqs. 5.12 and 5.13, we can obtain the quantity P_{out}, given that we make some assumptions related to the particular active region and metasurface design. We assume a fixed bias area of $AF = 0.3$ mm^2, which is close to that for the metasurface QC-VECSEL discussed in Section 5.4.2, and a fixed injection current density of $J = 600$ cm^2. The values of other involved parameters including $\sigma\tau_{eff}/eL_p = 0.64$ cm/A, $v = 3.4$ THz, $\eta_i = 0.43$, $\eta_u = 0.65$, $J_{leak} = 343$ A/cm^2 are inherited from the experimentally extracted values presented in Sec. 5.4.2. The number of periods N_p scales as the height h as we assume the microcavity is filled with the QC-laser active material. In Figure 5.6, we plot the normalized output power per period P_{out}/N_p as a function of output coupler reflectance R_2 for two cases of round-trip external cavity loss: an ideal external cavity without diffraction or transmission losses ($T^2 = 1$), and a more realistic with $T^2 = 0.82$ based on the estimate in Section 5.4.2.

As is standard, there is an optimum value for R_2 which maximizes the output power, which reflects the trade-off between the large slope efficiency for small values of R_2 and the low threshold for high R_2. The optimum value of R_2 varies for different metasurface designs and occurs at lower values for metasurfaces with higher ξ. In other words, a more transmissive output coupler is needed to achieve the optimum output coupling condition for a metasurface with high radiative Q-factor. Another observation is that in a lossless external cavity with $T^2 = 1$ the maximum power per period P_{max}/N_p is higher for metasurfaces with a lower g_{tr}; this is not surprising as more absorption loss is always bad. However, if external cavity loss is unavoidable (i.e. $T^2 < 1$), a higher ξ may be beneficial even if it is accompanied by higher transparency gain. Larger ξ corresponds to a larger field enhancement within the

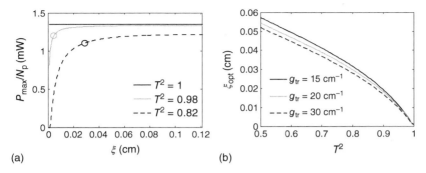

Figure 5.7 (a) Maximum power per period P_{max}/N_p change with ξ for different external cavity losses represented by T^2 and a fixed $g_{tr} = 15$ cm^{-1}. The red and green dots indicate where 90% of $P_{max}|_{\xi \to \infty}$ is reached. (b) ξ_{opt} change with T^2 for three values of g_{tr}. ξ_{opt} is where 90% of $P_{max}|_{\xi \to \infty}$ is achieved.

QC-active material, i.e. less field resides in the lossy open cavity. This finding implies that one should optimize the maximum power output from an actual QC-VECSEL in the presence of external cavity loss by optimizing both ξ and R_2.

The optimization of ξ depends on how lossy the external cavity is, which follows the trend that a higher ξ is preferred for a lossier cavity. Figure 5.7a plots the calculated maximum output power per period vs. ξ for three levels of cavity loss (represented by T^2) and a fixed $g_{tr} = 15$ cm^{-1}, which shows that P_{max}/N_p increases with a larger ξ. However, it is worth noticing that the increase of P_{max} tends to saturate at lower values of ξ for a less lossy cavity, with an extreme case being that for a lossless cavity with $T^2 = 1$, P_{max}/N_p has no dependence on ξ at all. We can reasonably define an "optimized" value of ξ as ξ_{opt} at which which 90% of $P_{max}|_{\xi \to \infty}$ is reached. The so-defined ξ_{opt} are indicated by circles in Figure 5.7a. ξ_{opt} is further calculated and plotted vs. T^2 for different g_{tr} (see Figure 5.7b), which suggests that a higher ξ is preferred for a lossier external cavity with lower T^2 to extract more power out of the VECSEL. In the actual design process, even if the values of T^2 and g_{tr} are typically estimated with an accuracy limit, an optimum range of ξ can still be inferred from the results shown in Figure 5.7b, which provides a guideline for metasurface design and optimization. The value of ξ is also readily engineerable for metasurfaces. While we have considered here the effects of varying the height of the microcavities, the radiative Q-factor can also be controlled using other techniques. For example, the polarimetric metasurfaces shown in Section 5.5 exhibit high-Q Fano resonances arising from interference between various coupled resonances [36].

5.4 THz QC-VECSEL Performance: Power, Efficiency, and Beam Quality

In this section, we will provide a snapshot for the current state of QC-VECSEL performance. With the exception of Section 5.4.4, the VECSELs discussed in this

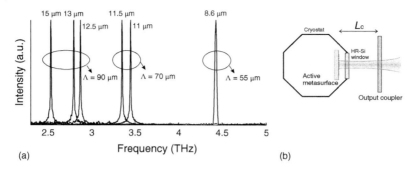

Figure 5.8 (a) Measured lasing spectra for some demonstrated QC-VECSELs based on various metasurfaces designed with different ridge widths (labels above each spectrum show the nominal width) and periods Λ fabricated on different active QC-laser materials. (b) Measurement set-up of plano-plano cavity where output coupler is external to cryostat. Source: ©2017 IEEE. Reprinted, with permission, from [34].

section were demonstrated using an external cavity configuration where the metasurface was mounted to the cold stage of a cryostat, and the output-coupler (OC) reflector was outside. This set-up is convenient for easy alignment. This set-up is shown schematically in Figure 5.8b. However, the cryostat window is an intracavity element, and one must take care to minimize the transmission loss. Best results have been obtained using a 3-mm thick high-resistivity (HR) silicon window, the estimated round-trip loss is 7% based upon literature values for the loss tangent for HR-silicon. Our first QC-VECSEL demonstration used an off-the-shelf wire grid polarizer as an output coupler [29], but since then we have more often used mesh couplers, i.e. a periodic subwavelength inductive or capacitive metal mesh pattern is defined by photolithography and evaporated on a crystal quartz substrate. Mesh couplers were developed for use with far-IR molecular gas lasers and allow ready design of the reflectance/transmittance by choice of the mesh dimensions [37]. Of course, using a mirror OC with a central hole drilled is an option; however, we have avoided this path as we wish to preserve the highest quality beam pattern.

5.4.1 Effect of Metasurface on Spectrum

So far, our group has demonstrated various QC-VECSELs lasing over a range from 2.5 to 4.4 THz. This is achieved by designing metasurfaces with different periods Λ and ridge widths w and pairing them with various resonant-phonon QC-laser active region designs. Figure 5.8a shows a family of lasing spectra from different THz QC-VECSELs, all of which were taken using the OC outside the cryostat (except the 4.4 THz demonstrated with an intra-cryostat cavity). As expected, there is an approximate inverse relationship between ridge width w and the lasing frequency. Typically, the lasing is in single-mode when the cavity length is optimized to achieve a maximum power output. This is attributed to the frequency dependence of the metasurface response, in conjunction with the etalon filter effect of the HR-Si cryostat window. As the cavity is tuned away from its optimum length, hopping between

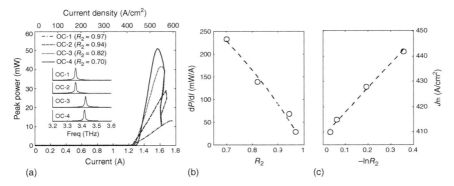

Figure 5.9 (a) Measured 77 K pulsed P–I characteristics for QC-VECSELs built with one metasurface and four different output couplers. The inset shows the lasing spectra for each VECSEL. (b) Slope efficiency change with the OC reflectance R_2. The black dashed line is the fitted curve. (c) Threshold current density J_{th} plotted vs. $-\ln R_2$. The black dashed line is the fitted curve according to Eqs. (5.13) and (5.12). Source: ©2017 IEEE. Reprinted, with permission, from [34].

longitudinal modes is occasionally observed. Temperature change can also induce a longitudinal mode hop.

5.4.2 Effect of Output Coupler

The ability to vary the output coupler transmittance to optimize the emitted power, threshold, and slope efficiency is one of the useful features of the QC-VECSEL. We fabricated a series of four mesh output couplers with varying reflectance values R_2. Due to the Fabry–Pérot fringes associated with the 100 μm quartz substrate, the transmittance varies slowly with frequency; FTIR spectroscopy was used to measure the transmittance at the relevant frequency, which gives 3.2, 5.8, 17.5, 30% for OC1–4, respectively. P–I measurements were performed on VECSELs constructed using the same uniform metasurface, with the output couplers placed external to the cryostat (shown in Figure 5.9a. Due to the dependence of the OC's reflectance on frequency, the VECSEL lasing frequencies are slightly different. Figure 5.9b and c shows the slope efficiency plotted as a function of R_2, and J_{th} plotted as a function of $\ln(R_2^{-1})$ (proportional to the output coupling loss coefficient α_m). As expected, as the transmittance increases, the slope efficiency increases as does the threshold current denstiy. The dashed lines are fits to the functions based upon Eqs. (5.12), (5.13) and (5.14), which show good qualitative agreement.

If certain assumptions are made, these fits can be further used for quantitative estimates. For example, we assume $T^2 = 0.82$ based upon 7% round trip loss in the HR-Si cryostat window, and 11% diffraction loss estimated using Fox-and-Li cavity modeling. Fitting the data yields values of $R_1 = 0.62$, $\eta_u \eta_i = 0.29$ and $J_{leak} = 343$ A/cm^2. This is close to a measured passive reflection of 0.76 (at room temperature) from a similar metasurface. If we assume $\eta_u = 0.65$, this implies that the internal quantum efficiency at 77 K is $\eta_i = 0.45$. These values are relatively insensitive to the estimated value of T_2. The bulk gain increase per injected current density was extracted as

$\sigma\tau_{\text{eff}}/eL_p = 0.64$ cm/A. This value was obtained assuming $\xi = 1.63 \times 10^{-2}$ cm from simulation data as in Section II; however, there is some uncertainty in ξ if the VECSEL is in fact lasing slightly detuned from the metasurface resonance. Direct measurements of the metasurface reflectance and gain will be necessary in the future to directly measure this value. Nonetheless, these measurements validate (at least qualitatively) the basic model for the VECSEL cavity. Furthermore, if the estimated value for $T^2 = 0.82$ is accurate, the simulation data shown in Figure 5.7 suggests that performance in this cavity configuration could be improved with a higher quality factor metasurface, as well as a more transmittive output coupler. This assumption is borne out by the data in Section 5.4.4, which shows a dramatic increase in performance when a fully intra-cryostat cavity is implemented, which eliminates the contribution of cryostat window losses to T^2.

5.4.3 Focusing Metasurface VECSEL

While the simplest QC-VECSELs are based upon spatially uniform metasurfaces, one has great flexibility to engineer inhomogeneous metasurfaces. This approach recalls the reflectarray antennas that were first developed for use in the millimeter-wave region, in which the reflection phase response of various curved reflectors can be mapped onto planar surfaces by varying some critical dimension of a subwavelength resonant antenna [30]. More recently, this concept has expanded across the spectrum even into the optical regime, where engineering of inhomogeneous metasurfaces are allowing the development of a wide variety of flat optical components [38]. Our approach allows the addition of gain via stimulated emission as an additional design variable.

As a demonstration of this concept, so-called focusing metasurface QC-VECSELs were demonstrated in Ref. [33], in which a parabolic reflection phase so as to mimic a parabolic mirror. Thus, the laser cavity acts as a hemispherical laser cavity, which exhibits improved geometric stability compared with a uniform metasurface cavity. Thanks to the focusing effect, we observe a significant improvement in cavity stability and output beam pattern compared to a nonfocusing metasurface configuration. This functionality is particularly useful in the THz, where partially transmitting concave output coupler mirrors are not readily available. Indeed, since both the active metasurface and mesh output couplers are fabricated via photolithography, they are naturally planar, which prevents realization of a hemispherical two-mirror cavity.

Phase modulation is obtained by varying the width w of the microcavity ridge antenna both in the transverse and longitudinal direction. Figure 5.10a shows that a phase change of 311° is achieved by varying w from 9 to 14 μm. The ridge width at the metasurface center is chosen to match the resonant frequency of the element to intersubband gain spectrum peak. The modulation in ridge width is designed to achieve the target parabolic phase profile (for paraxial focusing) of $2\pi r^2/R\lambda_0$, where r is the radial distance to the metasurface center and R is the effective radius of curvature (i.e. twice the desired focal length). As an example, a focusing metasurface designed with $R = 10$ mm at 3.4 THz has its transverse distribution of ridge width through the center, as shown in Figure 5.10b. Although this microcavity design provides

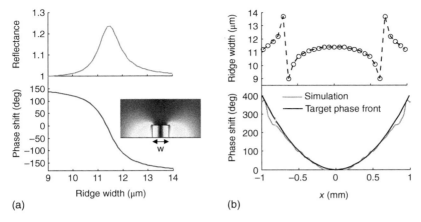

Figure 5.10 (a) Simulated reflectance and reflection phase from a uniform metasurface as a function of the ridge width w for $\Lambda = 70$ µm, at a fixed frequency of 3.4 THz. (b) Designed ridge width as a function of position (top) in order to achieve the target parabolic phase shift (bottom). Source: Adapted with permission from Ref. [33] [The Optical Society].

the desired phase profile only near a single frequency, the fact that the reflectance is highest near the resonance provides a "self-selection" to the correct frequency. Not only is the phase spatially modulated, but the gain is as well. Oxide isolation is used such that only a 1-mm diameter circular region in the center of the metasurface receives current injection. This is a form of integrated spatial filtering that encourages lasing of the fundamental Gaussian cavity mode, since the center of the beam has the largest transverse confinement factor Γ_t with the gain.

Experimental results demonstrated the efficacy of this approach in a cavity configuration where the mesh output coupler is mounted outside the cryostat. Upon testing, it was immediately apparent that the focusing designs were easier to align and more tolerant of misalignment compared to uniform metasurface designs. This was confirmed through a controlled set of measurements where the output coupler was deliberately misaligned. The focusing metasurface VECSEL cavities exhibited significantly less increase in threshold current compared to uniform metasurface VECSELs. Figure 5.11a presents measured P-I-V for a QC-VECSEL based on a 2-by-2 mm² focusing metasurface designed with an effective curvature radius of $R = 10$ mm. Two different metal mesh output couplers with different reflectances R_2 were used outside the cryostat with a cavity length of ~9 mm.

Beams were measured for VECSEL cavities based upon focusing metasurfaces with $R = 10$ mm and $R = 20$ mm and a cavity length of $L = 9$ mm. As expected from Gaussian mode theory, the cavity with $R = 20$ mm exhibited a less divergent beam than that with $R = 10$ mm. The beam quality was in general very high, with circular, near-diffraction limited beam patterns observed with divergence as narrow as 3.4°-by-3.3° FWHM. This is among the narrowest beam divergences directly produced by a THz QC-laser so far. The measured result agrees with the calculated FWHM divergence of 3.5° for an ideal hemispherical Gaussian cavity with a concave mirror of 20 mm curvature radius. The M^2 beam parameter is further measured

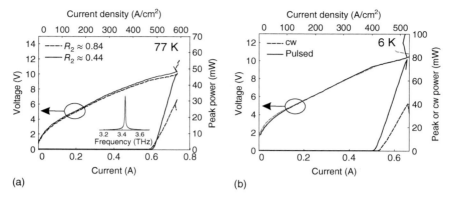

Figure 5.11 (a) Pulsed P–I–V curves for the R = 10 mm focusing metasurface QC-VECSEL designed for 3.4 THz at 77 K, paired with two different output couplers with different reflectance R_2. (b) Pulsed and cw P–I–V curves for the QC-VECSEL composed of the R = 10 mm focusing metasurface and the coupler with $R_2 = 0.44$ at 6 K. Source: Adapted with permission from Ref. [33] [The Optical Society].

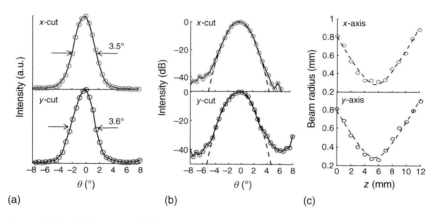

Figure 5.12 (a) Measured 1D beam pattern cuts in x and y directions for a QC-VECSEL based on a focusing metasurface with effective radius of curvature R = 20 mm. (b) The same beam pattern cuts in (a) plotted in logarithmic scale. (c) M^2 factor measurement results for the output beam radius measured using knife edge method measured in both x and y direction as a function of position along the optical (z) axis through a focused beam waist. Data is circles, with the curve fitting results plotted in black dashed line. Source: Adapted with permission from Ref. [33] [The Optical Society].

for the beams generated from some several focusing metasurface QC-VECSELs by measuring the Gaussian spot size through a beam focus, as shown in Figure 5.12. The best result was obtained from an R = 20 mm metasurface, which gave value of $M^2 = 1.3$ in both x- and y-direction while operating at peak power of 27 mW. This gives a value of brightness ($B_r = P/(M_x^2 M_y^2 \lambda^2)$) of 1.86×10^6 W sr^{-1} m^{-2}, the highest reported to date from a THz QC-laser. Achieving this relies on proper cavity alignment of course; we observed a degradation in beam quality with M^2 increased to 2.2–2.5 as the cavity length approaches the effective radius of curvature.

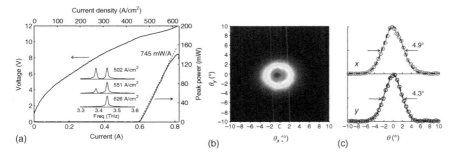

Figure 5.13 (a) Measured 77 K pulsed P–I–V for an intra-cryostat cavity QC-VECSEL based on a uniform metasurface. Paired with a metal mesh OC mounted inside the cryostat. The inset shows the spectra change with the current injection level. (b) Measured 2D beam pattern for the intra-cryostat cavity QC-VECSEL and 1D cuts of the beam in x and y directions. Black dashed lines are the Gaussian curve fits. Source: ©2017 IEEE. Reprinted, with permission, from [34].

5.4.4 Intra-cryostat Cavity QC-VECSEL

To further improve the compactness and performance of QC-VECSEL, it is possible to contain the entire cavity within the cryostat. This improves the laser performance, since there is no longer any loss associated with atmospheric absorption. Furthermore, removal of the cryostat window as an intracavity element reduces the transmission loss; it can also change the laser spectrum, since an intracavity etalon has been removed. The primary disadvantage is the inaccessibility of the optics within the cryostat, while the optical alignment is performed at room temperature, there is some risk that angular misalignment of the output coupler may occur during cooldown due to thermal contraction of the device mount. For this reason, it is helpful to use short cavity lengths of 2–3 mm in order to keep the walk-off loss low. Use of a focusing metasurface is also helpful, although not essential, to mitigate the effects of misalignment.

The intra-cryostat cavity configuration has proven to be very promising for improving the QC-VECSEL performance [34]. As an example, we present a device at ~3.4 THz uniform metasurface with area of 2×2 mm^2, and a circular bias area of diameter 1 mm, 11.5-μm ridge width and $\Lambda = 70$ μm. It was paired with an output coupler with $T_2 \approx 18 - 20\%$ (Figure 5.13). This device exhibited a high peak power of 140 mW in pulsed mode at 77 K. The slope efficiency is a record high 745 mW/A for a THz QC-laser operating at 77 K, which corresponds to roughly 0.33 photons emitted per electron per stage above threshold. The peak wall-plug efficiency was 1.5%. The measured spectra show lasing in two neighboring longitudinal modes at low biases, which gradually evolves to a dominant high frequency mode with a higher bias. At peak power, the lasing is single mode. The two modes are separated by 61 GHz (measured with a FTIR spectrometer with 7.5 GHz resolution), from which we can infer that the cavity length is 2.5 mm. Accompanied with the high power output is a near-Gaussian circular beam pattern with a FWHM divergence angle of 4.9°-by-4.3°.

In order to optimize the device for cw operation at 77 K, an intra-cryostat QC-VECSEL was demonstrated based upon a focusing metasurface with $R = 20$ mm

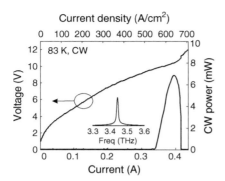

Figure 5.14 Measured 77 K cw P–I–V for an intra-cryostat cavity QC-VECSEL based on a uniform metasurface with a small bias area of 0.7-mm diameter.

with a reduced bias area of 0.7-mm diameter. The smaller bias area helps to reduce the total power consumption. As shown in Figure 5.14, the device exhibits single-mode cw lasing at 83 K with a power output as high as 7 mW. This is significantly higher than the previously reported values of 1–2 mW cw power at 77 K [16, 22, 39]. We expect that the cw performance can be further improved by using sparse and low-power metasurface designs, improved low threshold QC-active materials, and better thermal engineering.

5.5 Polarization Control in QC-VECSELs

In addition to phase and amplitude, the metasurface approach also allows one to engineer the polarization response. Indeed, the conventional metasurface made up of metal–metal ridge microcavities is already polarization sensitive, since it only responds to incident light polarized perpendicular to the ridges. This feature was leveraged in Ref. [29], when a wire grid polarizer was used as an output coupler, whose effective reflectance R_2 could be changed by rotating polarizer orientation with respect to the ridges. This allowed one to achieve the optimum coupling condition to maximize the output power "on-the-fly".

A more sophisticated control of polarization was reported in Ref. [36], in which a QC-VECSEL was demonstrated with electrically controlled switching of the output beam between linear polarization states. The enabling concept is a metasurface designed around two interleaved sets of cross-polarized "zigzag" antennas, each of which is designed to interact with an orthogonally polarized radiation. An image of such a metasurface is shown in Figure 5.15a. It is most intuitive to consider each "zigzag" antenna as a set of patch antennas that couple to the incident electric field polarized along the patch width (13 μm in this case). Sets of patches are rotated either at an angle of 45° or 135° from the x-axis; these patches are then connected by narrower segments needed to allow a continuous dc injection current path (as seen in Figure 5.15b). Patches of one orientation type are all electrically connected to one wire bonding area and thus can be biased separately from the other type. By switching the electrical bias between the two sets, we can select the polarization preference that the metasurface amplifies.

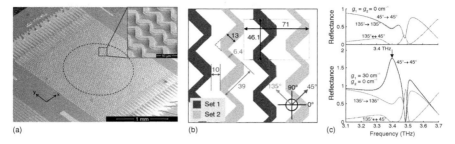

Figure 5.15 (a) SEM image of polarimetric metasurface covering an area of 2 × 2 mm². Only a center circular region of 1.5 mm diameter is biased, shown by the dashed circle. Antennas preferring one polarization direction are electrically connected together through the tapers on the top left of the metasurface, while others preferring the orthogonal polarization direction are connected together on the bottom right side. (b) Top view of a portion of the metasurface illustrated with dimensions given in microns. Set 1 of antennas – the ones interacting with radiation linearly polarized at 45° – is shown in dark gray, while the Set 2 of antennas, which interact with radiation linearly polarized at 135°, is shown in light gray. The region inside the dashed rectangular is one unit cell. (c) Top: Co- and cross-polarization reflectance of the metasurface when Set 1 and Set 2 are both passive. The 45°–45° reflectance $|\Gamma_{45°-45°}|^2$ designates the reflectance of light linearly polarized at 45° into light linearly polarized at 45°, and so on. Bottom: Co- and cross-polarization reflectance of the metasurface when a QC gain of $g_1 = 30$ cm^{-1} is assumed for Set 1 patches and Set 2 is kept passive. Source: Adapted with permission from Ref. [36] [The Optical Society].

Reflection mode simulations at normal incidence were performed for a unit cell using periodic boundary conditions to simulate infinite periodic arrays. The results are shown in Figure 5.15c. For clarity, we define gain coefficients g_1 and g_2 to represent the amount of gain supplied to Set 1 and Set 2 patches, respectively. The simulated copolarization and cross-polarization reflectance $|\Gamma_{ij}|^2$ are considered for two cases: first where the metasurface is passive ($g_1 = g_2 = 0$), and second where Set 1 patches only are supplied with a gain of $g_1 = 30$ cm^{-1} (emulating "turning on" Set 1 patches using bias current) and Set 2 patches kept passive ($g_2 = 0$). When gain is supplied to Set 1 only, net gain is observed for incident E-field polarized at 45° near the target frequency of 3.4 THz, while the orthogonal polarization 135° is almost unchanged compared to the fully passive case. Furthermore, the design shows high effectiveness in suppressing cross-polarization near the target frequency ($|\Gamma_{45°-135°}|^2$ and $|\Gamma_{135°-45°}|^2 < 0.01$ across a bandwidth of 51 GHz). The asymmetric reflectance lineshape is characteristic of a Fano resonance owing to the interactions and coupling paths between the complex set of resonances present within the metasurface lattice. By pairing such a metasurface with an output coupler that is insensitive to polarization, a QC-VECSEL is created with the ability to switch its output polarization with electrical control. Because the cavity mode profile does not depend upon the detailed antenna structure, high power, and excellent beam pattern can be consistently maintained as polarization is switched.

Results from a typical device are shown in Figure 5.16. The polarization of the output beam was analyzed by placing a wire-grid polarizer between the output coupler

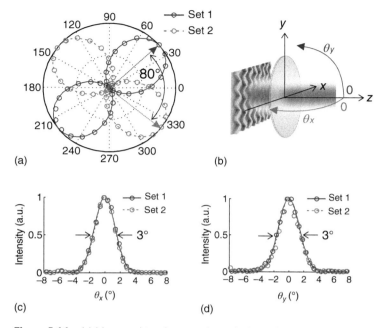

Figure 5.16 (a) Measured total power through the polarizer versus the polarizer angle when each set of antennas is separately biased. 80° linear polarization angle switching is shown in arrow. Circles are experimental data, and the solid lines in red and blue are fitting curves (to linear polarization). The schematic on the right of (a) shows the 2-axis far-field beam pattern measurement scheme. (b, c) Measured 1D beam pattern far-field cuts along x- and y-axes for bias of either Set 1 and Set 2. Circles are measured data, with fits to a Gaussian profile plotted in solid/dashed lines. Source: Adapted with permission from Ref. [36] [The Optical Society].

and the detector and measuring the power as the polarizer is rotated. This measurement was performed when either Set 1 and Set 2 antenna were biased. We observe switching between linear polarized states separated by 80°. The slight deviation from the ideal value of 90° is likely due to nonideal cross-polarized scattering from the metasurface, partially due to fabrication nonidealities. Far-field beam patterns were characterized at the same bias point near the maximum power output, with the axes defined with respect to the metasurface, as shown in Figure 5.16b. The measured beams are almost identical as the bias is switched between two sets; both exhibit a directive and narrow near-Gaussian pattern with FWHM angular divergence of ∼ 3° × 3° (see Figure 5.16). Although not shown, the emission spectrum is single mode, and no change is observed upon switching (within the resolution of the FTIR spectrometer). Polarization purity is observed to be greater than 13 dB for the total output, and the polarization state is mostly uniform across different positions within the beam pattern. It is likely this can be further improved by optimizing the VECSEL cavity to have the lasing frequency better match the metasurface resonance and exploration of advanced metasurface designs with improved broadband suppression of cross-polarized scattering.

This is a promising alternative approach for polarization control of a laser. Rapid polarization modulators that operate in the THz typically have large insertion losses, when available at all. Indeed, there does not appear to be a significant performance penalty of the polarimetric QC-VECSEL device compared to conventional QC-VECSELs. In principle, the switching speed is limited only by the build-up time of the laser oscillation (~nanosecond or less) and the RC time constant associated with the electrical bias. Such a high-performance QC-laser with electrically switchable polarization may find use within a multitude of terahertz applications, such as polarimetry, spectroscopy, and ellipsometry [40–42]. Furthermore, simply by adding a quarter waveplate in the path of the output beam, the laser can be converted into one which switches between circular polarizations of opposite handedness. Or alternately a future chiral metasurface design could selectively amplify specific circular polarization states directly.

5.6 Conclusion

The VECSEL cavity configuration holds the promise of addressing one of the most limiting problems for THz QC lasers – the difficulty of obtaining a good beam patterns with high (or even modest) power levels. We believe that the results so far on QC-VECSELs are just the "tip of the iceberg," with a great potential for further developments. First, there is still likely significant room for improvement in terms of raw performance. The pulsed powers can be increased by scaling up the metasurface size, and the cw power and efficiency can be increased by developing metasurfaces with reduced power dissipation density and improved heat sinking.

Second, another rich area for exploration is broadband tuning of single-mode lasers. This is particularly important, since spectroscopy is a major application for THz QC-lasers, and the various Bragg grating (i.e. DFB) strategies for THz QC-laser beam control are not suitable for broadband tuning. Prior to the QC-VECSEL, only a few demonstrations of THz QC-external cavity lasers had been reported, with the most successful demonstrations based upon surface-plasmon waveguides for either tuning [43] or detection via self-mixing [44]. In general though, edge-emitting metal–metal waveguide external cavities are nearly impossible to implement properly, since the strong facet reflectivity cannot be eliminated by using an antireflective coating. The metasurface approach opens up the wide array of tunable external cavity configurations (e.g. Littrow cavity ECLs) that were previously not available for metal–metal waveguides due to their high facet reflectivity – the metasurface eliminates the facet entirely! In principle, the metasurface can be designed with a large gain bandwidth, and when this is paired with a broadband QC-laser gain material extremely broad tuning may be possible.

Third, there is a wide open design space in the ability to engineer novel metasurfaces afforded by the ability to engineer the phase, amplitude, and polarization response, in both the spatial and spectral domains. This provides a platform to leverage extensive past research in reflectarray antennas and emerging research in metasurfaes and flat optical components [38]. The metasurface QC-VECSEL approach allows the addition of gain via stimulated emission and cavity feedback as an additional design variable. Some preliminary examples of that capability were shown here in the form of the focusing metasurface VECSEL and the polarization switchable VECSEL. In the future, this might include design of lasers that directly generate specific (or even arbitrary) beams, such as vector beams, vortex beams, multilobed beams. What is more, dynamic electrical control of gain on the unit cell level can be incorporated into the metasurface itself (as shown by the example of polarization switching).

Fourth, further work is warranted to explore the design-space that connects low-Q active metasurfaces intended for QC-VECSELs, with recent demonstrations of coherent arrays of microcavity lasers that phase lock without an external cavity but purely through mutual radiative coupling. Such devices have been demonstrated for THz QC-lasers [24] and plasmonic arrays in the near-IR [45]; they can be considered to be a manifestation of the so-called "lasing-spaser" [46]. This is ordinarily not the intended design space for metasurfaces intended for QC-VECSELs, which are deliberately designed to have a low radiative quality factor to prevent self-oscillation. The conventional wisdom is that such self-oscillation will occur in a high-Q "dark" mode, with low output power and/or an antisymmetric beam (such as in a second-order DFB or a bound-state-in-the-continuum laser [47, 48]). However, such collective self-oscillatory behavior has indeed been observed for the polarimetric metasurface in [36] when both antenna sets were biased. The metasurface exhibited self-oscillation with high power and a high-quality, narrow beam.

Finally, the scalability of the metasurface QC-VECSEL approach to shorter wavelengths remains an open question. Mid-IR QC-lasers may well benefit, but different metasuface designs will be needed to avoid the losses associated with metals at shorter wavelengths. We look forward to see what the future brings!

References

1 Faist, J., Capasso, F., Sivco, D.L. et al. (1994). Quantum cascade laser. *Science* 264: 553–556.
2 Köhler, R., Tredicucci, A., Beltram, F. et al. (2002). Terahertz semiconductor-heterostructure laser. *Nature* 417: 156–159.
3 Rochat, M., Ajili, L., Willenberg, H. et al. (2002). Low-threshold terahertz quantum-cascade lasers. *Appl. Phys. Lett.* 81: 1381–1383.
4 Williams, B.S., Callebaut, H., Kumar, S. et al. (2003). 3.4-THz quantum cascade laser based on longitudinal-optical-phonon scattering for depopulation. *Appl. Phys. Lett.* 82: 1015–1017.
5 Williams, B.S. (2007). Terahertz quantum cascade lasers. *Nature Photon.* 75: 517–525.
6 Kumar, S. (2011). Recent progress in terahertz quantum cascade lasers. *IEEE. J. Sel. Topics Quantum Electron.* 17: 38–47.

7 Vitiello, M.S., Scalari, G., Williams, B., and De Natale, P. (2015). Quantum cascade lasers: 20 years of challenges. *Opt. Express* 23: 5167–5182.

8 Fathololoumi, S., Dupont, E., Chan, C.W.I. et al. (2012). Terahertz quantum cascade lasers operating up to ~200 K with optimized oscillator strength and improved injection tunneling. *Opt. Express* 20: 3866–3876.

9 Wienold, M., Röben, B., Schrottke, L. et al. (2014). High-temperature, continuous-wave operation of terahertz quantum-cascade lasers with metal–metal waveguides and third-order distributed feedback. *Opt. Express* 22: 3334–3348.

10 Li, L.H., Chen, L., Freeman, J.R. et al. (2017). Multi-Watt high-power THz frequency quantum cascade lasers. *Electron. Lett.* 53 (12): 799–800.

11 Wang, X., Shen, C., Jiang, T. et al. (2016). High-power terahertz quantum cascade lasers with ~0.23 W in continuous wave mode. *AIP Adv.* 6: (075210).

12 Ulrich, J., Zobl, R., Finger, N. et al. Terahertz-electroluminescence in a quantum cascade structure. *Physica B* 272: 216–218.

13 Kohen, S., Williams, B.S., and Hu, Q. (2005). Electromagnetic modeling of terahertz quantum cascade laser waveguides and resonators. *J. Appl. Phys.* 97: 053106.

14 Williams, B.S., Kumar, S., Callebaut, H. et al. (2003). Terahertz quantum-cascade laser at $\lambda \approx 100$ μm using metal waveguide for mode confinement. *Appl. Phys. Lett.* 83: 2124–2126.

15 Unterrainer, K., Colombelli, R., Gmachl, C. et al. (2002). Quantum cascade lasers with double metal–semiconductor waveguide resonators. *Appl. Phys. Lett.* 80: 3060–3062.

16 Williams, B.S., Kumar, S., Hu, Q., and Reno, J.L. (2005). Operation of terahertz quantum-cascade lasers at 164 K in pulsed mode and at 117 K in continuous-wave mode. *Opt. Express* 13: 3331–3339.

17 Hon, P., Tavallaee, A., Chen, Q.S. et al. (2012). Radiation model for terahertz transmission-line metamaterial quantum-cascade lasers. *IEEE Trans. THz Sci. Technol.* 2: 323–332.

18 Tavallaee, A.A., Hon, P.W.C., Mehta, K., Itoh, T., and Williams, B.S. (2010). Zero-index terahertz quantum-cascade metamaterial lasers. *IEEE J. Quantum Electron.* 46: 1091–1098. http://dx.doi.org/10.1109/JQE.2010.2043642.

19 Adam, A.J.L., Kašalynas, I., Hovenier, J.N. et al. (2006). Beam patterns of terahertz quantum cascade lasers with subwavelength cavity dimensions. *Appl. Phys. Lett.* 88: 151105.

20 Lee, A.W.M., Qin, Q., Kumar, S. et al. (2007). High-power and high-temperature THz quantum-cascade lasers based on lens-coupled metal–metal waveguides. *Opt. Lett.* 32: 2840–2842.

21 Kumar, S., Williams, B.S., Qin, Q. et al. (2007). Surface-emitting distributed feedback terahertz quantum-cascade lasers in metal–metal waveguides. *Opt. Exp.* 15: 113–128.

22 Xu, G., Colombelli, R., Khanna, S.P. et al. (2012). Efficient power extraction in surface-emitting semiconductor lasers using graded photonic heterostructures. *Nature Commun.* 3: 952.

23 Chassagneux, Y., Colombelli, R., Maineult, W. et al. (2009). Electrically pumped photonic-crystal terahertz lasers controlled by boundary conditions. *Nature* 457: 174–178.
24 Kao, T.Y., Hu, Q., and Reno, J.L. (2010). Phase-locked arrays of surface-emitting terahertz quantum-cascade lasers. *Appl. Phys. Lett.* 96: 101106.
25 Amanti, M.I., Fischer, M., Scalari, G. et al. (2009). Low-divergence single-mode terahertz quantum cascade laser. *Nature Photon.* 3: 586–590.
26 Kao, T.Y., Hu, Q., and Reno, J.L. (2012). Perfectly phase-matched third-order distributed feedback terahertz quantum-cascade lasers. *Opt. Lett.* 37: 2070–2072.
27 Kao, T.Y., Cai, X., Lee, A.W.M. et al. (2015). Antenna coupled photonic wire lasers. *Opt. Express* 23: 17091.
28 Wu, C., Khanal, S., Reno, J.L., and Kumar, S. (2016). Terahertz plasmonic laser radiating in an ultra-narrow beam. *Optica* 3: 734–740.
29 Xu, L., Hon, P.W.C., Curwen, C. et al. (2015). Metasurface external cavity laser. *Appl. Phys. Lett.* 107: 221105.
30 Huang, J. and Encinar, J.A. Reflectarray Antennas. Wiley-IEEE.
31 Tavallaee, A.A., Williams, B.S., Hon, P.W.C. et al. (2011). Terahertz quantum-cascade laser with active leaky-wave antenna. *Appl. Phys. Lett.* 99: 141115.
32 Tavallaee, A.A., Hon, P.W.C., Chen, Q.S. et al. (2013). Active terahertz quantum-cascade composite right/left-handed metamaterial. *Appl. Phys. Lett* 102: 021 103.
33 Xu, L., Chen, D., Itoh, T. et al. (2016). Focusing metasurface quantum-cascade laser with a near diffraction-limited beam. *Opt. Express* 24: 24117–24128.
34 Xu, L., Curwen, C.A., Chen, D. et al. (2017). Terahertz metasurface quantum-cascade vecsels: theory and performance. *IEEE J. Sel. Topics Quantum Electron.* 23: 1200512. https://doi.org/10.1109/JSTQE.2017.2693024.
35 Faist, J. (2007). Wallplug efficiency of quantum cascade lasers: critical parameters and fundamental limits. *Appl. Phys. Lett.* 90: 253512.
36 Xu, L., Chen, D., Curwen, C.A. et al. (2017). Metasurface quantum-cascade laser with electrically switchable polarization. *Optica* 4: 468–475.
37 Densing, R., Erstling, A., Gogolbwski, M. et al. (1992). Effective far infrared laser operation with mesh couplers. *Infrared Phys.* 33: 219–226.
38 Yu, N., Genevet, P., Aieta, F. et al. (2013). Flat optics: controlling wavefronts with optical antenna metasurfaces. *IEEE J. Sel. Topics Quantum Electron.* 19: 4700423.
39 Amanti, M.I., Scalari, G., Castellano, F. et al. (2010). Low divergence terahertz photonic-wire laser. *Opt. Express* 18: 6390–6395.
40 Doradla, P., Alavi, K., Joseph, C., and Giles, R. (2014). Single-channel prototype terahertz endoscopic system. *J. Biomed. Opt.* 19 (8): 080501. https://doi.org/10.1117/1.JBO.19.8.080501.
41 Mochizuki, K., Aoki, M., Tripathi, S.R., and Hiromoto, N. (2009). Polarization-changeable THz time-domain spectroscopy system with a small incident-angle beam-splitter. *2009 34th Int. Conf. Infrared, Millimeter, Terahertz Waves* 1 (2): 690–691.

42 van der Valk, N.C.J., van der Marel, W.A.M., and Planken, P.C.M. (2005). Terahertz polarization imaging. *Opt. Lett.* 30: 2802–2804.

43 Lee, A.W.M., Qin, Q., Kumar, S. et al. (2006). Real-time terahertz imaging over a standoff distance (> 25 meters). *Appl. Phys. Lett.* 89: 141125.

44 Ren, Y., Wallis, R., Jessop, D.S. et al. (2015). Fast terahertz imaging using a quantum cascade amplifier. *Appl. Phys. Lett.* 107: 011107.

45 Zhou, W., Dridi, M., Suh, J.Y. et al. (2013). Lasing action in strongly coupled plasmonic nanocavity arrays. *Nature Nano.* 8: 506–511.

46 Zheludev, N.I., Prosvirnin, S.L., Papasimakis, N., and Fedotov, V.A. (2008). Lasing spaser. *Nature Photonics* 2: 351–354.

47 Kodigala, A., Lepetit, T., Gu, Q. et al. (2017). Lasing action from photonic bound states in continuum. *Nature* 541: 196–199. https://doi.org/10.1038/nature20799.

48 Noll, R.J. and Macomber, S.H. (1990). Analysis of grating surface emitting lasers. *IEEE J. Quantum Electron.* 26 (3): 456–466. doi: 10.1109/3.52121.

6

DBR-free Optically Pumped Semiconductor Disk Lasers

Alexander R. Albrecht, Zhou Yang, and Mansoor Sheik-Bahae

Department of Physics and Astronomy, The University of New Mexico, Albuquerque, NM, USA

6.1 Introduction

Since the first demonstration of Watt-level output powers from a vertical external-cavity surface-emitting laser (VECSEL) [1], the basic design has remained largely unchanged: an active region, typically consisting of several quantum wells (QWs), sits on top of a monolithically integrated semiconductor distributed Bragg reflector (DBR) mounted to a heatspreader and heatsink (Figure 6.1a), that together act as an active mirror of the laser's external cavity. This approach has been wildly successful, yielding continuous-wave (CW) output powers exceeding 100 W [2], ultrashort pulses [3], and spanning a wide wavelength range [4]. One of the shortcomings of this design is the need for a DBR, which requires lattice-matched, high index contrast materials to achieve sufficiently high reflectivity. While this requirement is easily met in the GaAs material system, InP, for example, does not lend itself to DBR growth [5]. As a solution, direct bonding or wafer fusion of the gain chip to GaAs-based DBRs has been implemented [5] and more recently achieved relatively high output power [6]. In this chapter, we discuss a new alternative that does away with DBRs altogether. Departing from the traditional VECSEL design, optically pumped semiconductor disk lasers without DBRs are investigated. In addition to eliminating the need for an epitaxial growth of a DBR, this architecture offers further advantages that include improved heat management, broad tunability, and no pump dissipation (excess heat generation) in the DBR.

With the enormous successes of VECSEL devices and applications, we often overlook the fact that the first optically pumped semiconductor disk laser did not employ semiconductor DBRs: it consisted of 2–4 μm thick CdSe platelets mounted to the facet of a GaAsP laser diode using vacuum grease, all at a temperature of 77 K [7]. Lasing operation was achieved in a cavity formed by the CdSe–air interface and the facet of the pump laser (including the vacuum grease). Pulsed operation in single or multiple modes (depending on distance between the CdSe platelet and pump laser) near 690 nm was observed. Shortly after, the same group realized CW operation [8], by mounting a similar CdSe platelet between two sapphire windows, one of which

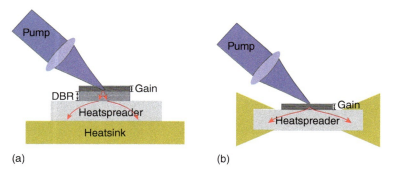

Figure 6.1 Comparison of sample, heatspreader, and heatsink geometries for (a) traditional VECSELs and (b) DBR-free SDLs.

was metalized with aluminum to increase reflectivity to approximately 90%, while the other reflector relied on the sapphire–air interface, to allow for higher transmission of the He-Ne pump laser. Around the same time, a "radiating mirror" concept [9] was proposed, involving a thin CdSe, GaAs, or GaSe disk mounted on top of a mirror and using an external output coupler, much like today's VECSEL.

In 1981 a group at Bell Labs mounted first GaAs [10] and later InP, InGaAs, and InGaAsP [11] films of micrometer thickness between two dielectric mirrors using epoxy. The resulting devices were pumped with picosecond pulses and achieved lasing at a variety of wavelengths from 0.77 to 1.59 μm. In the same year, another group demonstrated an external cavity CdS platelet laser having TEM_{00} output beam [12]. While it required cryogenic temperatures for operation, this laser was also shown to modelock and produced ps pulses at mW-level average output powers.

In 1991 the disk laser geometry for optically pumped semiconductor lasers was revisited in order to investigate power scaling while maintaining good beam quality [13]. Pumped by a Ti:sapphire laser, a 135 μm thick GaAs wafer in a V-shaped cavity produced 20 ns pulses with approximately 500 W peak power at room temperature with good beam quality.

In 2009 AlGaInAs/InP QW-based semiconductor disk lasers (SDLs) without semiconductor DBRs at 1.3 μm [14] and 1.5 μm [15] were demonstrated where the active regions on the pump-transparent growth substrates were pumped with Q-switched lasers. Later, the average output power was improved by adding heatspreaders, though still under pulsed operation [16].

6.2 DBR-free Semiconductor Disk Lasers

In its simplest form, as first demonstrated by Yang et al. [17], a DBR-free SDL consists of an optically pumped semiconductor active region (Figure 6.1b) in an external cavity, typically formed by conventional dielectric mirrors as shown schematically in Figure 6.2. The active region of choice is a QW resonant periodic gain structure, although bulk (double heterostructure [DHS]) lasers have also been demonstrated. A GaAs-based active membrane is usually about 2–3 μm thick and

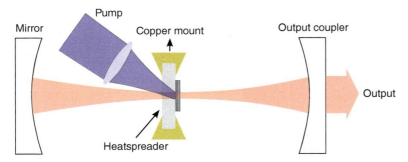

Figure 6.2 Schematic diagram of a typical linear DBR-free SDL cavity.

thus would be difficult to handle free standing unless bonded to a transparent substrate (window). More importantly, to facilitate heat removal, this transparent substrate must have high thermal conductivity and act as a heatspreader to the thermal ground (heatsink). Typical candidate window materials are sapphire [18], silicon carbide [19], or diamond [20]. Such DBR-free structures have also been referred to as membrane external-cavity surface-emitting lasers or MECSELs [21].

Needless to say, DBR-free SDLs are not limited to the linear cavity geometry shown in Figure 6.2. Depending on the application, various configurations like V- or Z-shaped and ring cavities can be implemented as will be discussed later.

Another potential advantage of DBR-free structures is the intriguing possibility of exploiting total internal reflection (TIR) [17], in particular in monolithic ring cavities. These configurations and other novel concepts will be discussed in more detail in Section 6.5.

In the following sections, we present certain advantages of DBR-free SDLs over more traditional VECSEL geometries.

6.2.1 Opportunities and Advantages

As mentioned earlier, the most obvious advantage of DBR-free SDLs is that they do not rely on an epitaxially grown DBR. Since AlAs and GaAs are almost perfectly lattice-matched, growth of an Al(Ga)As-GaAs-based DBR on a GaAs substrate is not too challenging, and the high refractive index contrast allows for reflectivities well above 99% with 20–30 DBR pairs. This is part of the reason why GaAs-based VECSELs (and VCSELs) have been so successful in the wavelength range near 1 μm. Nonetheless, it proves advantageous to avoid altogether the epitaxial growth of several micrometer DBRs in order to reduce the growth time and related complexities and costs. The problem is exacerbated when approaching visible wavelengths in the GaInP/AlGaInP system on GaAs, where AlAs/AlGaAs DBRs of reduced refractive index step require close to 60 DBR pairs [22].

The requirements for a lattice-matched DBR are also hard to fulfill in other material systems: On InP substrate, very common for the 1.5 μm wavelength range, the index contrast between InP and lattice-matched InGaAsP is much lower, practically doubling the number of required DBR pairs when compared to GaAs. This is not only impractical to grow but also decreases the thermal conductivity, at least for the case of heat extraction through the DBR [23]. The same is true for II-VI material systems in the visible, also grown on InP [24].

Another important factor is the interaction of the pump laser with the DBR in a standard VECSEL structure: in the case of a GaAs-based VECSEL, a very common pump laser configuration involves an 808 nm laser diode, which is readily absorbed in a GaAs(P) barrier. When combined with an AlAs/GaAs DBR, any pump power transmitted by the active region could be absorbed by the DBR, or the solder layers [25], causing additional heating (see Figure 6.1a). One can design a pump-transparent AlAs/AlGaAs DBR [26, 27], but this reduces the index step, resulting in more DBR pairs. Both the increased thickness and the ternary alloy used will lead to an increased thermal resistance of the device. An alternative approach is to design a DBR that reflects not only the VECSEL wavelength but also the pump wavelength [25]. This helps to maximize pump absorption but again yields a thicker, more complicated, and less thermally conductive DBR.

There have also been studies that use hybrid mirrors in VECSELs to expedite the heat-removal process, including metallic-semiconductor-DBR [23, 28] and even metallic-dielectric-semiconductor-DBR [29] arrangements. Even though hybrid mirrors loosen some restrictions of semiconductor DBRs, they could introduce other issues, for example, phase mismatch at higher temperatures [28].

6.2.2 Thermal Analysis

Similar to other lasers, the lasing process in SDLs is accompanied by heat generation, due to the quantum defect and nonradiative recombination processes. Overheating of the active region could degrade laser beam quality, decrease material gain, and eventually lead to catastrophic damage of the gain medium [30]. Therefore, thermal management is most critical to SDLs' performance when moderate to high average powers are desired. In traditional VECSELs (Figure 6.1a), by heatsinking the backside of the DBR, an approximately one-dimensional heat flow is ensued, which controls the temperature gradients in the transverse plane and, as a result, reduces the beam distortion. However, due to the high thermal resistance of semiconductor DBRs [31, 32], these VECSELs face practical limitations in their power scalability [33]. Without the integrated DBR, DBR-free geometries could potentially facilitate heat extraction from the active region and outperform traditional VECSELs in power scaling.

Following similar numerical procedure in thermal analysis of VECSELs reported earlier [34], we developed a finite-element model for DBR-free SDLs, utilizing the commercial software COMSOL Multiphysics 4.4 [17]. We compare the thermal performance of the traditional VECSEL and DBR-free SDL geometries subject to various heat loads. To simplify the problem, we make the following assumptions: (i) Cylindrical symmetry: both the VECSEL or the DBR-free gain chip and the heatspreader are taken to be circular, and the pump lasers are assumed to have a fundamental Gaussian profile centered on the sample. (ii) The thermal conductivity

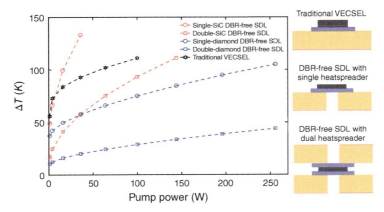

Figure 6.3 Maximum active region temperature rise with incident pump power for five SDL geometries. The maximum pump laser intensity is kept constant, and the pump beam size is scaled up with the square root of the incident pump power. All heatspreaders in the simulation are 0.5 mm in thickness. The red lines in schematics represent where the constant temperature boundary conditions are applied.

coefficients of all materials are taken to be temperature independent. (iii) We assume a constant proportion (40%) of the absorbed pump light as the heat load in the active region; any pump light not absorbed in the active region is assumed to be absorbed in the DBR for traditional VECSELs. (iv) Traditional VECSELs are treated as three equivalent layers: DBR, active region, and cap layers with equivalent parameters, and the thermal inhomogeneity in these equivalent layers is not considered here. (v) For DBR-free geometries, perfect bonding (i.e. perfect thermal contact) is assumed between semiconductor and heatspreader, a reasonable assumption for van der Waals bonding. The heatspreaders are taken to be 500 μm in thickness as it is often the case in practice. Other parameters used in the simulation are detailed in [17].

Under the same pumping condition, we compare the maximum temperature rise in active regions among five geometries, which are shown in Figure 6.3, including traditional VECSELs with one extra-cavity diamond heatspreader only, and DBR-free geometries with single- and dual-intracavity diamond or silicon carbide (SiC) heatspreaders. To mimic the power scaling process, the maximum pump intensity is kept constant. Geometries with lower temperature rise could allow for higher power operation.

None of these five geometries are power scalable, for which the maximum temperature rise is constant with pump power when the maximum pump intensity is constant. With a single diamond heatspreader, the DBR-free geometry exhibits lower temperature rise in comparison to the VECSEL geometry due to the absence of high thermal resistance DBRs. The dual-intracavity-diamond-heatspreader configuration has less than half of the temperature rise of the single diamond DBR-free geometry since active cooling can be applied from both sides. Such a dual-intracavity-heatspreader cooling scheme has also been proposed in [35] with the high-contrast-gratings concept.

It is worth emphasizing that even though the thermal conductivity of SiC (370 W/m·K) is much lower than that of diamond (2000 W/m·K), the dual-SiC-intracavity-heatspreader geometry still outperforms the traditional VECSEL geometry in heat dissipation, which is due to the high thermal resistance of the semiconductor DBR [31, 32]. With good surface quality, cost-effectiveness, and wafer-size availability, SiC is a very promising heatspreader material for a myriad of SDL applications.

6.2.3 Longitudinal Mode Structure and Broadband Tunability

Typical SDLs adopt the periodic gain structure, in which the discrete QW or quantum dot (QD) gain layers are positioned at the peak of the cavity's standing wave field to enhance the integrated gain or gain bandwidth. Therefore, the position of gain layers is critical. For typical VECSELs, these gain layers are close to the cavity antinodes, forming a so-called resonant periodic gain structure [36].

Without the integrated DBR attached to the active region acting as a global or local cavity node, the mode structure for DBR-free SDLs will be different from traditional VECSELs, especially for the transmission geometry. In addition, with much lower reflectivity (than unity reflectivity from DBR), the integrated gain from the bonded gain element is also very different. Such disparity is obvious from the gain perspective. Here we extend the traditional longitudinal mode confinement factor [37] or integrated gain factor concept [36] to the DBR-free case and introduce its gain position dependence [38].

To simplify the analytical calculation, we assume a free-standing periodic gain film as the active region and neglect refractive index difference among materials and the Fabry–Perot effects, which could modulate the integrated gain. The stable standing wave resonator is formed with a pair of curved mirrors with a separation of L. For a small incident signal at normalized intensity, the integrated gain or the amplification for a given mode at wavelength λ is [36]

$$g(\lambda, z) = \int_0^L \gamma(\lambda, z' - z) \sin^2\left[\frac{2\pi n z'}{\lambda}\right] dz', \tag{6.1}$$

where z is the distance between the global cavity node (i.e. the end mirror) to the closest gain layer, and $\gamma(\lambda, z)$ is the longitudinal gain profile (per unit length) of the active region having an index of refraction n. Considering that typical gain layers (QWs or QDs) are much thinner than the periodicity (half design wavelength in optical path length), with one gain layer per period, the longitudinal gain profile can be approximated as a series of Dirac delta functions separated by half-wavelengths:

$$\gamma(\lambda, z) = \sum_{i=1}^{N} \delta\left[z - \frac{(i-1)\lambda_0}{2n}\right] g_m(\lambda), \tag{6.2}$$

where N is the number of gain periods, λ_0 is the periodic-gain design wavelength, and $g_m(\lambda)$ is the integrated gain of each layer. Here we assumed one QW per layer, having the same gain. By combining Eqs. (6.1) and (6.2), the integrated gain of the system can be expressed as:

$$g(\lambda, z) = G(\lambda, z) g_m(\lambda), \quad (6.3)$$

where an integrated gain factor G is defined as:

$$G(\lambda, z) = \frac{N}{2} - \frac{\sin\left(\frac{N\pi\lambda_0}{\lambda}\right)}{2\sin\left(\frac{\pi\lambda_0}{\lambda}\right)} \cos\left[\frac{\pi\lambda_0}{\lambda}\left(N - 1 + \frac{4nz}{\lambda_0}\right)\right] \quad (6.4)$$

which effectively describes the overlap between the periodic gain structure and the longitudinal modes. Within typical semiconductor material gain bandwidth, $\lambda_0/\lambda \approx 1$, this further simplifies to:

$$G(\lambda, z) \approx \frac{N}{2}\left[1 + (-1)^N \mathrm{sinc}(N\pi(\lambda_0/\lambda - 1))\cos\left[\frac{\pi\lambda_0}{\lambda}\left(N - 1 + \frac{4nz}{\lambda_0}\right)\right]\right] \quad (6.5)$$

For a traditional VECSEL, ignoring the penetration of the field into the DBR, and assuming $z = m\lambda_0/4n$ (m=odd, but small integer; typically 1 or 3), after minor mathematical manipulations yields:

$$G_{VECSEL}(\lambda, z) = \frac{N}{2}[1 - \mathrm{sinc}(2N\pi(\lambda_0/\lambda - 1))] \quad (6.6)$$

For a DBR-free SDL (MECSEL) in transmission geometry where $z \gg \lambda_0$, the cosine term in Eq. (6.5) represents a fast modulation, and the overall envelope of the integrated gain function can be expressed as

$$G_{MECSEL}(\lambda, z) \approx \frac{N}{2}[1 \pm \mathrm{sinc}(N\pi(\lambda_0/\lambda - 1))] \quad (6.7)$$

Comparing Eq. (6.6) and Eq. (6.7), the DBR-free SDL can offer nearly twice the integrated gain bandwidth ($\approx \lambda_0/N$) as the traditional VECSEL geometry ($\approx \lambda_0/2N$), assuming the gain bandwidth $g_m(\lambda)$ is not a limitation. As an example, for a gain structure consisting of $N = 12$ periodically distributed QWs, the integrated gain factors are compared in Figure 6.4 for three z positions. When $z = 0.75\lambda_0/n$ (Figure 6.4a), the MQW is adjacent to the global cavity node, which is typical of a traditional VECSEL. In these plots, the integrated gain factor is the product of the envelope and a slow modulation. By moving the MQW further away from the cavity node, that modulation period rapidly decreases as shown in Figure 6.4b for $z = 100.25\lambda_0/n$. With $z = 1$ cm, as in a typical DBR-free SDL geometry, the modulation becomes even finer, beyond the resolution of the figure, with an effective gain bandwidth that is significantly broader than the traditional-VECSEL. Another important implication of these results is that the integrated gain of neighboring longitudinal cavity modes in a DBR-free VECSEL vary significantly in magnitude compared to smooth changes in the traditional VECSEL case. These variations in turn have implications for mode-locking applications that need further investigation.

Besides broader gain bandwidth, DBR-free active regions are also more robust to growth errors than traditional VECSEL. Following the previous derivation, the integrated gain of traditional VECSEL and DBR-free active regions with and without

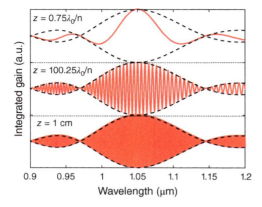

Figure 6.4 Position dependent (z) integrated gain for a periodic MQW (N = 12) in a linear cavity. z is the distance from the global cavity node to the closet QW. The cavity length is 10 cm. The MQW is designed at $\lambda_0 = 1050$ nm and the QW separation is $\lambda_0/2n$. The dashed lines are the envelopes.

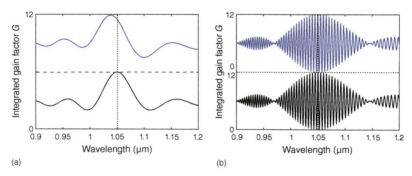

Figure 6.5 Integrated gain factors for (a) VECSEL and (b) DBR-free geometry in perfect resonant periodic gain design (black) and with a 15 nm thinner window layer (blue). The dashed lines at 1050 nm represent the design wavelength.

a growth error are shown in Figure 6.5. With the window layer thinner by 15 nm than optimal design, the integrated gain peak of the traditional VECSEL shifts to a shorter wavelength with a 2.0% drop in amplitude, while the integrated gain of DBR-free geometry exhibits little change in center wavelength and 0.5% decrease in amplitude.

6.3 Device Fabrication

Due to the high thermal resistance and optical losses of the growth substrates, the fabrication of DBR-free SDLs typically requires substrate transfer and adhesive-free bonding. Depending on the fabrication techniques, the bonding process could be performed before [39] or after the substrate-removal process [40]. To release the active region from the growth substrate, there are two approaches: undercutting a release layer [40] and etching off the growth substrate [41]. To bond the released active regions onto destination substrates, we adopt the direct bonding or van der Waals bonding technique [40]. We will focus on the fabrication of DBR-free SDLs near 1 μm based on the GaAs material system.

Before detailing the fabrication process, we would like to describe the structure of our typical wafers. After the typical GaAs buffer layer, there is a 100 nm thick AlAs layer, which serves as the sacrificial layer for undercutting process or the etch-stop layer for the substrate-etching process. This is followed by the MQW gain structure surrounded by two InGaP window layers for carrier confinement. Apart from the typical VECSEL design, because of much lower reflectivity or weaker Fabry–Perot effect in DBR-free gain structure, the InGaP layer thickness is less critical but is kept at $\lambda/2$ in optical thickness for our designs. For DBR-free SDLs designed at 1020 nm, there are about 12 periodic InGaAs/GaAsP QW gain layers, each with $\lambda/2$ in optical thickness. For DBR-free SDLs designed at 1178 nm, due to challenges in strain balancing, there are only eight QWs. Typical barrier-pumped DBR-free active regions designed near 1 μm are about 2 μm thick.

In the fabrication, a wafer is cleaved into small pieces, usually 3 mm by 3 mm, limited by diamond size. A larger sample-to-pump area ratio helps to suppress the possibility of unwanted lateral lasing [42], which we have observed for large pump spot sizes on small samples (or samples with cracks). Before releasing the MQW samples from the growth substrate, the epitaxial surface is coated with about 0.5 mm of black wax (Apiezon Wax W) at 100 °C. This serves as mechanical support for the few-micrometer thick free-standing gain film, which is very fragile and difficult to handle on its own. Besides, black wax is resistant to etchants and protects the epitaxial surface. Our experimental study shows that after bonding, the sample surface protected by black wax is consistently of higher quality with brighter photoluminescence than the one adjacent to the sacrificial or etch-stop layer.

The next step is etching. For the undercutting process [43], the wax-coated sample is attached to a U-shaped Teflon mount from the substrate side via black wax, with the wax-coated epitaxial surface facing up. Then the mount is flipped upside down, positioned in a Teflon beaker with the sample facing down, and immersed into 49% concentration HF acid. Once the sacrificial AlAs layer has been etched away (about 2.5 hours for our typical 3 mm by 3 mm chip), the gain membrane covered in wax should fall down due to gravity, preventing it from rebonding to the substrate. For jet etching, the wax-coated sample is mounted onto a microscope slide with crystal bonding wax, with the substrate facing up. The etchant is a mixture of hydrogen peroxide and ammonium hydroxide at a 33 : 1 ratio, which is pumped onto the sample substrate at a constant flow. Mechanical thinning could accelerate the process. For DBR-free devices, direct bonding without adhesive is unavoidable, and the sample and heatspreader surface quality is critical to both optical and thermal properties of the bonded structure. The bond quality can be characterized with a Nomarski microscope, as shown in Figure 6.6. On the left is a bonding result with many voids and right is a good bonding result with only a few tiny dust particles on top.

To evaluate the potential degradation after the fabrication process, carrier lifetime is characterized before and after the device fabrication process. Time-resolved photoluminescence measurements are performed before and after the fabrication process. Employing a 910 nm pulsed laser diode as pump source, carriers are excited only in QWs for an active region designed at 1040 nm. As shown in Figure 6.7, no significant change is observed from the late-time carrier lifetime, which suggests material

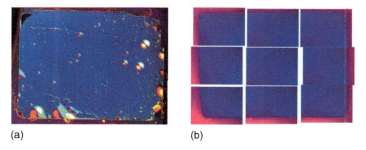

Figure 6.6 Nomarski micrographs of bonding examples on single-crystalline chemical vapour deposition (CVD) diamond substrates: (a) Sample with several bonding issues, including possible contamination, air bubbles, and partial delamination (sample size 3.5 mm by 2.5 mm); (b) well-bonded sample (3 mm by 2.5 mm).

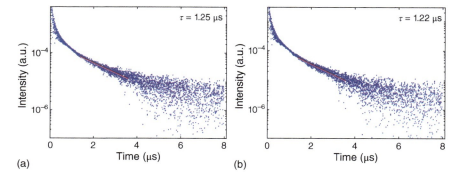

Figure 6.7 Lifetime characterization of an active region before (a) and after (b) substrate transfer. The MQW active region is transferred from growth substrate to a diamond heatspreader via the undercutting technique. The InGaAs/GaAsP MQW is designed at 1040 nm. A diode laser at 910 nm is employed as the pump source. The pump pulses are 10 ns in length with a repetition rate of 6 kHz. The photoluminescence is collected through a telescope system and detected with a photomultiplier tube (PMT). Estimated lifetime is 1.25 µs before and 1.22 µs after the device fabrication process.

quality is preserved in the epitaxial lift-off process, in contrast to [44]. This may be due to reduced bending of the thin film during the fabrication process because of the mechanical support provided by the black wax.

A thermal resistance measurement of bonded DBR-free gain elements could also shine light on the bonding quality. Investigations are ongoing using techniques based on temperature-dependent wavelength shift of the laser emission [45] and fluorescence of the cap layers [46].

As mentioned earlier, another promising geometry utilizes heatspreaders on both sides of the gain membrane. The active region is etched off from the growth substrate and transferred onto one of the destination substrates (heatspreader) via direct bonding. The second substrate is attached in the same manner.

6.4 DBR-free SDL Implementation

In the DBR-free transmission geometry, to date, CW laser operation has been demonstrated with sapphire, silicon carbide, and single-crystalline diamond substrates. Among the three heatspreader materials, sapphire has the lowest thermal conductivity at room temperature (27 W/m·K) and, as a result, lowest thermal rollover pump power under the same operation conditions. Even though diamond has much higher thermal conductivity than the other two materials, it is of limited supply, has worse surface quality, and higher optical losses (between 0.01 and 0.1 cm^{-1} absorption coefficient and potential losses due to birefringence [47, 48]). Silicon carbide (SiC) and sapphire have much better optical quality and lower absorption at common laser wavelengths, but they also have much lower thermal conductivities. The thermal expansion coefficient of SiC is close to that of GaAs, which might improve reliability of bonded devices. Here, we will focus the discussion on DBR-free SDL geometries with single-crystalline diamond and SiC heatspreaders. Recently, Mirkhanov et al. has collected more than 10 W of output power from a DBR-free SDL with a SiC heatspreader [49].

For thermal management, the bonded gain element is mounted onto a custom-made copper heatsink, which is water cooled to about 10 °C. To achieve a good thermal contact, an indium foil layer of 50 μm thickness has been applied between the copper mount and substrate or sample. In [49], even silver foil is employed to minimize the thermal resistance between the heatspreader and heatsink.

6.4.1 High Power Operation

To verify the prediction from the thermal model, a direct comparison on laser performance has been made between single- and dual-heatspreader configurations, as shown in Figure 6.8a. Low optical losses, double side polished 4H-SiC windows have been employed as intracavity heatspreaders. Compared to single-crystalline diamond heatspreaders, SiC wafers have the advantages of better surface quality, low cost, and easier, wafer-size availability. The SiC chips used [50] are 0.37 mm thick, have been diced into 5 mm by 5 mm squares, with surface roughness (RMS) less than 0.2 mm [50].

Two DBR-free gain chips from one wafer are used for the bonding. The multi-quantum-well gain structure is etched off from the GaAs growth substrate and transferred to an uncoated SiC heatspreader with direct bonding technique. The laser performances of both active regions are shown in Figure 6.8 under the same cavity conditions. At high pump powers, both configurations deviate from their initial linear trends. The pump power for such deviation in the dual-SiC configuration is almost twice of single SiC, which agrees with our previous thermal simulation results [17]. The second heatspreader significantly accelerates the heat dissipation process. A wafer-scale fabrication result will be shown later.

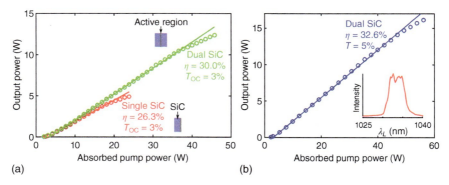

Figure 6.8 (a) Laser performance comparison between gain structures with single-SiC heatspreader and dual-SiC heatspreader. The circles represent the experiment results and curves are linear fits. The inset is the laser setup. (b) Laser output power as a function of absorbed pump power with a dual-SiC-heatspreader geometry. The inset laser spectrum is collected at 24 W of absorbed pump power.

By optimizing the laser cavity and employing a 5% transmission output coupler, 16.1 W of output power is collected with the dual-SiC configuration at 10.5 °C coolant temperature [50]. We believe this to be limited by the thermal contact between heatspreader and heatsink, which could be further optimized but utilizing an improved design and/or larger area heatspreader to increase contact area. Considering the absorbed pump power, a slope efficiency of 32.6% is achieved. A broad emission spectrum (FWHM ≈ 5 nm) is observed at high power, as shown in Figure 6.8b inset, which may be due to the large spatial temperature variation across the active region [51].

Compared to SiC heatspreaders, DBR-free SDLs with diamond heatspreaders have reported lower output powers [17, 38, 52]. As shown in Figure 6.9, to date only 6 W has been reported near 1 μm, with a significantly lower slope efficiency of 21.8%. This may be due to worse surface quality or higher absorption for diamond heatspreaders compared to SiC.

Without the integrated semiconductor DBR and adopting the transmission geometry, the subcavity effect for DBR-free SDLs is much weaker than the traditional VECSEL configuration. Same as the intracavity heatspreaders in traditional VECSEL

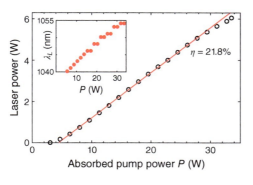

Figure 6.9 Laser performance for a gain chip bonded onto a single-crystalline diamond. The inset shows the lasing wavelength, starting at 1040 nm near threshold.

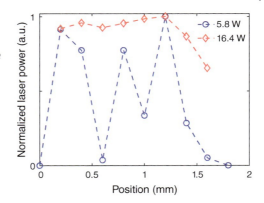

Figure 6.10 Active region uniformity check at different pump powers. With 5.8 W (blue) and 16.4 W (red) pump power, the maximum output powers are 10, 140, and 810 mW, respectively. The sample is bonded onto a 3×3 mm^2 diamond.

geometry, the heatspreader could modulate laser spectra [53]. In the ideal scenario, there could be a spectral modulation from the coupled cavity of the gain membrane and heatspreader(s) [53, 54]. Simple analysis shows that a symmetric dual heatspreader structure (e.g. as shown in Figure 6.3) provides an additional advantage due to its spectrally broader subcavity and null reflectivity at peak (gain) transmission. This arrangement is therefore less sensitive to angular deviations from normal incidence. Alternatively, antireflection-coated surfaces could minimize the subcavity modulations (at the cost of reducing the gain per pass). This may prove necessary if imperfect parallelism of heatspreaders gives rise to unwanted optical losses. Similarly, a Brewster's angle geometry can be used to eliminate the subcavity modulations, thus leading to laser oscillation in the p-polarization, provided that the subcavity gain enhancements for the s-polarization are suppressed by additional intracavity polarizers.

Under the same pump and cavity conditions, the spatial dependent laser output power could give us a hint. An uncoated diamond is used as the bonding substrate. Before the experiments, both the cavity (including the sample orientation) and pump parameters are optimized at 16.4 W pump power for maximum output power with near TEM$_{00}$ mode. Then the pump condition is kept constant, and the active region is scanned transversely, with the position dependent normalized output power graphs as shown in Figure 6.10. Even though there are power fluctuations at pump powers near threshold, the output power is stable over a 1 mm window for this 3 mm by 3 mm size sample.

6.4.2 Broad Tunability

Inserting a quartz birefringent filter (BRF) into the cavity at Brewster's angle, the lasing wavelength can be tuned via rotating the BRF. Approximately 80 nm of tuning range is observed in two gain elements designed for operation at a wavelength of 1040 nm [38] and 1160 nm [17], respectively, as shown in Figure 6.11. Such broad tunability is consistent with the previously discussed integrated gain picture.

To verify the predicted unique mode structure in the DBR-free geometry, we study the longitudinal mode beating signal. The cavity is formed with two concave mirrors

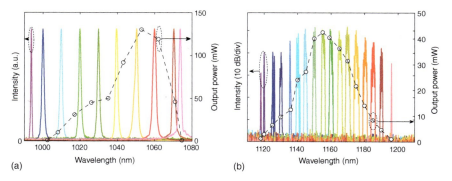

Figure 6.11 Laser tuning spectra of DBR-free SDLs with active regions designed at 1040 nm (a) and 1160 nm (b).

with 100 mm radius of curvature, separated by about 200 mm. In experiment, to keep the pumping condition the same, the two mirrors are translated parallel to the laser path while the active region and pump optics stay unchanged. RF spectra are obtained by Fourier transform of the laser output trace collected by a high-speed detector.

When the active region is 5 mm away from the cavity center, the RF spectrum contains the components of integer times of cavity repetition rate f_{rep}, as shown in Figure 6.12. When the active region is closer to the cavity center, however, the odd harmonics of f_{rep} component disappear. This agrees with our integrated gain picture, previously discussed in Section 6.2.3. As shown in Figure 6.12c, d, under the two scenarios, the integrated gain for these selected longitudinal modes could be significantly different. When it is closer to the cavity center point, half of the longitudinal modes experience near-zero gain. Therefore, the longitudinal mode spacing doubles.

As a proof of concept, a DBR-free SDL based on bulk GaAs as the gain medium in a GaAs/InGaP DHS has also been demonstrated [54]. In this implementation, the DHS (having 0.75 μm thickness for each layer) was directly bonded to the backside of a dielectric mirror on a fused silica substrate. The structure was pumped by a CW Ti:sapphire laser at 810 nm. However, since thermal conductivity of the fused silica substrate is poor, the pump laser was mechanically chopped (1% duty cycle) in order to avoid substantial thermal loading. Pulsed operation near $\lambda = 900$ nm was observed with good mode quality. For a sample bonded onto SiC in a straight cavity with two external dielectric high-reflecting mirrors, threshold could be achieved under CW pumping with an 808 nm diode, but no significant output power was recorded. Switching to a 1% output coupler increased the threshold to the limit of thermal contact between SiC and the water-cooled heatsink. Experiments are under way to more efficiently heatsink the heatspreader. Although there are known shortcomings for bulk gain materials, such as higher laser threshold and limited material choices, such gain media may offer advantages for high-power SDLs and amplifiers due to their inherently large power density. Also, in contrast to in-well pumping, a DHS can easily be pumped with very little quantum defect.

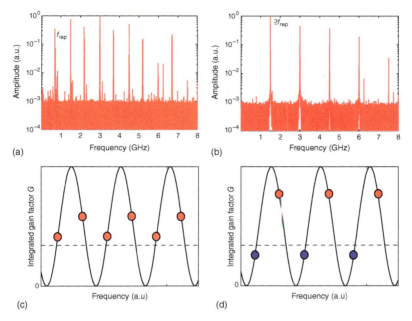

Figure 6.12 RF spectra for DBR-free geometry with near-symmetric (a) and perfectly symmetric cavity configuration (b). (c, d) show the integrated gain factor for both cases, with circles representing the longitudinal modes supported by the cavity; red circles indicate modes that are above threshold (dashed line) and blue circles are below threshold.

6.4.3 Wafer-scale Processing

The highest thermal conductivity heatspreader – single-crystalline diamond – is only available in relatively small pieces, requiring separate processing and bonding of every laser chip. SiC, on the other hand, is available in relatively large wafers with excellent surface quality and, despite the smaller thermal conductivity, has produced the highest CW output powers reported to date. This promises the potential of wafer-scale processing and bonding, followed by dicing of the individual laser chips as shown in Figure 6.13 for a GaAs membrane, which is of great interest for commercial use of DBR-free SDLs.

6.5 Novel Concepts

With the van der Waals bonding technique, the gain membrane can be transferred onto arbitrary smooth surfaces, which enables many novel concepts. By bonding onto a nonplanar optical element, such as a prism or a cube, TIR could be utilized to form the optical cavity, which could allow a multitude of monolithic geometries. Two examples are presented in Figure 6.14. When using only planar surfaces, the thermal lens effect in the active region might help stabilize the cavity. In addition, good surface parallelism is required for both configurations. For the quadrilateral

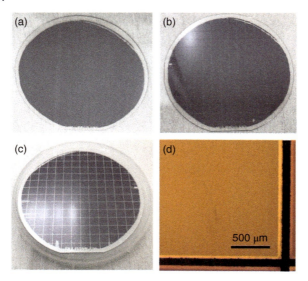

Figure 6.13 Wafer-scale manufacturing process for a dual-SiC-heatspreader (SiC/epi/SiC) structure, demonstrating a cost-efficient path for generating high-power SDL gain media (or DBR-free VECSEL or MECSEL gain chips, etc.). The key fabrication steps include (a) direct fusion bonding of an epitaxial GaAs layer to a single-crystal SiC wafer with subsequent GaAs substrate removal, (b) optical lithography and wet-chemical etching of dicing lanes into the epi-GaAs, (c) and a second fusion bonding step to finish the SiC/epi/SiC structure. A dicing saw was used to cut 5 mm × 5 mm chips out of the 3″ (76.2 mm) diameter wafer. A close-up image of the dicing lane is shown in (d) and demonstrates minimal chip-out in the sawing process. Source: Photos courtesy Garrett D. Cole, Thorlabs Crystalline Solutions, Santa Barbara, CA, USA.

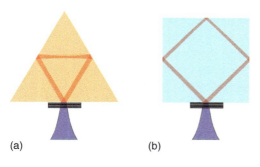

Figure 6.14 Schematic of the TIR-based monolithic ring DBR-free SDLs. (a) Equilateral prism with refractive index $n > 2$ material, such as ZnSe, ZnS, or diamond; (b) quadrilateral prism with $n > 1.42$.

prism geometry, extra geometrical conditions must be satisfied, as discussed in [54]. Variable output coupling could be achieved via an adjustable coupling prism. These monolithic geometries have the advantage of being compact, alignment free, mechanically robust, and offer wavelength flexibility, since they do not rely on any dielectric mirrors or DBRs. Other functional layers, like a saturable absorber (for example, semiconductor QW or graphene), can be applied onto other surfaces for high repetition rate, short pulse generation, similar to what has been demonstrated for a Nd:YVO$_4$ laser in [55], but with the inherent wavelength flexibility of semiconductor materials.

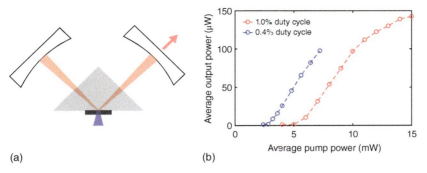

Figure 6.15 (a) Schematic of the total internal-reflection-based V-shaped cavity. (b) Power conversion graph with 0.4% (blue) and 1.0% (red) duty cycle for the TIR-based DBR-free SDL.

In a proof-of-concept experiment, a TIR-based DBR-free laser was demonstrated in an InGaAs/GaAs MQW active region (designed for $\lambda = 1040$ nm) bonded onto the hypotenuse side of a fused silica right-angle prism [54]. As shown by the schematic in Figure 6.15, two curved mirrors of 100 mm radii of curvature were employed to form the V-shaped cavity. Because of the low thermal conductivity of fused silica and the lack of active cooling, the active region was pumped at $\lambda_P = 810$ nm with a mechanically chopped Titanium–Sapphire laser (0.4–1% duty cycle) to reduce the average heat load.

In addition to flat substrates, active regions can also be bonded onto curved surfaces, like a sphere or cylinder. Smaller spheres have less optical losses, including scattering and absorption losses, and higher Q factors, but the highly curved surfaces are more challenging for bonding. Smaller samples are easier to handle, since less deformation is needed, but lateral lasing may be involved. Bonding has been attempted with 6 and 9 mm diameter sapphire spheres (Edmund Optics, Inc.), but no high-quality bonding has been achieved yet [54].

Another promising DBR-free concept that has been recently proposed involves using subwavelength gratings in an active mirror structure [54]. Termed gain-embedded meta-mirror or GEMM, in such a structure the gain medium is embedded inside a subwavelength grating that is bonded to a heatspreader, as shown in Figure 6.16 (left). The grating period (Λ), height (h), and fill factor (f) are carefully selected to ensure that most (> 98%) of the incident light is diffracted only to the first order inside the high index (n_H) grating medium, and that they are total internally reflected from the heatspreader (n_L) interface. While the grating pitch constraints $1 < \Lambda n_H/\lambda_0 < n_H/n_L$ are the same as for broadband mirrors [56], Yang and coworkers [57] found that the fill factor f plays a critical factor when identifying the optimum grating parameters that will satisfy the TIR condition for desirable heatspreaders, such as diamond or SiC, with a relatively high index of refraction ($n_L \approx 2.5$).

The calculations based on rigorous coupled wave analysis (RCWA) found that, for GaAs-based gain structure ($n_H \approx 3.4$), a fill factor of $0.55 < f < 0.60$ is required if it is to be bonded to a diamond heatspreader [57]. Yang et al. also performed thermal

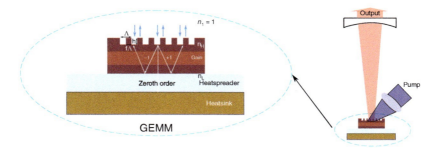

Figure 6.16 The GEMM structure as an active mirror in a VECSEL. Source: Reprinted with permission from Yang et al. [57], © 2019 The Optical Society.

analysis on these structures and showed that a GEMM/diamond VECSEL can potentially dissipate multikilowatts of heat load, thus vastly outperforming the standard DBR-based VECSELs.

6.6 Conclusions

Since the first demonstration in 2015 [17], DBR-free SDLs, or MECSELs have already made tremendous progress, with several groups reporting CW laser operation from the visible (following chapter) to the infrared, the latter with output powers in excess of 20 W [58]. Thermal analysis presented in this chapter even suggests that this design might help overcome the heatsinking and power limitations of traditional VECSELs. DBR-free SDLs offer great promise for several material systems that currently lag behind due to a lack of a convenient monolithically integrated DBR. The current technology already relies on wafer bonding – but to a conventional DBR, which brings with it all the limitations in design flexibility, heatsinking ability, and difficult growth process. InP-based devices for emission near 1550 nm are currently being investigated, but the same approach could certainly be utilized for other materials as well.

While CVD diamond has long been the heatspreader of choice for VECSELs, DBR-free SDLs typically require single-crystalline diamond for successful bonding, and for low intracavity parasitic losses. Although such diamond-based lasers have been implemented with good results, the supply of good-quality single-crystalline diamond is very limited, and the cost is high. Another promising heatspreader material is SiC: it is readily available in large sizes at a fraction of the cost of diamond, with similar optical losses in the infrared, but much better surface quality, facilitating reliable bonding. This also opens up the possibility of wafer-scale processing of single or double heatspreader DBR-free SDLs and MECSELs, which might be of great interest for commercial applications.

Finally, novel concepts, like monolithic designs relying on cavities formed by TIR inside the heatspreader might lead to compact and robust laser sources for applications not typically suited for a VECSEL. The proposed grating-based GEMM concept finally combines the advantages of DBR-free SDLs in a design that could easily replace conventional VECSEL chips, without the need to switch to a transmission geometry cavity design.

References

1 Kuznetsov, M., Hakimi, F., Sprague, R., and Mooradian, A. (1997). High-power (>0.5-W CW) diode-pumped vertical-external-cavity surface-emitting semiconductor lasers with circular TEM00 beams. *IEEE Photon. Technol. Lett.* 9 (8): 1063–1065.
2 Heinen, B., Wang, T.-L., Sparenberg, M. et al. (2012). 106 W continuous-wave output power from vertical-external-cavity surface-emitting laser. *Electron. Lett.* 48 (9): 516.
3 Tilma, B.W., Mangold, M., Zaugg, C. et al. (2015) Recent advances in ultrafast semiconductor disk lasers. *Light Sci. Appl.* 4 (7): e310.
4 Guina, M., Rantamaki, A., and Harkonen, A. (2017). Optically pumped VECSELs: review of technology and progress. *J. Phys. D Appl. Phys.* 50: 383001.
5 Sirbu, A., Volet, N., Mereuta, A. et al. (2011). Wafer-fused optically pumped VECSELs emitting in the 1310-nm and 1550-nm wavebands. *Adv. Opt. Technol.* 2011: 209093, pp. 1–8.
6 Leinonen, T., Iakovlev, V., Sirbu, A. et al. (2017). 33 W continuous output power semiconductor disk laser emitting at 1275 nm. *Opt. Express* 25 (6): 516–517.
7 Stillman, G.E., Sirkis, M.D., Rossi, J.A. et al. (1966). Volume excitation of an ultrathin single-mode CdSe laser. *Appl. Phys. Lett.* 9 (7): 268–269.
8 Johnson, M.R., Holonyak, N., Sirkis, M.D., and Boose, E.D. (1967). Volume excitation of an ultrathin continuous-wave cdse laser at 6900 Å output. *Appl. Phys. Lett.* 10 (10): 281–282.
9 Basov, N., Bogdankevich, O., and Grasyuk, A. (1966). Semiconductor lasers with radiating mirrors. *IEEE J. Quantum Electron.* 2 (9): 594–597.
10 Damen, T.C., Duguay, M.A., Shah, J. et al. (1981). Broadband tunable picosecond semiconductor lasers. *Appl. Phys. Lett.* 39 (2): 142–145.
11 Stone, J., Wiesenfeld, J., Dentai, A. et al. (1981). Optically pumped ultrashort cavity $In_{1-x}Ga_xAs_yP_{1-y}$ lasers: picosecond operation between 083 and 159 μm. *Opt. Lett.* 6 (11): 534.
12 Roxlo, C.B. and Salour, M.M. (1981). Synchronously pumped mode-locked CdS platelet laser. *Appl. Phys. Lett.* 38 (10): 738–740.
13 Le, H.Q., Di Cecca, S., and Mooradian, A. (1991). Scalable high-power optically pumped GaAs laser. *Appl. Phys. Lett.* 58 (18): 1967–1969.
14 Su, K.W., Huang, S.C., Li, A. et al. (2009). High-peak-power AlGaInAs quantum-well 1.3 μm laser pumped by a diode-pumped actively Q-switched solid-state laser. *Opt. Lett.* 31 (13): 2009–2011.
15 Huang, S.C., Chang, H., Su, K. et al. (2009). AlGaInAs/InP eye-safe laser pumped by a Q-switched Nd:GdVO4 laser. *Appl. Phys. B Lasers Opt.* 94 (3): 483–487.

16 Wen, C.P., Tuan, P., Liang, H. et al. (2015). High-peak-power optically-pumped AlGaInAs eye-safe laser with a silicon wafer as an output coupler: comparison between the stack cavity and the separate cavity. *Opt. Express* 23 (24): 20800–20805.

17 Yang, Z., Albrecht, A.R., Cederberg, J.G., and Sheik-Bahae, M. (2015). Optically pumped DBR-free semiconductor disk lasers. *Opt. Express* 23 (26): 33164.

18 Alford, W.J., Raymond, T.D., and Allerman, A.a. (2002). High power and good beam quality at 980 nm from a vertical external-cavity surface-emitting laser. *J. Opt. Soc. Am. B* 19 (4): 663.

19 Hastie, J.E., Jeon, C., Burns, D. et al. (2003). A 0.5-W 850-nm $Al_x Ga_{1-x} As$ VECSEL with intracavity silicon carbide heatspreader. *SPIE Proc.* 5137: 201–206.

20 Kim, J.-Y., Cho, S., Lee, S.-M. et al. (2007). Highly efficient green VECSEL with intra-cavity diamond heat spreader. *Electron. Lett.* 43 (2): 105–107.

21 Kahle, H., Mateo, C.M.N., Brauch, U. et al. (2016). Semiconductor membrane external-cavity surface-emitting laser (MECSEL). *Optica* 3 (12): 1506–1512.

22 Mateo, C.M.N., Brauch, U., Khale, H. et al. (2016). 2.5 W continuous wave output at 665 nm from a multipass and quantum-well-pumped AlGaInP vertical-external-cavity surface-emitting laser. *Opt. Lett.* 41 (6): 1245.

23 Tourrenc, J.P., Bouchoule, B., Khadour, A. et al. (2007). High power single-longitudinal-mode OP-VECSEL at 1.55 μm with hybrid metal-metamorphic Bragg mirror. *Electron. Lett.* 43 (14): 754–755.

24 Jones, B.E., Schlosser, P.J., De Jesus, J. et al. (2015). Processing and characterisation of II–VI ZnCdMgSe thin film gain structures. *Thin Solid Films* 590: 84–89.

25 Hader, J., Wang, T.-L., Yarborough, J. et al. (2011). VECSEL optimization using microscopic many-body physics. *IEEE J. Sel. Top. Quantum Electron.* 17 (6): 1753–1762.

26 Keller, S.T., Sirbu, A., Iakovlev, V. et al. (2015). 85 W VECSEL output at 1270 nm with conversion efficiency of 59%. *Opt. Express* 23 (13): 17437.

27 Wang, T.-L., Kaneda, Y., Hader, J. et al. (2012). Strategies for power scaling VECSELs. *Proc. SPIE* 8242: 824209.

28 Gbele, K., Laurain, A., Hader, J. et al. (2015). Design and fabrication of hybrid metal semiconductor mirror for high power VECSEL. *IEEE Photon. Technol. Lett.* 28 (7): 732–735.

29 Rantamäki, A., Saarinen, E.J., Lyytikäinen, J. et al. (2014). High power semiconductor disk laser with a semiconductor-dielectric-metal compound mirror. *Appl. Phys. Lett.* 104 (10): 101110.

30 Bedford, R.G., Kolesik, M., Chilla, J.L.A. et al. (2005). Power-limiting mechanisms in VECSELs. *Proc. SPIE* 5814: 199–208.

31 Piprek, J., Tröger, T., Schröter, B. et al. (1998). Thermal conductivity reduction in GaAs-AlAs distributed Bragg reflectors. *IEEE Photonics Technol. Lett.* 10 (1): 81–83.

32 Borca-Tasciuc, T.W., D Song, J., Meyer, R. et al. (2002). Thermal conductivity of and alloys and digital-alloy superlattices. *J. Appl. Phys.* 4994: 92.

33 Chernikov, A., Herrmann, J., Koch, M. et al. (2011). Heat management in high-power vertical-external-cavity surface-emitting lasers. *IEEE J. Sel. Top. Quantum Electron.* 17 (6): 1772–1778.

34 Kemp, A.J., Valentine, G., Hopkins, J. et al. (2005). Thermal management in vertical-external-cavity surface-emitting lasers: Finite-element analysis of a heatspreader approach. *IEEE J. Quantum Electron.* 41 (2): 148–155.

35 Iakovlev, V., Walczak, J., Gebski, M. et al. (2014). Double-diamond high-contrast-gratings vertical external cavity surface emitting laser. *J. Phys. D. Appl. Phys.* 6: 47.

36 Raja, M.Y.A., Brueck, S., Osinski, M. et al. (1989). Resonant periodic gain surface-emitting semiconductor lasers. *IEEE J. Quantum Electron.* 25 (6): 1500–1512.

37 Corzine, S.W., Geels, R.S., Scott, J.W. et al. (1989). Design of Fabry–Perot surface-emitting lasers with a periodic gain structure. *IEEE J. Quantum Electron.* 25 (6): 1513–1524.

38 Yang, Z., Albrecht, A.R., Cederberg, J.G., and Sheik-Bahae, M. (2016). 80 nm tunable DBR-free semiconductor disk laser. *Appl. Phys. Lett.* 109 (2): 022101.

39 Liau, Z.L. (2000). Semiconductor wafer bonding via liquid capillarity. *Appl. Phys. Lett.* 77 (5): 651–653.

40 Yablonovitch, E., Hwang, D.M., Gmitter, T.J. et al. (1990). Van der Waals bonding of GaAs epitaxial liftoff films onto arbitrary substrates. *Appl. Phys. Lett.* 56 (24): 2419–2421.

41 DeSalvo, G.C. (1996). Wet chemical digital etching of GaAs at room temperature. *J. Electrochem. Soc.* 143 (11): 3652.

42 Wang, C., Malloy, K., Sheik-Bahae, M. et al. (2015). Influence of coulomb screening on lateral lasing in VECSELs. *Opt. Express* 23 (25): 32548–32554.

43 Wang, C. (2015). Precise characterization and investigation of laser cooling in III–V compound semiconductors. Dissertation. University of New Mexico.

44 Imangholi, B., Hasselbeck, M.P., Sheik-Bahae, M. et al. (2005). Effects of epitaxial lift-off on interface recombination and laser cooling in GaInPGaAs heterostructures. *Appl. Phys. Lett.* 86 (8): 1–3.

45 Heinen, B., Zhang, F., Sparenberg, M. et al. (2012). On the measurement of the thermal resistance. *IEEE J. Quantum Electron.* 48 (7): 934–940.

46 Yang, Z., Follman, D., Albrecht, A.R. et al. (2018). High-power DBR-free membrane semiconductor disk lasers. *CLEO* JTu2A.14: 1–2.

47 van Loon, F., Kemp, A.J., Maclean, A. et al. (2006). Intracavity diamond heatspreaders in lasers: the effects of birefringence. *Opt. Express* 14 (20): 9250–9260.

48 Kim, J.Y., Yoo, J., Cho, S. et al. (2007). Effect of the properties of an intracavity heat spreader on second harmonic generation in vertical-external-cavity surface-emitting laser. *J. Appl. Phys.* 101 (7): 1–5.

49 Mirkhanov, S., Quarterman, A., Kahle, H. et al. (2017). DBR-free semiconductor disc laser on SiC heatspreader emitting 10.1 W at 1007 nm. *Electron. Lett.* 53 (23): 1537–1539.

50 Yang, Z., Follman, D., Albrecht, A.R. et al. (2018). 16 W DBR-free membrane semiconductor disk laser with dual-SiC heatspreader. *Electron. Lett.* 54 (7): 430–432.

51 Chernikov, A., Hermann, J., Scheller, M. et al. (2010). Influence of the spatial pump distribution on the performance of high power vertical-external-cavity surface-emitting lasers. *Appl. Phys. Lett.* 97: 191110.

52 Broda, A., Kuzmicz, A., Rychlik, G. et al. (2017). Highly efficient heat extraction by double diamond heat-spreaders applied to a vertical external cavity surface-emitting laser. *Opt. Quantum Electron.* 49 (9): 1–7.

53 Giet, S., Kemp, A.J., Burns, D. et al. (2008). Comparison of thermal management techniques for semiconductor disk lasers. *SPIE Proc.* 6871: 687115.

54 Yang, Z. (2016). Novel concepts in semiconductor disk lasers. Dissertation. University of New Mexico.

55 Krainer, L., Paschotta, R., Lecomte, S. et al. (2002). Compact Nd:YVO$_4$ lasers with pulse repetition rates up to 160 GHz. *IEEE J. Quantum Electron.* 38 (10): 1331–1338.

56 Brückner, F., Clausnitzer, T., Burmeister, O. et al. (2008). Monolithic dielectric surfaces as new low-loss light-matter interfaces. *Opt. Lett.* 33 (3): 264–266.

57 Yang, Z., Lidsky, D., and Sheik-Bahae, M. (2019). Gain-embedded meta mirrors for optically pumped semiconductor disk lasers. *Opt. Express* 27 (20): 27882–27890.

58 Priante, D., Zhang, M., Albrecht, A.R. et al. (2021). Demonstration of a 20-W membrane-external-cavity surface-emitting laser for sodium guide star applications. *Electron. Lett.* 57(8): 337–338.

7

Optically Pumped Red-Emitting AlGaInP-VECSELs and the MECSEL Concept

Hermann Kahle[1,2], Michael Jetter[1], and Peter Michler[1]

[1] *Institut für Halbleiteroptik und Funktionelle Grenzflächen (IHFG), Universität Stuttgart, Stuttgart, Germany*
[2] *Optoelectronics Research Centre (ORC), Physics Unit/Photonics, Tampere University, Tampere, Finland*

7.1 Introduction

The first part of this chapter deals with the development of optically pumped and directly red-emitting VECSELs (see Section 7.2). This development began in 2002. Gallium-indium-phosphide (GaInP) quantum wells (QWs) embedded in alloys of aluminum-gallium-indium-phosphide (AlGaInP) were used as active laser material. Lasers based on the optically pumped semiconductor VECSEL (OPS-VECSEL) scheme have not been shown in the visible spectral range at that time. Several restrictions like the limited charge carrier confinement, low thermal conductivity of the used semiconductor materials, and therefore poor heat removal out of the laser-active region (also discussed in Section 7.3) hampered early progress. Nevertheless, red-emitting AlGaInP-based VECSELs were successfully developed and improved continuously during the last years. The following lines will briefly summarize this development (Figure 7.1).

In a first approach, disk laser architectures with active regions containing GaInP QWs that were arranged in groups of four QWs each were grown as a resonant periodic gain (RPG) structure. Gain elements with three, five, and eight groups of QWs, each group separated by a distance of $\lambda/2$ to each other, have been compared. With the 8×4 QW design, which turned out to be the optimum structure then [1], more than 200-mW peak power at a wavelength of 660 nm and at a heat sink temperature of $-30\,°C$ could be shown with a pulsed pump mode with 514-nm pump wavelength [2]. Heat incorporation due to the large quantum defect made continuous-wave (CW) operation at the beginning completely impossible. With the use of a dye laser adjusted to emit 630 nm as pump source and by that pumping only the QWs, CW operation of the VECSEL with a maximum output power of 55 mW at a heat sink temperature of $-35\,°C$ was possible [3]. By in-well pumping, the temperature increase in the active region could be drastically reduced but cannot totally be abandoned. This leads to thermally limited performance, as is the case for all semiconductor devices. Also, poor absorption of the 630-nm pump light in the

Vertical External Cavity Surface Emitting Lasers: VECSEL Technology and Applications, First Edition.
Edited by Michael Jetter and Peter Michler.
© 2022 WILEY-VCH GmbH. Published 2022 by WILEY-VCH GmbH.

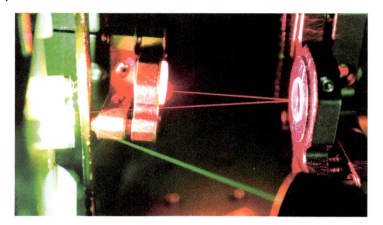

Figure 7.1 Close-up view of a red-emitting AlGaInP-VECSEL in a V-shaped cavity without any intracavity elements, pumped with a 532-nm laser. Source: Photographed by Hermann Kahle.

GaInP QWs hampered further progress here. For the given semiconductor material (AlGaInP), charge carrier confinement and material gain are given parameters. Therefore, investigations and improvements, mainly conducted on the thermal management, seem to be the most promising path for future work.

Continuing research in the following years, also with a GaInP QW-based structure with 20 QWs and arranged in pairs in a RPG design, aimed also toward this goal. An intracavity heat spreader, in this case a 250-μm-thick single-crystal diamond platelet, was liquid capillary bonded to the VECSEL chip to remove heat from the surface as well as the whole active region [4, 5]. In that way, 1.1 W of output power at around 675 nm at a heat sink temperature of −10 °C [6] was possible and the use of intracavity heat spreaders became a standard method for thermal management on VECSELs with low charge carrier confinement gain materials.

The shortest emission wavelength from a GaInP-based direct emitting VECSEL in the red spectral range was reported in 2009. This VECSEL emitted 625-nm light in a pulsed pumping mode with a repetition rate of 6 kHz and a pulsed output peak power of up to 3 W [7]. At the same time, CW emission around 646 nm was reported [8], but with a relatively low power of around 1 mW. 1.2 W of laser output at 670 nm emission wavelength and at a heat sink temperature of −31 °C was reported [9] in 2011. The used chip architecture was a 5 × 4 QW distribution. A wide tuning range of 21 nm (661–682 nm) of laser emission was shown. Investigations on strain compensation of the gain structure comparing the 5 × 4 with a 10 × 2 QW designs found an influence of the number of QWs per package as well as the consequences of stacking them to build up an active region in 2013 [10, 11]. Mode-locking of directly red-emitting AlGaInP-VECSELs, which were also investigated in recent years [12, 13], is described in detail in Chapter 10.

Another way to create red light emission from semiconductor-based structures is the use of InP quantum dot (QD) layers instead of QWs. A structure comprising 7 × 1 QD layers delivered 1.39 W at a heat sink temperature of −25 °C and at 655 nm emission wavelength [14]. The superior charge carrier confinement of QDs compared to QWs leads to a further improvement of the output power despite a slight blue shift of the emission wavelength.

It is also possible to cover parts of the red spectral range by frequency-doubling infrared-emitting VECSELs. 650-nm emission wavelength and up to 3-W output power were shown in 2010, where an InP-based active medium was fused with a GaAs (gallium arsenide)/AlGaAs (aluminum gallium arsenide) DBR (distributed Bragg reflector) resulting in an integrated monolithic gain mirror [15]. Furthermore, second-harmonic generation of GaInNAs-based semiconductor disk lasers performed with a periodically poled Mg:SLT crystal, which covered the short wavelength red-orange area. 610-nm emission wavelength with 730-mW stable output power was presented [16], later up to 4.6 W was reached [17], including frequency quadrupling to 305 nm with more than 10 mW output power. 615-nm emission wavelength finally exceeding 10-W output power [18–20] was also reached with GaInNAs QW-based disk laser structures designed for emission around 1230 nm. An optical-to-optical conversion efficiency of 17.5% was achieved here via frequency-doubling. Actively stabilized VECSELs in the red spectral range were also under investigation during the last years. A relative line width of 200 kHz could be realized when locking the laser to a reference cavity [21]. Later, 16 kHz in the fundamental and even 50 kHz in the UV were shown [22]. The very latest results on stabilized red VECSELs emitting at around 690 nm aim on atom cooling applications for strontium [23].

The second part of this chapter, Section 7.3, describes the latest breakthrough to a new semiconductor-based laser technology – the membrane external-cavity surface-emitting laser (MECSEL [24]). In a MECSEL, a semiconductor membrane consisting only of a laser-active region with micron or submicron thickness sandwiched between transparent heat spreaders represents the compact gain element of the laser. The absence of a substrate and a DBR and the fact that the growth of a monolithically integrated DBR is not at all necessary avoid a lot of growth-related problems and furthermore extend the choice of gain materials. Furthermore, the MECSEL concept aims on creating a thermally perfect environment for the semiconductor gain membrane and thereby optimizing the heat dissipation out of the semiconductor gain region.

7.2 Direct Red-Emitting AlGaInP-VECSELs and Second-Harmonic Generation

In this section, the latest developments in the field of direct red-emitting VECSELs are presented as well as achievements on intracavity frequency-doubling to reach the UV-A spectral range.

7.2.1 GaInP Quantum Wells and the AlGaInP Material System

QWs have been routinely used in optoelectronic devices since many years. They can be grown with great precision by molecular beam epitaxy (MBE) or metal organic vapor phase epitaxy (MOVPE). These techniques allow the easy production

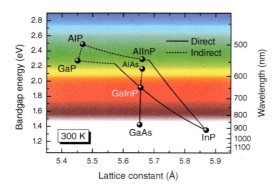

Figure 7.2 Epitaxy map of the AlGaInP material system.

of layered structures containing different semiconductor materials with precise control of the layer thickness down to the atomic level (Figure 7.2).

QWs are formed by growing a layer of a semiconductor of thickness L_w between layers of another semiconductor with a larger bandgap. The typical material system to create light emission in the red spectral range is > $Ga_y In_{1-y})_{0.5} P_{0.5}$ (GaInP). GaInP QWs are typically embedded in $[(Al_x Ga_{1-x})_y In_{1-y}]_{0.5} P_{0.5}$ (AlGaInP) where a carefully selected composition of x and y enables lattice-matched growth on GaAs substrates. The use of QWs as gain material compared to semiconductor bulk material delivers several benefits [25, 26]; the keyword here is bandgap engineering. The bandgap can be adjusted just by the thickness L_w of the QW, which changes the quantization energy. This adjustment is independent from the material composition [27]. Below a critical layer thickness [28, 29], which is about 14 nm for $(Ga_{0.4} In_{0.6})P$ [30], one can independently select the $(Ga_y In_{1-y})_{0.5} P_{0.5}$ composition where the phosphorus fraction is always 50%. Thereby, the bandgap energy can be adjusted. A gallium fraction selected to be $y < 0.516$ leads to a compressive strained layer compared to the GaAs substrate due to the larger radius of the indium ions. For a fraction of $y > 0.516$, the layer is tensile-strained and reveals pseudomorphic growth. The material composition can be determined by measuring lattice constants with X-ray diffraction, which allows one the derivation of the strain situation in the material [11, 31]. The effective emission wavelength is further influenced by the strain present in the QW. The bandgap reduction caused by the composition is partly compensated via compressive strain [30–32]. One has to point out here that the biaxial strain in the layer plane defines the strain situation (compressive, lattice-matched, and tensile, see Figure 7.3). The strain inside the QW causes a bandgap renormalization, and the light hole and the heavy hole have again different energy levels, which reduces the bandgap [29, 32–36]. This causes a reduction of the density of states. Therefore, population inversion can be reached at a lower charge carrier density, which leads to reduction of the threshold current in electrically driven devices [25, 26, 37–40]. We assume here a similar effect as the electric current producing charge carriers can be replaced by photon flux, the incident optical pump power. Looking at the material gain of compressively strained layers compared to tensile-strained ones, a slightly lower transparency charge carrier density and a slightly increased differential efficiency can be expected [26, 32, 41, 42].

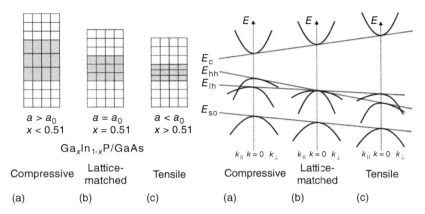

Figure 7.3 (a–c) Schematic of the gallium to indium composition influencing the internal strain and the band structure in the GaInP material system. Source: Schematic adapted from [32].

Due to the beneficial properties of compressively strained GaInP QWs as gain material in laser devices, we use a composition of $Ga_{0.4}InP$ to realize GaInP-based VECSELs. Nevertheless, it is still a challenge to realize short wavelength emitting VECSELs in the AlGaInP material system due to the charge carrier confinement. The latter is comparatively low when looking at other material systems [34, 43]. As an example, we compare $Ga_{0.4}InP$ QWs in $Al_{0.5}GaInP$ barriers with $In_{0.2}GaAs$ in $Al_{0.4}GaAs$ barriers [44]: the electrons in the conduction band in the GaInP wells reveal a confinement energy of $\Delta E_c = 225$ meV, and the holes in the valence band are trapped by $\Delta E_v = 150$ meV, which is the situation for a emission wavelength of 670 nm [43]. In the near-infrared spectral range for $In_{0.2}GaAs/Al_{0.4}GaAs$, we have $\Delta E_c = 480$ meV and $\Delta E_v = 290$ meV. This means a by far better confinement in $In_{0.2}GaAs/Al_{0.4}GaAs$ as the thermal activation for charge carriers is proportional to $e^{(-\Delta E_{c/v})/(k_B \cdot T)}$. Therefore, a loss of charge carriers due to thermal escape back into the barrier lowers the radiative efficiency of the system. This is especially a problem for high internal temperatures or high operating temperatures and becomes a real challenge if one aims on power scaling of these systems. For shorter emission wavelengths, the situation of carrier confinement gets even worse [44]. Assuming a relation of $\Delta E_c/\Delta E_v = 0.6/0.4$ and starting the calculation at 670 nm, the confinement energy changes by -8% when going to 660 nm and even by -16% for 650 nm. Otherwise for larger wavelengths, for 680 nm, we get an improvement of the confinement energy by 7%, for 690 nm by even 16%. Therefore, evaluating measured results with respect to the wavelength is indispensable for VECSELs in the AlGaInP material system.

7.2.2 GaInP Quantum Well VECSELs: A Comparison

A comparison of epitaxial designs for barrier pumped VECSELs in the red spectral range (see Section 7.2.2.1–7.2.2.4) is presented here. The compared VECSEL structures are grown by MOVPE as GaInP/AlGaInP multi-QW structures with 20 and 21 compressively strained QWs, respectively. The QWs are placed in various packages in a separate confinement heterostructure with quaternary AlGaInP barrier and cladding layers. We compare three different QW distributions: the standard 10×2 QW design includes 20 QWs arranged in 10 pairs [45]. The second gain structure to compare contains also 20 QWs arranged in 10 pairs, but with tensile-strained barriers to compensate the compressive strain of the QWs. The third gain structure has an inhomogeneous or, more accurately, an exponential distribution of the QWs in tensile-strained barriers. Laser parameters such as laser emission wavelength, differential efficiency, optical output power, and absorption of the pump laser were measured or calculated for the different designs.

7.2.2.1 Architecture of the Semiconductor Structures

In all three design cases (10×2 QWs, 10×2 QWs strain-compensated, and inhomogeneous QW distribution strain-compensated shown in Figure 7.4), the VECSEL chips were fabricated by MOVPE and they consist of the standard monolithic integrated DBR [9] and the active region providing the gain. In the standard 10×2 active region design [45], 20 QWs are contained, arranged in 10 pairs embedded in barrier layers of $[(Al_{0.33}Ga_{0.67})_{0.51}In_{0.49}]_{0.5}P_{0.5}$. Compressively strained GaInP QWs with an approximate thickness of 5 nm are used to achieve a maximum PL emission at a wavelength of ~665 nm. The QW packages are embedded in $Al_{0.55}Ga_{0.45}InP$ cladding layers on top of the DBR. The standard samples were grown at a temperature of 750 °C.

The strain-compensated design with the same QW distribution as the standard design, but slightly different compositions (see [11]), was grown with tensile-strained $[(Al_{0.33}Ga_{0.67})_{0.52}In_{0.48}]_{0.50}P_{0.5}$ barriers and tensile-strained cladding layers to compensate the compressive strain of the QW. In order to achieve high gain, it is important to locate the QW packages in the antinodes of the electric field component of the optical standing wave field, which represents a RPG structure [46, 47]. The sample with the inhomogeneous QW distribution (see Figure 7.4) has the same length of the active region and also tensile-strained barriers and cladding layers for strain compensation. For an optimized charge carrier distribution due to

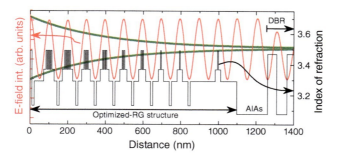

Figure 7.4 Index of refraction (black curve) and electric field intensity (red curve) in the VECSEL chip simulated with the transfer matrix method. The QW packages are located in the antinodes of the electric field in the exponentially distributed QW design. Also, a scheme of the pump light intensity (green curve) as it travels through the structure under absorption is plotted.

absorption, the QWs were rearranged in a pattern of 4-4-3-3-2-2-1-1-0-1 QWs per antinode to fit to the exponential absorption profile of the pump power P_{pump} (see the schematically included green curve in Figure 7.4) in the active region, which follows the Beer-Lambert-law:

$$P_{trans} = P_{pump} \cdot e^{-\alpha \cdot d_{abs}} \qquad (7.1)$$

P_{trans} represents the non-absorbed transmitted pump power (which is then absorbed in the DBR), α is the average absorption coefficient of the semiconductor material, and d_{abs} is the thickness of the active region seen as absorbing path.

According to the number of QWs per antinode/QW-package, a higher number than four QWs would not be useful because the overlap of the electric standing wave with the outer QWs in the package would be poor and so the enhancement factor Γ_{enh}, which quantifies this effect (for more details, see [46–48]), would be reduced. This would finally have negative effects on the performance of the laser. The overall value of Γ_{enh} for the active region with the inhomogeneous QW distribution was calculated to be $\Gamma_{enh} \approx 1.93$ ($\Gamma_{enh} \approx 1.96$ for the 10×2 structures). This means the individual QW number per package plays a minor role in the overall performance of the whole structure as the difference between the Γ_{enh}-factors is rather small.

Nevertheless, due to the absorption-adapted QW distribution, we expected a more equal carrier concentration in the QWs and therefore a more uniform gain over the whole active region. Additionally, AlInP layers were positioned between the QW packages to enhance the charge carrier confinement and suppress thermal escape and diffusion. To achieve CW laser operation, the heat removal from the active region is improved by bonding a 0.5-mm-thick intracavity diamond heat spreader onto the chip surface via capillary forces [49].

7.2.2.2 Experimental Setup

The VECSEL chip is mounted with thermal grease on a copper plate cooled by a thermoelectric cooler (TEC), which is itself water cooled. A downholder made of polyethylene with a small aperture in the middle is used to apply additional gentle pressure onto the diamond heat spreader to improve the thermal contact between diamond and chip and therefore increase the heat removal. The chip is optically pumped by a frequency-doubled Nd:YAG laser emitting at 532 nm. The pump beam is focused with a lens (focal length of 100 mm) and irradiated on the chip with an angle of incidence of about 40° resulting in an elliptical pump spot. The pump spot size on the chip can be varied by changing the distance between the chip and the lens. It is adjusted for a maximum output power of the VECSEL [50], which results in mode-matching of the pump spot and the mode diameter on the chip. We use a V-shaped cavity to measure a set of power transfer curves of the three different VECSEL designs for the derivation of the transmission losses T_{loss} with the Findlay-Clay method [51] and for the comparison of internal parameters due to Kuznetsov's model [52]. The concave folding mirror has a radius of curvature of 50 mm and a reflectivity of $R_{fold} \approx 99.9\%$. The plane outcoupling mirror was varied for the test series and therefore had different reflectivities (R_{out} = 99.9, 99.8, 99.5, 99.0, 98.0, 97.0 and 95.0%). The individual cavity arms of the compact setup have a length of ~68 mm and ~38 mm, respectively. The measurements for a maximum output power were performed in a linear cavity with an outcoupling mirror with a radius of curvature of 50 mm and a distance of ~49 mm to the chip.

7.2.2.3 Characterization Results

The spectra of the laser emission of the 10×2 QW design and the optimized design with the exponential QW distribution emitting at 661.5 and 665.5 nm, respectively, are shown in Figure 7.5a. The 10×2 strain-compensated design emits slightly blue-shifted at 654.7 nm. This blue shift can either be explained by a slight variation in the QW growth process or by a strain overcompensation [11] due to the tensile-strained barrier and cladding layers.

The measurements of the output power shown in Figure 7.5b reveals a maximum output power of 1.6 W for the absorption-optimized structure with exponentially distributed QWs, which is an enhancement to the values published [14] even at a higher heat sink temperature. The 1.6 W compared to the maximum output of 1.34 W from the 10×2 standard design shows an increase in output power of 20% due to the arrangement of the QWs in the active region. The 10×2 strain-compensated design has a lower output power of about 1.0 W here, which can be explained by the significantly shorter wavelength, and therefore a reduced confinement of the charge carriers, which increases the thermal sensitivity leading to an earlier thermal rollover.

7.2.2.4 Internal Efficiency

To get a relative comparison of the three different VECSEL structures, one has to look at the efficiencies in a VECSEL introduced by Kuznetsov et al. [53]:

A VECSEL's output power is then given by

$$P_{out} = (P_{pump} - P_{th}) \cdot \eta_{diff}, \tag{7.2}$$

where the VECSEL's differential efficiency η_{diff} is

$$\eta_{diff} = \eta_{abs} \cdot \eta_{geo} \cdot \eta_{quant} \cdot \eta_{out} \cdot \eta_{int}. \tag{7.3}$$

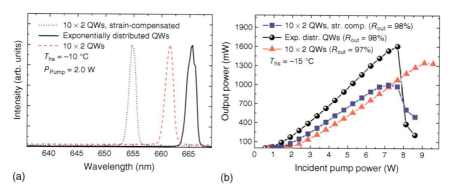

Figure 7.5 (a) Laser spectra of the different epitaxial designs. (b) Power transfer measurements for the different epitaxial designs. All measurements were done with the best-performing outcoupler.

7.2 Direct Red-Emitting AlGaInP-VECSELs and Second-Harmonic Generation

The components of the differential efficiency η_{diff} are the absorption efficiency η_{abs}:

$$\eta_{\text{abs}} = \frac{P_{\text{abs}}}{P_{\text{pump}}} = \frac{P_{\text{pump}} - P_{\text{refl}} - P_{\text{trans}}}{P_{\text{pump}}}, \tag{7.4}$$

where P_{refl} is the reflected pump light on all present interfaces (e.g. air/heat spreader or heat spreader/semiconductor) and P_{trans} is the transmitted pump power which is neither reflected at nor absorbed in the laser-active region (but probably absorbed in the DBR, where it creates excess heat); and the geometrical efficiency η_{geo} [3]:

$$\eta_{\text{geo}} = \frac{A_{\text{mode}}}{A_{\text{pump}}} = \frac{\pi \cdot r_{\text{mode}}^2}{\pi \cdot a_{\text{pump}} \cdot b_{\text{pump}}} = \frac{r_{\text{mode}}^2}{a_{\text{pump}} \cdot b_{\text{pump}}}, \tag{7.5}$$

with the mode area A_{mode} of the VECSEL and the pumped area A_{pump}, which can be of elliptical shape and therefore consists next to π of a short and a long radius a_{pump} and b_{pump}; and the quantum efficiency η_{quant}:

$$\eta_{\text{quant}} = \frac{E_{\text{phot. laser}}}{E_{\text{phot. pump}}} = \frac{h \cdot \left(\frac{c}{\lambda_{\text{laser}}}\right)}{h \cdot \left(\frac{c}{\lambda_{\text{pump}}}\right)} = \frac{\lambda_{\text{pump}}}{\lambda_{\text{laser}}} \tag{7.6}$$

with h as Planck's constant, c represents the speed of light. $E_{\text{phot. laser}}$ and $E_{\text{phot. pump}}$ are the energies of the laser and pump photons, and λ_{laser} and λ_{pump} are the corresponding wavelengths; and the outcoupling efficiency η_{out}:

$$\eta_{\text{out}} = \frac{\ln(R_{\text{out}})}{\ln(R_{\text{out}} \cdot R_{\text{DBR}} \cdot R_{\text{fold 1}} \cdot \ldots \cdot R_{\text{fold } n} \cdot T_{\text{loss}})} \tag{7.7}$$

with the ratio of the reflectivity of the outcoupling mirror R_{out} with the cavity transparency factor T_{loss} and all other loss channels (reflectivity of the DBR R_{DBR} and the reflectivity of all other cavity folding mirrors $R_{\text{fold 1...}n}$ including outcoupling mirror reflectivity; and the internal efficiency η_{int}, which can be derived from all other efficiencies, if η_{diff} can be taken from a measurement:

$$\eta_{\text{int}} = \frac{\eta_{\text{diff}}}{\eta_{\text{abs}} \cdot \eta_{\text{geo}} \cdot \eta_{\text{quant}} \cdot \eta_{\text{out}}}. \tag{7.8}$$

The internal efficiency η_{int} shows us the percentage of charge carriers originating from absorbed pump photons (η_{abs}) within the laser mode area (η_{geo}) that are actually transferred into laser photons considering the differential efficiency (η_{diff}), the quantum efficiency (η_{quant}), and the outcoupling efficiency (η_{out}).

For a relative comparison of the three given GaInP/AlGaInP-VECSEL gain structures, one can neglect η_{abs} because all of the three investigated and here compared structures contain a very similar active region. Especially, the overall thickness of the active region is important here and only varies in the range of a few nanometers. Also the difference in architecture, namely an additional GaInP layer on the surface of the strain-compensated 10 × 2 design and the exponentially distributed QW design and the additional electron blocking layers in the latter, should not have a significant difference in absorption at the used pump wavelength and therefore also have been neglected in further considerations. For the geometrical efficiency η_{geo}, the situation is very similar. As for all three test series, the pump conditions and the resonator were kept unchanged, so the geometrical efficiency η_{geo} should be identical for all test series and can therefore also be neglected in a relative comparison. On the other hand, the quantum efficiency η_{quant} remains important because the different VECSEL structures emit at different wavelengths. The outcoupling efficiency η_{out} contains the constant T_{loss} as well as the product of all other reflectivities of mirrors used in the resonator ($R_{DBR}, R_{fold}, R_{out}$) from which the measurable differential efficiency η_{diff} significantly depends. We have then left the equation:

$$\eta_{diff}(R_{out}) = \frac{\lambda_{pump}}{\lambda_{VECSEL}} \cdot \frac{\ln(R_{out})}{\ln(R_{DBR} \cdot R_{fold} \cdot R_{out} \cdot T_{loss})} \cdot \eta_{int} \tag{7.9}$$

with η_{int} as the only unknown parameter, which can now be derived and is named as $\eta_{rel.int.eff.}$ because it represents only a relative comparing factor. By fitting this equation to the differential efficiencies $\eta_{diff}(R_{out})$ plotted versus outcoupler reflectivity R_{out} (see Figure 7.6) for each of the three test series, the relative internal efficiencies $\eta_{rel.int.eff.}$ can be determined and are noted in Table 7.1.

In a relative comparison, these results allow the following conclusions:

(I) Comparing the standard 10 × 2 QW structure with the strain-compensated, one it seems obvious that an increase in the internal efficiency η_{int} by 13.6% can only be connected to the reduced internal strain inside the structure. Earlier investigations clearly confirm this improvement [11]. It is also accompanied with a narrowing of the PL and with this probably a narrowing as well as an

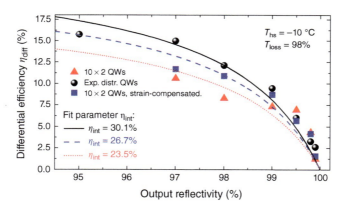

Figure 7.6 Differential efficiency η_{diff} plotted versus outcoupler reflectivity R_{out} and via the fit function Equation 7.9 determined internal efficiencies η_{int}.

Table 7.1 From the data plotted in Figure 7.6 with the fit Equation 7.9 derived relative internal efficiencies $\eta_{rel.int.eff.}$ expressed as absolute value and in percent.

	10 × 2 QWs	10 × 2 QWs, strain-compensated	Exp. distr. QWs
$\eta_{rel.int.eff.}$	0.235	0.267	0.301
$\eta_{rel.int.eff.}$ in %	100.0%	113.6%	128.1%

increase of the overall gain of the VECSEL chip due to better spectral overlap of the single gain curves of each QW or QW package (see Figure 7.7).

The relative internal efficiency $\eta_{\text{rel.int.eff.}}$ can also be identified with the radiative efficiency η_{rad}. In this case, strain compensation can also result in a more homogeneous growth and therefore in a lower density of nonradiative decay channels like defects and dislocations in the semiconductor crystal.

(II) If we now compare the VECSEL structure with the inhomogeneously distributed QWs, which was also grown with strain-compensating barriers and cladding layers, to the strain-compensated 10 × 2 QW structure, we find again an additional improvement on the relative internal efficiency $\eta_{\text{rel.int.eff.}}$ of 14.5 percentage points. In order to determine the reasons for this, one needs a close look at the differences between the two designs. One difference is the distribution of the QWs within the structures. It is 4-4-3-3-2-2-1-1-0-1 QWs per QW package or per antinode of the standing electric field. This distribution is approximately adapted to the absorption of the pump light in the active region. Therefore, the probability of fully saturated QWs, which mean a waste of further present charge carriers, or on the other hand not saturated QWs, which even act as loss channel because they have the potential to absorb laser light until they reach transparency charge carrier density, is significantly reduced. It is well known that gain in QWs is charge carrier density-dependent [1, 54] and its maximum shifts to shorter wavelengths with increasing charge carrier density.

The situation of inhomogeneously distributed charge carriers per QW package is schematically plotted in Figure 7.7a. A more homogeneous charge carrier density in all present QWs or QW packages in the active region of the VECSEL implies a much better spectral overlap of the gain in each single QW or QW package (see Figure 7.7b), which leads to a further overall increase in the gain and therefore probably to the increase in the relative internal efficiency $\eta_{\text{rel.int.eff.}}$.

Further, charge carrier diffusion would also be hindered by the presence of the AlInP electron blocking layers separating the QW packages. The reason for the implementation of electron blocking layers here is not to suppress charge carrier diffusion in general but to hamper thermal escape of charge carriers out of the QWs at high pump power densities. Due to the fact that in the test series of power transfer measurements high pump powers have been avoided because only the differential efficiency and the laser threshold were necessary for this evaluation, the impact of the electron blocking layers can be neglected here. This can surely not be assumed for high-power measurements. Finally, one can state that the reason for the improvement of the relative internal efficiency $\eta_{\text{rel.int.eff.}}$ is the pump power-adapted QW distribution.

(III) Another hint that the new VECSEL structure with the exponentially distributed QWs is capable to receive more losses than the others is the fact that laser operation was still possible with an outcoupler with a reflectivity of $R_{\text{out}} = 95\%$, while with the other structures measurements only with outcouplers up to $R_{\text{out}} = 97\%$ were possible.

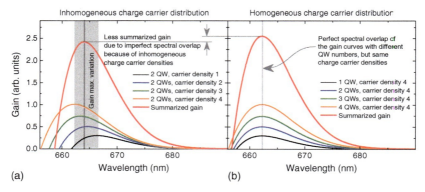

Figure 7.7 Gain schematics, gain (arbitrary units) is plotted versus wavelength: (a) Black, blue, green, and orange curves represent gain schemes of QW packages with a different number of QWs and at the same time different charge carrier densities. The red curve represents the sum of these four curves. The different charge carrier densities lead to spectrally shifted gain. (b) The consequence of homogeneously distributed charge carriers leads to a perfect spectral overlap of the gain curves with different numbers of QWs.

7.2.3 Power Scaling via Quantum Well and Multi-Pass Pumping

At the beginning, AlGaInP-VECSELs have in most cases been pumped into the barriers (except the first QW pumping approach by Müller et al. [3]). The typical pump source is a frequency-doubled 532-nm-emitting DPSS laser. Also, barrier pumping with blue-emitting GaN (gallium nitride) diodes [55, 56] was under investigation in recent years. Such short-pumping wavelengths simplify the pumping configuration due to high absorption within the active region. Looking at the quantum defect:

$$\eta_{\text{defect}} = 1 - \eta_{\text{quant}}, \tag{7.10}$$

which includes the quantum efficiency η_{quant} (see Eq. 7.6), one problem becomes obvious – the huge amount of heat created inside the active region itself due to the excess energy of the pump photons. For an emission wavelength of 665 nm and a pump wavelength of 532 nm, the quantum defect is 20%. Compared to InGaAs-based VECSELs emitting around 1060 nm, which are typically pumped with 808 nm laser diodes, η_{quant} is quite the same, but due to the worse charge carrier confinement (see Section 7.2.1) the incorporated thermal energy has a by far larger impact on the performance of GaInP/AlGaInP structures. This results with respect to output power in a difference of two orders of magnitude ranging from 1.6 W [57] with the AlGaInP-VECSEL to 106 W [58] with an InGaAs VECSEL. In both cases, the structures were barrier-pumped resulting in $\eta_{\text{defect}} \approx 20\%$. In the following section, we describe a way to scale up efficiency [59] and power [60, 61] of AlGaInP-based VECSELs.

7.2.3.1 Quantum Well Pumping

In order to determine the difference between barrier pumping and QW pumping, a standard 5 × 4 GaInP QW structure was used. It was optimized for emission at 665 nm, and its design is described in detail elsewhere [9]. In order to dissipate most of the introduced heat, an uncoated single-crystal diamond ($\sim$ 500 m in thickness) was liquid capillary bonded onto the surface of a cleaved 2.5 2.5 mm² gain chip. After bonding, both, the diamond and the disk, were fixed to a water-cooled brass heat sink. The laser cavity was a standard I-cavity and formed by the semiconductor's DBR as one end mirror and an output coupler with a transmittance of 1% on the other end. The radius of curvature of the output coupler was 100 mm and the cavity length was about 98 mm. The VECSEL was pumped with a dye laser adjusted to an emission wavelength of 640 nm and a frequency-doubled Nd:YAG laser emitting at 532 nm for QW and barrier pumping configurations, respectively. Because of the limited output power of the dye laser (2 W) and the very low pump absorption of the QWs at the pump wavelength of 640 nm, the pump spot size was set to be 40 μm in order to have a high pump power density. For the barrier pumping, the same spot size was used.

Output versus input power (see Figure 7.8a, top) was compared at a heat sink temperature of 15 °C for 532 and 640 nm pump wavelengths. With 532-nm-pumping, a differential efficiency of $\eta_{diff} = 12.5\%$ was achieved as well as a threshold pump power of 130 mW. The latter corresponds to a threshold input power density of 10.3 kW cm^{-2}. The output of the laser, however, was limited by the rollover occurring just a few hundreds of milliwatts above the threshold. This is an indication that the temperature rise in the gain region is high. A similar laser device [9], but using a larger pump spot at a heat sink temperature of −31 °C, exhibited a differential efficiency η_{diff} of 18% and a threshold power of 960 mW corresponding to a threshold power density of 8.5 kW/cm². The differential efficiency with respect to the absorbed pump power (see Figure 7.8a, bottom) for the 640-nm-pumped VECSEL was 60%, which is 3.5 times higher than that of the VECSEL pumped at 532 nm. A maximum output power achieved was only limited by the available pump power, since no thermal rollover was observed. At the laser threshold, the absorbed pump power at 640 nm was 40% less than via 532-nm pumping, which resulted in a higher threshold. This

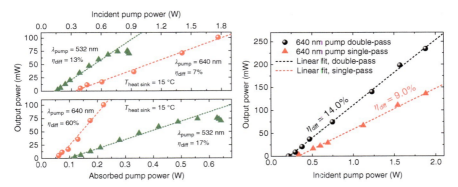

Figure 7.8 (a) Optical output power versus (top) incident and (bottom) absorbed pump power for the GaInP-VECSEL operated at 15 °C for 640-nm pumping (spheres) and 532-nm pumping (triangles). (b) Output power versus incident pump power of the QW pumped VECSEL at 15 °C heat sink temperature for both operation with (spheres) and without (triangles) pump light recycling. Source: Adapted with permission from [59]. ©The Optical Society.

comparison shows a clear improvement on the efficiency of the laser itself with respect to the absorbed power. The drawback that occurs here is the lower absorption of the pump light when pumping in the QWs. A good way to compensate the lower absorption is to recycle the pump light several times through the active region. The influence of this method and its influence on the differential efficiency and the laser threshold can be directly seen in Figure 7.8b. This so-called multi-pass pumping represents the common pumping scheme in solid-state thin disk lasers [62].

7.2.3.2 Multi-Pass Pumping

In this section, we give an exemplary overview on a multi-pass and QW-pumped AlGaInP-VECSEL using a fiber-coupled diode laser as a pump source. The same gain structure as mentioned in Section 7.2.3.1 was used in this study. The used diode laser delivered 100 W at ~638 nm out of a fiber with a core diameter of 400 µm and a numerical aperture of 0.22. More details about the pump optics can be found elsewhere [60]. Figure 7.9 shows the scheme of the setup used in the experiments. Multi-pass pump optics, well known from solid-state disk lasers [62], allow in the present case eight double passes of the pump light. It consists of two prism pairs for pump beam displacement, and a parabolic mirror with a focal length of $f = 32.5$ mm for pump beam focusing. Due to the comparatively low brightness of available diode pump lasers, the minimum pump spot size realized was between 600 and 700 µm. Smaller pumping spots could be achieved by inserting apertures in an intermediate image plane while maintaining maximum power density. The laser cavity was formed by the semiconductor's DBR as one end mirror and an output coupler with a transmissivity of 0.5% on the other end. The radius of curvature of the output coupler was 100 mm and the cavity length was (98 ± 0.5) mm. The performance of the diode-pumped VECSEL operated at a heat sink temperature of 10 °C and with different pump spot sizes is shown in Figure 7.10a. The differential efficiency with respect to the incident pump power was 19% for pump spot diameters of 150 and 240 µm and 17% for the pump spot diameter of 300 µm. The maximum output power of 2.5 W in fundamental TEM_{00} operation was achieved with a pump spot size of

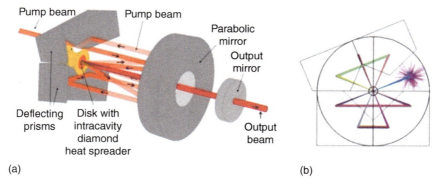

Figure 7.9 (a) Mirror and laser beam schematic of a linear VECSEL with multi-pass pump optic. (b) Simulated pump beam path viewed through the parabolic mirror. Source: Adapted with permission from [60]. ©The Optical Society.

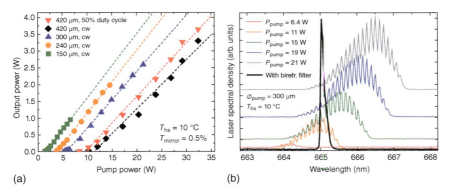

Figure 7.10 (a) Laser performance with 0.5% of output coupling and a heat sink temperature of 10 °C. The pump power was measured at the entrance of the pumping optics, and losses in the multi-pass pumping optics were not subtracted.
(b) Power-dependent laser emission spectra of the AlGaInP-VECSEL with intracavity diamond. For the subsequent nonlinear conversion into the UV, a birefringent filter was inserted into the resonator to narrow down the laser spectrum. Source: Adapted with permission from [60]. ©The Optical Society.

300 μm. With larger pump spots, the output power could be further increased, however with higher order mode operation and unstable output. Thermal rollover was not observed in any configuration which indicates that the output power was only limited by the pump power.

7.2.4 Second-Harmonic Generation into the UV-A Spectral Range

Frequency-doubling of directly red-emitting VECSELs is very attractive because it enables the creation of wavelengths in the UV-A spectral range with decent output power. Continuous research on improving the performance of red VECSELs to upscale second-harmonic generation into the UV during the last years enabled laser light emission in the 325–340-nm wavelength range. The latest results exceed the output power of earlier laser systems (e.g. HeCd gas lasers) by a multitude.

The first results from a frequency-doubled red-emitting VECSEL have been reported in 2006 [63]. The use of a β-barium-borate (BBO) crystal as second-harmonic element inside the laser resonator lead to 120 mW output at around 338 nm [64]. A 10 × 2 QW-VECSEL was used to be frequency-doubled [45] in a compact V-shaped cavity. The short fundamental emission wavelength lead to an also very short UV emission at 328 nm. A tuning range in the UV of about 7 nm (328–335 nm) at a heat sink temperature of −25 °C was reported. Further consequent improvements of the GaInP-active region like the application of strain compensation strategies [11] and a redesign of the active region in total via adapting the QW distribution to the pump light absorption [57] lead to a slight blue shift [45] as well as improved output power [65] in the UV. As a direct result of further power-scaling methods applied to the fundamental laser (QW and multi-pass pumping, see Section 7.2.3), the emission power in the UV-A spectral region could be scaled up again and reached up to 820 mW (see Figure 7.11) at a wavelength of

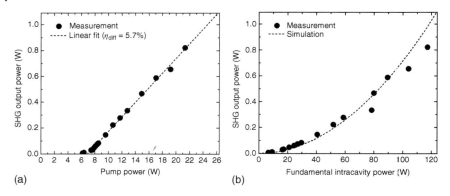

Figure 7.11 (a) Frequency-doubled UV output power (330 nm) plotted over pump power a heat sink temperature of 10 °C. The pump power was measured at the entrance of the pumping optics, and losses in the multi-pass pumping optics were not subtracted. (b) Frequency-doubled UV output power plotted over fundamental intracavity power. Source: Adapted with permission from [60]. ©The Optical Society.

333 nm. These results imply that future development of improved AlGaInP-based gain structures will enable the Watt level in the UV-A spectral region to be reached.

7.3 The Membrane External-Cavity Surface-Emitting Laser (MECSEL)

In this section, a new laser concept is introduced; the MECSEL (see Figure 7.12).

Since the work of Basov et al. in 1966 [66] about the search for strategies to make semiconductor-based lasers more powerful, the driving idea to reach this goal was mainly thermal management. Additionally, using thin semiconductor material vertical to emission direction would lead to a nearly perfect beam quality due to a by far less beam distortion by the gain medium itself. Despite huge improvements during the last decades, the main limitation of nowadays high-power OPS-VECSELs is still heat incorporation into the active region. This leads to a strongly temperature-dependent performance [67] due to the interplay of gain and cavity resonance and the limited charge carrier confinement. The latter is especially a challenge for the AlGaInP material system [68, 69] in which the charge carrier confinement is rather low compared to, e.g. the AlGaInAs material system. In addition to that, the laser structure is based on a thick DBR. The thermal conductivity of this DBR is one order of magnitude inferior compared to well-conducting metals which are often used as backside heat sink, and two orders of magnitude worse compared to diamond used as common backside or intracavity heat spreader [70–72]. The semiconductor structure itself with a thickness of several microns (active region plus DBR) and the substrate with a typical thickness of 350–500 m impede the heat flow out of the active region. To overcome this, numerous strategies for thermal management like heat spreader arrangement [73], substrate removing [52, 74], flip-chip processes [75], or the insertion of compound mirrors [76, 77] improved the performance of VECSELs perpetually. Following this path further, the corollary would be finally abandoning each semiconductor part of a VECSEL not

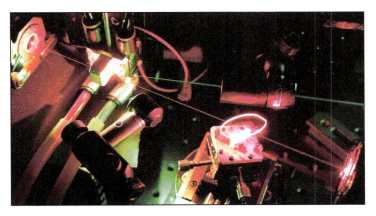

Figure 7.12 Photograph of the operating semiconductor membrane external-cavity surface-emitting laser [24] in an asymmetric linear resonator including a birefringent filter for wavelength selection. Additionally, the optics for the pump beam as well as the green pump beam itself can be seen. Source: Adapted with permission from [24]. ©The Optical Society.

essentially needed to build up the whole laser. Still years before the first optically pumped VCSEL [78, 79] and the first high-power VECSEL [52] were operating, the idea of sandwiching thin semiconductor gain elements between transparent heat spreaders was proposed in a patent by Aram Mooradian in 1992 [80]. This can be achieved by growing the active region directly onto the substrate without the DBR, removing the substrate and finally embedding the released active region membrane in-between diamond heat spreaders (or any other highly transparent material of good thermal conductivity) to create a compact gain device with superior cooling. Such a sandwiched active region membrane configuration would additionally allow the growth of semiconductor structures that are otherwise impossible to grow due to limitations imposed by the need to lattice-match the DBR to the substrate or the active region to the DBR.

This concept to improve the cooling in a VECSEL was theoretically studied and simulated by Iakovlev et al. [81] in 2014. A "DBR-free VECSEL" was realized with an GaInAs-based and released active region bonded to one side of an intracavity heat spreader by Yang et al. [82, 83] in 2015. Simulations were performed, which show the superiority of this system. In this work, it is also mentioned that this can be significantly improved by placing heat spreaders on both sides of the active region [83]. Aiming on other applications, membrane processes with, for example, the AlGaInP material system were already successfully performed [84]. Especially due to the low confinement energy in AlGaInP, the possibility of double-sided cooling promises significant improvements. The first realization of this new laser has been developed during the last years and is presented in this Section [24].

7.3.1 The Semiconductor Active Region Membrane

In the case of the conventional VECSEL, a ~ 5-µm-thick DBR (for details see [9]) is grown instead of the 200-nm-thick AlAs layer. The whole structure of the MECSEL shown here was designed to depict a 3λ-cavity for a wavelength of 665 nm,

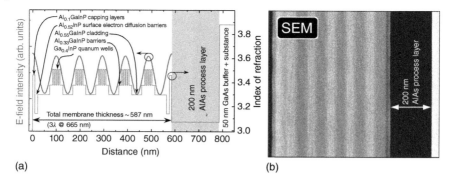

Figure 7.13 (a) Electric field intensity and index of refraction in the RPG structure of the MECSEL as it is grown plotted over the distance in the chip. The five QW packages, each consisting of four QWs, are located at the antinodes of the electric field standing wave and (b) the corresponding SEM (scanning electron microscope) picture of the whole active region with the 200 nm thick AlAs process layer. Source: Adapted from [85].

which leads to a designed thickness of 587 nm. The gain region (detailed scheme shown in Figure 7.13a, the corresponding SEM picture of the whole active region as grown is shown in Figure 7.13b), is built up very similar to the gain region of a standard 5 × 4 QW-VECSEL [9]. The only differences are lattice-matched 12-nm-thick AlInP electron or charge carrier diffusion blocking layers followed by a 10-nm thick $Al_{0.1}GaInP$ capping layer on each side of the active region enclosing the structure. The $Al_{0.1}GaInP$ capping layer acts as protection layer to prevent oxidation as well as an etch stop layer for the selective etching of a 200-nm-thick AlAs layer, which separates the active region from the substrate. Stopping a wet-chemically HF acid etching process, a non-aluminum-containing material depicts the best choice. This is represented here by GaInP. Due to the fact that the gain delivering QWs consist of GaInP, capping layers of the same material act as absorbers for our aimed emission wavelength. To prevent absorption, $Al_{0.1}GaInP$ with a bandgap of around 640 nm is chosen as capping and etch stop material. Its aluminum content of 2.59% is still low enough to be resistive against a short exposure of process acids and to act as oxidation protection layer in regular laboratory atmosphere. A wet chemical process (more detailed descriptions can be found elsewhere [85, 86]) is applied to remove the substrate and to finally isolate the semiconductor active region membrane. A free-standing pice of the gain membrane can be seen in Figure 7.14. A small isolated piece of the membrane (∼ 0.25 mm²) is then transferred to one of the diamond heat spreaders. The other diamond heat spreader is placed onto the membrane, which is sticking already to the first diamond. The whole package is mechanically squeezed in a membrane holder made of brass shown in Figure 7.15b. This device is also designed to hold the heat spreader sandwiched active region during laser operation and to act as the heat sink. The mechanical squeezing leads to enforced good thermal contact between the membrane and the heat spreaders. A close look at the photograph reveals two important details. No damages aside from some cracks in the upper left corner and a small damaged area in the center top are visible. Furthermore, the surface of the membrane seems to be totally flat. This represents the most important

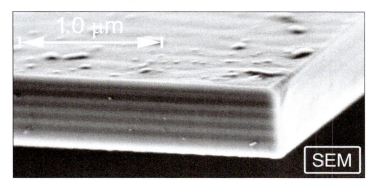

Figure 7.14 SEM picture of the gain membrane, taken from a free-standing piece sticking to a sample carrier. Dirt particles are visible on the sample surface. The QW packages appearing as lighter stripes are clearly visible. The thickness of the membrane is 590 nm, measured at an unprocessed cross section (Figure 7.13b) of the sample at several positions. Source: Adapted with permission from [24]. ©The Optical Society.

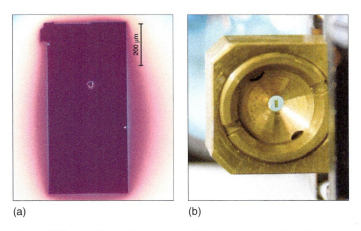

(a) (b)

Figure 7.15 (a) Microscope picture of the semiconductor gain membrane squeezed between two diamonds. The gain membrane is around 380 μm × 840 μm in size. Source: Adapted with permission from [24]. ©The Optical Society. (b) A photograph of the gain element holder. The membrane, shown in (a), can be seen as dark stripe in the center of the aperture. The diameter of the aperture is 1.5 mm. In the background, the bluish shimmering outcoupling mirror is visible. Source: Adapted from [85].

precondition to realize a good bonding. Secondly, the color gradient, induced by the Fabry-Pérot effect between the two diamond heat spreaders, which surrounds the whole membrane, allows the estimation that, if always the same color is found very close to the membrane, it is very homogeneous in flatness.

7.3.2 MECSEL Setup

The laser experiments are performed in a linear concentric resonator (Figure 7.16). The laser mirrors can be adjusted in all degrees of freedom and the sample holder

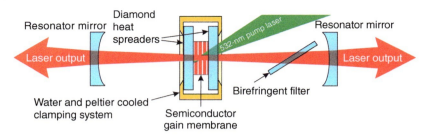

Figure 7.16 Schematic drawing of the MECSEL setup: A linear resonator with a birefringent filter in the long arm of the cavity as it can be seen in the photograph Figure 7.12 (relative dimensions not to scale). Source: Adapted with permission from [24]. ©The Optical Society.

can be shifted in the sample's plane and also tilted in two axes. A 532-nm laser is used as pump irradiating the sample under an angle of 15° to its normal. The two diamond heat spreaders enclosing the semiconductor gain membrane are anti-reflexion coated for the laser emission wavelength on the outer facets. Looking a bit more in detail into the the pump process reveals another major advantage of the MECSEL versus the conventional VECSEL: usually, the DBR has to be designed in such a way that the emitted laser light as well as the unabsorbed pump light are reflected. If, due to epitaxial restrictions (material parameters, strain, etc.), the DBR cannot be fabricated in such a manner, all residual pump light will be absorbed in the DBR. Together with the quantum defect, which is roughly 20% in the AlGaInP material system with a barrier pumping scheme at 532 nm, unwanted heat is produced inside but also close to the active region. This strongly restricts the performance of conventional VECSELs. In the MECSEL, the pump light which is not directly absorbed is only transmitted without disturbing the gain system or it is partly reflected back into the gain membrane on the first diamond–air interface. For characterization purposes, the absorption can be exactly determined by subtracting the transmitted and the reflected pump power from the irradiated pump power (Figure 7.17). About 6.5% of the incident pump power is reflected and about 11.5% is transmitted resulting in an absorption of about 82% of incident pump light in the present example. So, the absorption efficiency η_{abs}, which is a part of the differential efficiency η_{diff} of the laser, can be set to 1 when absorbed power is used for calculations. Therefore, the internal parameters of the MECSEL (efficiencies, etc.) can be determined more accurately. These values are also valid for conventional VECSELs if the different refractive index transition between the DBR and active region instead of diamond and active region is taken into account.

7.3.3 MECSEL Characterization

7.3.3.1 Output Power Measurements

For output power measurements, an output coupler with a reflectivity of $R_{out} = 96\%$ (radius of curvature $r_c = 50$ mm) and a highly reflective mirror ($R > 99.9\%$ for

Figure 7.17 High dynamic range photograph of the gain membrane holder, also shown in Figure 7.15b, while laser operation. The visible beams are named and percentages of the measured transmitted and reflected pump powers are given. Source: Adapted with permission from [24]. ©The Optical Society.

500–760 nm, $r_c = 100$ mm) were chosen. The cavity length was adjusted to roughly $L_{cav.} = 148.5$ mm, which leads to a mode diameter of approximately 76 μm at the beam waist of the resonator. There, the heat spreader sandwiched gain membrane was placed. In terms of mode-matching, the pump spot diameter was adjusted by adapting the distance of the pump lens to 103 mm corresponding to a pump spot diameter of approximately 80 μm in the short axis. The system was operated at a heat sink temperature of 10 °C. Figure 7.18a shows the output power of the MECSEL as a function of the irradiated pump power. A linear input-to-output behavior with a differential efficiency of $\eta_{diff} = 22.3\%$ can be clearly seen. The maximum output power of 595 mW was reached at an incident pump power of 3.67 W with the laser threshold at 1.0 W. A corresponding VECSEL comprising a DBR with an identical active region was tested under the same conditions with respect to pump spot size, mode diameter, outcoupling mirror, and heat sink temperature. The measurement is also plotted in Figure 7.18b. The best differential efficiency achieved here with the green-pumped VECSEL was 18.8% with a threshold pump power of 0.8 W. While the threshold of the MECSEL is slightly higher, probably due to additional losses caused by the additional intracavity heat spreader, η_{diff} is significantly exceeding any differential efficiency published before [9, 14, 64] with green-pumped conventional VECSELs in the AlGaInP material system at the elevated heat sink temperature of $T_{hs} = 10$ °C. Furthermore, the maximum output power for the VECSEL of 570 mW was achieved at 4.6 W of incident pump power where the thermal rollover already led to a saturation of the output power. For the MECSEL, a maximum of 595 mW was already reached at 3.7 W incident pump power before an output power breakdown was visible. The not-prefigured breakdown of the MECSEL indicates a sudden loss of thermal contact between the diamond heat spreaders and the semiconductor membrane. This is conceivable because a professional "bonding" procedure [87–89] was not applied here.

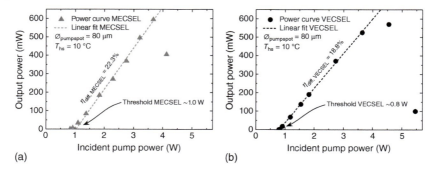

Figure 7.18 (a) Output power plotted over incident pump power of the MECSEL and (b) the corresponding VECSEL. The heat sink temperature was 10 °C and the pump spot diameter was adjusted to approximately 80 μm. Source: Adapted with permission from [24]. ©The Optical Society.

7.3.3.2 Beam Profile and Beam Quality Factor

Figure 7.19a shows the beam profile, recorded with a CMOS camera at a distance of 20 cm behind the $r_c = 100$ mm highly reflective mirror. The carefully adjusted resonator delivers a fundamental Gaussian TEM_{00} mode and the cross sections are plotted in Figure 7.19b. The beam quality factor of $M^2 < 1.06$ was measured at 160 mW output power (see Figure 7.20). The measurement device was a Coherent ModeMaster™ with a given inaccuracy of ±5%.

7.3.3.3 Spectra

Figure 7.21a (upper spectrum) shows a typical spectrum of the free-running (no birefringent filter or etalon in the cavity) MECSEL recorded during the power measurement. The spectral width of laser emission of the corresponding free-running VECSEL is below 2 nm and can be seen in Figure 7.21a (lower spectrum). For the MECSEL, we observed simultaneous laser emission in a range of more than 6.5 nm. This can be explained by the reduction of the subcavity effect (DBR and

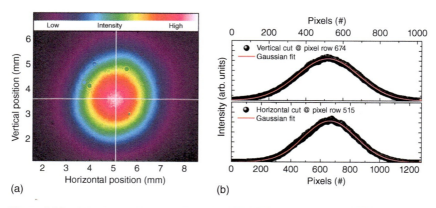

Figure 7.19 (a) A typical beam profile of the MECSEL demonstrating a TEM_{00} mode (Source: adapted with permission from [24]. ©The Optical Society.) and (b) a vertical and horizontal cut through the intensity (a) plotted over to the CMOS camera chip position corresponding pixels with an overlying Gaussian fit curve.

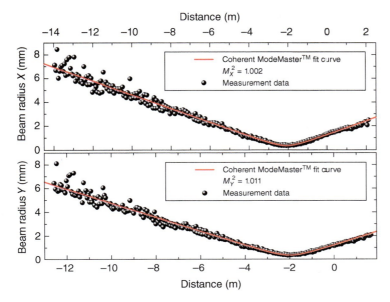

Figure 7.20 Beam propagation plot (beam radii versus distance) of the Coherent ModeMaster for the external beam after collimation with a 300-mm lens. Source: Adapted with permission from [24]. ©The Optical Society.

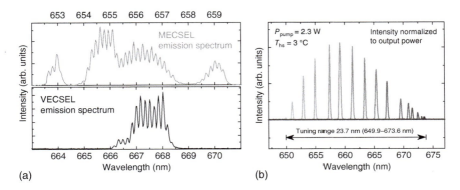

Figure 7.21 (a) A typical spectrum of the free-running MECSEL (upper graph) at around 3.2 W of incident pump power and a typical spectrum of the corresponding and free-running VECSEL (lower graph) at around 3.2 W of incident pump power. (b) Several exemplary laser emission spectra plotted versus wavelength using a 1-mm-thick birefringent filter for tuning. The measurements were performed with two highly reflective resonator mirrors. The intensities of the laser spectra are normalized to the measured output powers with a maximum of 2 mW around 660 nm. Source: Adapted with permission from [24]. ©The Optical Society.

semiconductor–heat spreader interface) [8], which is indicated by the typical cavity dip occurring in reflectivity measurements of conventional VECSELs [2]. In a MECSEL, this narrowing and frequency preselective part is missing and the gain bandwidth, if properly pumped, can show more of its potential. A closer inspection of the spectrum reveals further details. One is the Fabry-Pérot oscillation, visible in both MECSEL and VECSEL spectra in Figure 7.21a with a spacing of $\Delta\lambda_{FSR} \approx 0.16$ nm, which is impressed onto the whole emission spectrum due to the two approximately 550-µm-thick intracavity diamond heat spreaders. Although the diamonds used for the MECSEL are anti-reflexion coated on one side each, and in contact with the semiconductor membrane on the other, their impact is still large enough to also show the diamond introduced Fabry-Pérot oscillation. Another point is the beat note that can be identified in the MECSEL's spectrum in Figure 7.21a (upper graph). It originates from slightly different thicknesses of the two diamond heat spreaders. One is measured to be (540 ± 1) µm and the other is (554 ± 1) µm in thickness. The Fabry-Pérot effect and therefore the beat note can be avoided by the use of wedged heat spreaders or a tilt of the whole gain package preventing an overlap of internal reflections. To obtain information about the spectral range of the gain delivered by the membrane, a wavelength tuning measurement is of interest. Especially, the reduced active region subcavity enhancement [8] of the MECSEL (two times diamond/active region interface instead of the combination of diamond/active region and active region/DBR) possibly supports a larger spectral width of amplification. Therefore, a set of spectra was taken at a heat sink temperature of 3 °C and at an absorbed pump power of $P_{abs} = 1.73$ W (see Figure 7.21). Broadband highly reflective mirrors ($R > 99.95\%$ for 640–700 nm) in a 50–150 mm-cavity (see Figure 7.16) were used. For wavelength tuning, a 1-mm birefringent filter was adjusted at Brewster's angle inside the cavity and rotated around the normal axis of its surface to perform the spectral shift. The measurements reveal a tuning range of nearly 24 nm (649.9–673.6 nm), which is the highest value achieved in this spectral range by semiconductor lasers to date. The conventional VECSEL showed a tuning range of ~22 nm (656–678 nm) under the same conditions but at a heat sink temperature of $T_{hs} = 10$ °C. The tuning measurements revealed that the spectral range of the MECSEL is about 5 nm blue-shifted compared to the conventional VECSEL, which cannot be completely explained by the 7 °C lower heat sink temperature. The spectral shift with changing heat sink temperature was measured for the MECSEL to be 0.125 nm/K via measuring the shift of the long wavelength edge of the PL curve at the half maximum. This is in good accordance to earlier work on VECSELs (0.17 nm/K [44]) as the spectral shift with temperature is a material intrinsic parameter. From a change in the heat sink temperature of 7 °C, a spectral shift of only ~0.875 nm can be expected. According to earlier work [11], there is also the possibility that strain, which sums up when stacking the QW packages during growth and therefore causing a red shift of several nanometers, relaxes again when the gain membrane is released from its substrate. Further investigations are necessary here to fully resolve the effects connected to strain relaxation after processing and external stress [90] applied by the transparent diamond heat spreaders.

7.4 Conclusions

In the first part of this chapter, AlGaInP-based VECSELs are introduced and a chronological overview of research and development on red-emitting semiconductor disk lasers is given. AlGaInP-VECSELs have undergone a continuous process of improvement in recent years. The in overall improvement lead to more than a doubling of continuous-wave output power, namely 2.5 W, mainly driven by research on optimizing the QW arrangement and the reduction of heat deposition via QW and multi-pass pumping. The latter is also responsible for a drastic improvement of intracavity second-harmonic generation.

In the second part of this chapter, a novel laser system is presented – the semiconductor MECSEL. In this first approach, the concept shows its potential such as nearly room temperature operation of an AlGaInP-based gain membrane with improved slope efficiency at the same time. High output power of nearly 600 mW was achieved at a heat sink temperature of $T_{hs} = 10\,°C$. The MECSEL concept of sandwiching a semiconductor gain membrane between two transparent heat spreaders, while maintaining excellent beam properties, complements the achievements described in the first part of this chapter by a further aspect – the improvement of heat dissipation. To fully exploit the indubitable huge potential of this new semiconductor laser concept and to push its limits, further development must be pursued. Independent of the applied material system, this is the case as latest results show. Other heat spreader materials such as sapphire and SiC [91–93] are also under investigation, and a low thermal resistance [94] confirms the potential for power scaling of such systems. The possibility of applying pump light from both sides of the gain membrane element [95] leads to further potential for power scaling and brings the MECSEL system technically very close to classical solid-state lasers.

References

1 Müller, M.I. (2004). Optisch gepumpte oberflächenemittierende Halbleiterlaser mit externem Resonator für 650 nm auf der basis von InGaAlP. Ph.D. thesis, Friedrich-Alexander-Universität Ernangen-Nürnberg.

2 Müller, M., Linder, N., Karnutsch, C. et al. (2002). Optically pumped semiconductor thin-disk laser with external cavity operating at 660 nm. *Proceedings of SPIE* 4649: 265–271. doi: 10.1117/12.469242.

3 Müller, M., Karnutsch, C., Luft, J. et al. (2002). Optically pumped vertical external cavity semiconductor thin-disk laser with CW operation at 660 nm. In: *Compound Semiconductors 2002*, vol. 174 (eds. M. Ilegems, G. Weimann, and J. Wagner), pp. 427–430. Institute of Physics.

4 Hastie, J.E., Calvez, S., Sun, H. et al. (2005). High power, continuous wave operation of a vertical external cavity surface emitting laser at 674 nm. In: Advanced Solid-State Photonics, p. WC2. Optical Society of America. doi: 10.1364/assp.2005.117.

5 Hastie, J.E., Calvez, S., Dawson, M. et al. (2005). High power CW red VECSEL with linearly polarized TEM_{00} output beam. *Optics Express* 13 (1): 77–81. doi: 10.1364/opex.13.000077.

6 Morton, L.G., Hastie, J.E., Dawson, M.D. et al. (2006). 1 W CW red VECSEL frequency-doubled to generate 60 mW in the ultraviolet. In: *Conference on Lasers and Electro-Optics, 2006 and 2006 Quantum Electronics and Laser Science Conference. CLEO/QELS 2006*, pp. 1–2. doi: 10.1109/CLEO.2006.4628732.

7 Kozlovskii, V.I., Lavrushin, B.M., Skasyrsky, Y.K., and Tiberi, M.D. (2009). Vertical-external-cavity surface-emitting 625-nm laser upon optical pumping of an InGaP/AlGaInP nanostructure with a Bragg mirror. *Quantum Electronics* 39 (8): 731. doi: 10.1070/qe2009v039n08abeh014066.

8 Calvez, S., Hastie, J.E., Guina, M. et al. (2009). Semiconductor disk lasers for the generation of visible and ultraviolet radiation. *Laser & Photonics Review* 3 (5): 407–434. doi: 10.1002/lpor.200810042.

9 Schwarzbäck, T., Eichfelder, M., Schulz, W.M. et al. (2011). Short wavelength red-emitting AlGaInP-VECSEL exceeds 1.2 W continuous-wave output power. *Applied Physics B: Lasers and Optics* 102 (4): 789. doi: 10.1007/s00340-010-4213-5.

10 Schwarzbäck, T., Kahle, H., Jetter, M., and Michler, P. (2012). Red AlGaInP-VECSEL emitting at around 665 nm: strain compensation and performance comparison of different epitaxial designs. *Proceedings of SPIE* 8432: 843 209–843 209-7. doi: 10.1117/12.922593.

11 Schwarzbäck, T., Kahle, H., Jetter, M., and Michler, P. (2013). Strain compensation techniques for red AlGaInP-VECSELs: performance comparison of epitaxial designs. *Journal of Crystal Growth* 370: 208–211. doi: 10.1016/j.jcrysgro.2012.09.051.

12 Ranta, S., Härkönen, A., Leinonen, T. et al. (2013). Mode-locked VECSEL emitting 5 ps pulses at 675 nm. *Optics Letters* 38 (13): 2289–2291. doi: 10.1364/OL.38.002289.

13 Bek, R., Kahle, H., Schwarzbäck, T. et al. (2013). Mode-locked red-emitting semiconductor disk laser with sub-250 fs pulses. *Applied Physics Letters* 103 (24): 242101. doi: 10.1063/1.4835855.

14 Schwarzbäck, T., Bek, R., Hargart, F. et al. (2013). High-power InP quantum dot based semiconductor disk laser exceeding 1.3 w. *Applied Physics Letters* 102 (9): 092101. doi: 10.1063/1.4793299.

15 Rantamäki, A., Sirbu, A., Mereuta, A. et al. (2010). 3 W of 650 nm red emission by frequency doubling of wafer-fused semiconductor disk laser. *Optics Express* 18 (21): 21 645–21 650. doi: 10.1364/oe.18.021645.

16 Rautiainen, J., Okhotnikov, O.G., Eger, D. et al. (2009). Intracavity generation of 610 nm light by periodically poled near-stoichiometric lithium tantalate. *Electronics Letters* 45 (3): 177–179. doi: 10.1049/el:20093123.

17 Rautiainen, J., Härkönen, A., Korpijärvi, V.-M. et al. (2009). Red and UV generation using frequency-converted GaInNAs-based semiconductor disk laser. In: *Conference on Lasers and Electro-Optics, 2009 and 2009 Conference on Quantum electronics and Laser Science Conference (CLEO/QELS 2009)*, pp. 1–2.

18 Härkönen, A., Rautiainen, J., Guina, M. et al. (2007). High power frequency doubled GaInNAs semiconductor disk laser emitting at 615 nm. *Optics Express* 15 (6): 3224–3229. doi: 10.1364/oe.15.003224.

19 Leinonen, T., Penttinen, J.-P., Korpijärvi, V.-M. et al. (2015). > 8 W GaInNAs VECSEL emitting at 615 nm. *Proceedings of SPIE* 9349: 9349–9349-6. doi: 10.1117/12.2079162.

20 Kantola, E., Leinonen, T., Penttinen, J.-P. et al. (2015). 615 nm GaInNAs VECSEL with output power above 10 W. *Optics Express* 23 (16): 20 280–20 287. doi: 10.1364/OE.23.020280.

21 Morton, L.G., Foreman, H.D., Hastie, J.E. et al. (2007). Actively stabilised single-frequency red VECSEL. In: *Advanced Solid-State Photonics*, p. WB7. Optical Society of America. doi: 10.1364/assp.2007.wb7.

22 Paboeuf, D., Schlosser, P.J., and Hastie, J.E. (2013). Frequency stabilization of an ultraviolet semiconductor disk laser. *Optics Letters* 38 (10): 1736–1738. doi: 10.1364/OL.38.001736.

23 Paboeuf, D. and Hastie, J.E. (2016). Tunable narrow linewidth AlGaInP semiconductor disk laser for Sr atom cooling applications. *Applied Optics* 55 (19): 4980–4984. doi: 10.1364/AO.55.004980.

24 Kahle, H., Mateo, C.M.N., Brauch, U. et al. (2016). Semiconductor membrane external-cavity surface-emitting laser (MECSEL). *Optica* 3 (12): 1506–1512. doi: 10.1364/OPTICA.3.001506.

25 O'Reilly, E.P. and Adams, A.R. (1994). Band-structure engineering in strained semiconductor lasers. *IEEE Journal of Quantum Electronics* 30 (2): 366–379. doi: 10.1109/3.283784.

26 Adams, A.R. (2011). Strained-layer quantum-well lasers. *IEEE Journal of Selected Topics in Quantum Electronics* 17 (5): 1364–1373. doi: 10.1109/JSTQE.2011.2108995.

27 Rosencher, E. and Vinter, B. (2002). *Optoelectronics*. Cambridge: Cambridge University Press.

28 Matthews, J.W. and Blakeslee, A.E. (1974). Defects in epitaxial multilayers: I. misfit dislocations. *Journal of Crystal Growth* 27: 118–125. doi: 10.1016/S0022-0248(74)80055-2.

29 Winterhoff, R. (1998). Kurzwellige GaInP Quantenfilmlaser. Ph.D. thesis. University of Stuttgart.

30 Butendeich, R. (2003). Epitaxie von vertikal emittierenden (Ga,In)P-Lasern. Ph.D. thesis. Universität Stuttgart.

31 Koroknay, E. (2013). Epitaxial processes for low density quantum dots in III/V semiconductors. Ph.D. thesis. University of Stuttgart, Institite for Semiconductor Optics and Functional Interfaces.

32 Barth, F.H. (1995). GaInP Quantenfilmlaser für den sichtbaren Spektralbereich. Ph.D. thesis. University of Stuttgart.

33 Krijn, M.P.C.M. (1991). Heterojunction band offsets and effective masses in III-V quaternary alloys. *Semiconductor Science and Technology* 6 (1): 27.

34 Bour, D.P., Geels, R.S., Treat, D.W. et al. (1994). Strained $Ga_xIn_{1-x}P/(AlGa)_{0.5}In_{0.5}P$ heterostructures and quantum-well laser diodes. *IEEE Journal of Quantum Electronics* 30 (2): 593–607. doi: 10.1109/3.283808.

35 Novák, J., Hasenöhrl, S., Alonso, M.I., and Garriga, M. (2001). Influence of tensile and compressive strain on the band gap energy of ordered InGaP. *Applied Physics Letters* 79 (17): 2758–2760. doi: 10.1063/1.1413725.

36 Numai, T. (2015). *Fundamentals of Semiconductor Lasers*, vol. 93. Japan: Springer. doi: 10.1007/978-4-431-55148-5.

37 Yablonovitch, E. and Kane, E. (1986). Reduction of lasing threshold current density by the lowering of valence band effective mass. *Journal of Lightwave Technology* 4 (5): 504–506. doi: 10.1109/JLT.1986.1074751.

38 Adams, A.R. (1986). Band-structure engineering for low-threshold high-efficiency semiconductor lasers. *Electronics Letters* 22 (5): 249–250. doi: 10.1049/el:19860171.

39 Suemune, I. (1991). Theoretical study of differential gain in strained quantum well structures. *IEEE Journal of Quantum Electronics* 27 (5): 1149–1159. doi: 10.1109/3.83371.

40 Kamiyama, S., Uenoyama, T., Mannoh, M. et al. (1994). Analysis of GaInP/AlGaInP compressive strained multiple-quantum-well laser. *IEEE Journal of Quantum Electronics* 30 (6): 1363–1369. doi: 10.1109/3.299458.

41 Moritz, A. and Hangleiter, A. (1995). Optical gain in ordered GaInP/AlGaInP quantum wells. *Applied Physics Letters* 66 (24): 3340–3342. doi: 10.1063/1.113750.

42 Moritz, A., Wirth, R., Heppel, S. et al. (1997). Intrinsic modulation bandwidth of strained GaInP/AlGaInP quantum well lasers. *Applied Physics Letters*, 71 (5): 650–652. doi: 10.1063/1.119818.

43 Chow, W.W., Choquette, K.D., Crawford, M.H. et al. (1997). Design, fabrication, and performance of infrared and visible vertical-cavity surface-emitting lasers. *IEEE Journal of Quantum Electronics* 33 (10): 1810–1824. doi: 10.1109/3.631287.

44 Schwarzbäck,T. (2013). Epitaxie AlGaInP-basierter Halbleiterscheibenlaser: Dauerstrichbetrieb, Frequenzverdopplung und Modenkopplung. Ph.D. thesis, University of Stuttgart, Institite for Semiconductor Optics and Functional Interfaces.

45 Schwarzbäck, T., Kahle, H., Eichfelder, M. et al. (2011). Wavelength tunable ultraviolet laser emission via intra-cavity frequency doubling of AlGaInP vertical external-cavity surface-emitting laser down to 328 nm. *Applied Physics Letters* 99 (26): 261101. doi: 10.1063/1.3660243.

46 Raja, M.Y.A., Brueck, S.R.J., Osinski, M. et al. (1989). Resonant periodic gain surface-emitting semiconductor lasers. *IEEE Journal of Quantum Electronics* 25 (6): 1500–1512. doi: 10.1109/3.29287.

47 Corzine, S.W., Geels, R.S., Scott, J.W. et al. (1989). Design of fabry-perot surface-emitting lasers with a periodic gain structure. *IEEE Journal of Quantum Electronics* 25 (6): 1513–1524. doi: 10.1109/3.29288.

48 Coldren, L.A. and Corzine, S.W. (1995). *Diode Lasers and Photonic Integrated Circuits*, 1e. New York: Wiley.

49 Liau, Z.L. (2000). Semiconductor wafer bonding via liquid capillarity. *Applied Physics Letters* 77 (5): 651–653. doi: 10.1063/1.127074.

50 Laurain, A., Hader, J., and Moloney, J.V. (2019). Modeling and optimization of transverse modes in vertical-external-cavity surface-emitting lasers. *Journal of the Optical Society of America B* 36 (4): 847–854. doi: 10.1364/JOSAB.36.000847.

51 Findlay, D. and Clay, R. (1966). The measurement of internal losses in 4-level lasers. *Physics Letters* 20 (3): 277–278. doi: 10.1016/0031-9163(66)90363-5.

52 Kuznetsov, M., Hakimi, F., Sprague, R., and Mooradian, A. (1997). High-power (> 0.5-W CW) diode-pumped vertical-external-cavity surface-emitting semiconductor lasers with circular TEM_{00} beams. *IEEE Photonics Technology Letters* 9 (8): 1063–1065. doi: 10.1109/68.605500.

53 Kuznetsov, M., Hakimi, F., Sprague, R., and Mooradian, A. (1999). Design and characteristics of high-power (> 0.5-W CW) diode-pumped vertical-external-cavity surface-emitting semiconductor lasers with circular TEM_{00} beams. *IEEE Journal of Selected Topics in Quantum Electronics* 5 (3): 561–573. doi: 10.1109/2944.788419.

54 Fu, L. (2002). Degradation and reliability of GaInP/AlGaInP quantum well lasers. Ph.D. thesis, University of Stuttgart.

55 Smith, A., Hastie, J.E., Foreman, H.D. et al. (2008). GaN diode-pumping of red semiconductor disk laser. *Electronics Letters* 44 (20): 1195–1196. doi: 10.1049/el:20081435.

56 Smith, A., Hastie, J.E., Kemp, A.J. et al. (2008). GaN diode-pumping of a red semiconductor disk laser. In: *LEOS 2008 – 21st Annual Meeting of the IEEE Lasers and Electro-Optics Society*, pp. 404–405. doi: 10.1109/LEOS.2008.4688661.

57 Baumgärtner, S., Kahle, H., Bek, R. et al. (2015). Comparison of AlGaInP-vecsel gain structures. *Journal of Crystal Growth* 414: 219–222. doi: 10.1016/j.jcrysgro.2014.10.016.

58 Heinen, B., Wang, T.L., Sparenberg, M. et al. (2012). 106 W continuous-wave output power from vertical-external-cavity surface-emitting laser. *Electronics Letters* 48 (9): 516–517. doi: 10.1049/el.2012.0531.

59 Mateo, C.M.N., Brauch, U., Schwarzbäck, T. et al. (2015). Enhanced efficiency of AlGaInP disk laser by in-well pumping. *Optics Express* 23 (3): 2472–2486. doi: 10.1364/OE.23.002472.

60 Mateo, C.M.N., Brauch, U., Kahle, H. et al. (2016). 2.5 W continuous wave output at 665 nm from a multipass and quantum-well-pumped AlGaInP vertical-external-cavity surface-emitting laser. *Optics Letters* 41 (6): 1245–1248. doi: 10.1364/OL.41.001245.

61 Mateo, C.M.N., Brauch, U., Kahle, H. et al. (2016). Efficiency and power scaling of in-well and multi-pass pumped algainp-vecsels. *Proceedings of SPIE* 9734: 973410–973410-7. doi: 10.1117/12.2212162.

62 Giesen, A., Hügel, H., Voss, A. et al. (1994). Scalable concept for diode-pumped high-power solid-state lasers. *Applied Physics B: Lasers and Optics* 58 (5): 365–372. doi: 10.1007/BF01081875.

63 Hastie, J.E., Morton, L.G., Kemp, A.J. et al. (2006). High power ultraviolet vecsel through intra-cavity frequency-doubling in bbo. In: *Conference Digest.*

IEEE 20th International Semiconductor Laser Conference, pp. 109–110. doi: 10.1109/ISLC.2006.1708110.

64 Hastie, J.E., Morton, L.G., Kemp, A.J. et al. (2006). Tunable ultraviolet output from an intracavity frequency-doubled red vertical-external-cavity surface-emitting laser. *Applied Physics Letters* 89 (6): 061114. doi: 10.1063/1.2236108.

65 Kahle, H., Baumgärtner, S., Sauter, F. et al. (2015). High-power (>400 mW) laser emission at 332 nm of frequency-doubled, optically pumped AlGaInP disk laser with an optimized quantum well structure. In: *2015 European Conference on Lasers and Electro-Optics – European Quantum Electronics Conference*, p. CB_3_4. Optical Society of America.

66 Basov, N., Bogdankevich, O., and Grasyuk, A. (1966). Semiconductor lasers with radiating mirrors. *IEEE Journal of Quantum Electronics* 2 (9): 594–597. doi: 10.1109/JQE.1966.1074111.

67 Maclean, A.J., Birch, R.B., Roth, P.W. et al. (2009). Limits on efficiency and power scaling in semiconductor disk lasers with diamond heatspreaders. *Journal of the Optical Society of America B*. 26 (12): 2228–2236. doi: 10.1364/JOSAB.26.002228.

68 Piskorski, Ł., Sarzała, R.P., and Nakwaski, W. (2008). Analysis of anticipated performance of 650-nm GaInP/AlGaInP quantum-well GaAs-based VCSELs at elevated temperatures. *Opto-Electronics Review* 16: 34–41. doi: 10.2478/s11772-007-0027-3.

69 Afromowitz, M.A. (1973). Thermal conductivity of $Ga_{1-x}Al_xAs$ alloys. *Journal of Applied Physics* 44 (3): 1292–1294. doi: 10.1063/1.1662342.

70 Kemp, A.J., Valentine, G.J., Hopkins, J.M. et al. (2005). Thermal management in vertical-external-cavity surface-emitting lasers: finite-element analysis of a heat-spreader approach. *IEEE Journal of Quantum Electronics* 41 (2): 148–155. doi: 10.1109/JQE.2004.839706.

71 Zhang, P., Jiang, M., Zhue, R. et al. (2017). Thermal conductivity of GaAs/AlAs distributed Bragg reflectors in semiconductor disk laser: comparison of molecular dynamics simulation and analytic methods. *Applied Optics* 56 (15): 4537–4542. doi: 10.1364/AO.56.004537.

72 Huo, Y., Cho, C.Y., Huang, K.F. et al. (2019). Exploring the DBR superlattice effect on the thermal performance of a VECSEL with the finite element method. *Optics Letters* 44 (2): 327–330. doi: 10.1364/OL.44.000327.

73 Holl, P., Rattunde, M., Adler, S. et al. (2015). Recent advances in power scaling of GaSb-based semiconductor disk lasers. *IEEE Journal of Selected Topics in Quantum Electronics* 21 (6): 324–335. doi: 10.1109/JSTQE.2015.2414919.

74 Gerster, E., Ecker, I., Lorch, S. et al. (2003). Orange-emitting frequency-doubled GaAsSb/GaAs semiconductor disk laser. *Journal of Applied Physics* 94 (12): 7397–7401. doi: 10.1063/1.1625784.

75 Rantamäki, A., Saarinen, E., Lyytikäinen, J. et al. (2015). Thermal management in long-wavelength flip-chip semiconductor disk lasers. *IEEE Journal of Selected Topics in Quantum Electronics* 21 (6): 1–7. doi: 10.1109/JSTQE.2015.2420599.

76 Rantamäki, A., Saarinen, E.J., Lyytikäinen, J. et al. (2014). High power semiconductor disk laser with a semiconductor-dielectric-metal compound mirror. *Applied Physics Letters* 104 (10): 101110. doi: 10.1063/1.4868535.
77 Gbele, K., Laurain, A., Hader, J. et al. (2016). Design and fabrication of hybrid metal semiconductor mirror for high-power vecsel. *IEEE Photonics Technology Letters* 28 (7): 732–735. doi: 10.1109/LPT.2015.2507059.
78 Sandusky, J.V. and Brueck, S.R.J. (1996). A CW external-cavity surface-emitting laser. *IEEE Photonics Technology Letters* 8 (3): 313–315. doi: 10.1109/68.481101.
79 Brueck, S.R.J. and Sandusky, J.V. (1996). Correction to "A CW external-cavity surface-emitting laser." *IEEE Photonics Technology Letters* 8 (9): 1277. doi: 10.1109/LPT.1996.531862.
80 Mooradian, A. (1992). External cavity semiconductor laser system. *US Patent*, **Appl. No.: 654,798** (Patent No.: 5,131,002).
81 Iakovlev, V., Walczak, J., Gębski, M. et al. (2014). Double-diamond high-contrast-gratings vertical external cavity surface emitting laser. *Journal of Physics D: Applied Physics* 47 (6): 065 104. doi: 10.1088/0022-3727/47/6/065104.
82 Yang, Z., Albrecht, A.R., Cederberg, J.G., and Sheik-Bahae, M. (2015). DBR-free optically pumped semiconductor disk lasers. *Proceedings of SPIE* 9349: 934 905–934 905-6. doi: 10.1117/12.2079696.
83 Yang, Z., Albrecht, A.R., Cederberg, J.G., and Sheik-Bahae, M. (2015). Optically pumped DBR-free semiconductor disk lasers. *Optics Express* 23 (26): 33 164–33 169. doi: 10.1364/OE.23.033164.
84 Santos, J.M.M., Watson, S., Guilhabert, B. et al. (2015). MQW nanomembrane assemblies for visible light communications. In: *2015 IEEE Photonics Conference (IPC)*, pp. 523–524. doi: 10.1109/IPCon.2015.7323489.
85 Kahle, H., Mateo, C.M.N., Brauch, U. et al. (2017). The optically pumped semiconductor membrane external-cavity surface-emitting laser (MECSEL): a concept based on a diamond-sandwiched active region. *Proceedings of SPIE* 10087: 10 087–10 087-13. doi: 10.1117/12.2252182.
86 Cederberg, J., Albrecht, A., Ghasemkhani, M. et al. (2014). Growth and testing of vertical external cavity surface emitting lasers (VECSELs) for intracavity cooling of Yb:YLF. *Journal of Crystal Growth* 393: 28–31. doi: 10.1016/j.jcrysgro.2013.09.042.
87 Plößl, A. and Kräuter, G. (1999). Wafer direct bonding: tailoring adhesion between brittle materials. *Materials Science and Engineering R: Reports* 25 (1–2): 1–88. doi: 10.1016/S0927-796X(98)00017-5.
88 Gösele, U., Tong, Q.Y., Schumacher, A. et al. (1999). Wafer bonding for microsystems technologies. *Sensors and Actuators A: Physical* 74 (1–3): 161–168. doi: 10.1016/S0924-4247(98)00310-0.
89 Alexe, M. and Gösele, U. (eds.) (2004). *Wafer Bonding: Applications and Technology*, no. 75 in Springer Series in Materials Science, 1e. Berlin, Heidelberg: Springer-Verlag. doi: 10.1007/978-3-662-10827-7.
90 Kowalski, O.P., Cockburn, J.W., Mowbray, D.J. et al. (1995). GaInP-AlGaInP band offsets determined from hydrostatic pressure measurements. *Applied Physics Letters* 66 (5): 619–621. doi: 10.1063/1.114032.

91 Mirkhanov, S., Quarterman, A.H., Kahle, H. et al. (2017). DBR-free semiconductor disc laser on SiC heatspreader emitting 10.1 W at 1007 nm. *Electronics Letters* 53 (23): 1537–1539. doi: 10.1049/el.2017.2689.

92 Yang, Z., Follman, D., Albrecht, A.R. et al. (2018). 16 W DBR-free membrane semiconductor disk laser with dual-SiC heatspreader. *Electronics Letters* 54 (7): 430–432. doi: 10.1049/el.2018.0101.

93 Kahle, H., Penttinen, J.-P., Phung, H.-M. et al. (2019). MECSELs with direct emission in the 760 nm to 810 nm spectral range: a single- and double-side pumping comparison and high-power continuous-wave operation. *Proceedings of SPIE* 10901: 109 010D. doi: 10.1117/12.2512111.

94 Broda, A., Kuźmicz, A., Rychlik, G. et al. (2017). Highly efficient heat extraction by double diamond heat-spreaders applied to a vertical external cavity surface-emitting laser. *Optical and Quantum Electronics* 49 (9): 287. doi: 10.1007/s11082-017-1129-x.

95 Kahle, H., Penttinen, J.-P., Phung, H.-M. et al. (2019). Comparison of single-side and double-side pumping of membrane external-cavity surface-emitting lasers. *Optics Letters* 44 (5): 1146–1149. doi: 10.1364/OL.44.001146.

Part II

Mode-Locked VECSEL

8

Recent Advances in Mode-Locked Vertical-External-Cavity Surface-Emitting Lasers

Anne C. Tropper

Physics and Astronomy, Quantum, Light and Matter Group, University of Southampton, Southampton, UK

8.1 Introduction

The flourishing young field of ultrafast semiconductor disk lasers was comprehensively reviewed in 2010 by Thomas Südmeyer and his colleagues [1]. They noted that "tremendous progress" had already been made and defined the essential research task for this field; to push the performance of these lasers toward the high levels of peak power needed for applications such as frequency metrology and biomedical microscopy. Seven years on, significant new milestones on the road to application have indeed been passed by the mode-locked vertical-external-cavity semiconductor laser, or ML-VECSEL, and its compact cousin the mode-locked integrated external-cavity surface-emitting laser, or MIXSEL. At the time of the 2010 review, there were just two reports of sub-300-fs pulse generation from an ML-VECSEL in the literature [2, 3]; with peak pulse powers limited to a few 100 W. Since then we have seen pulse durations reach the 100-fs mark [4, 5]; average power scale up to the few-watt regime [6–8] (although not yet with the shortest pulse durations); and peak power attain levels of several kW [6, 7, 9]. It is now, indeed, possible to purchase a commercial ultrafast VECSEL system [10].

In this review, I shall outline the advances of the past seven years. For a description of the early development of the field, the first demonstration of passive mode-locking of a VECSEL [11] and the first report of sub-500-fs pulse generation [12], the reader is referred to previous review articles [1, 13–15]. Important new fields that have emerged over these years include the demonstration of red femtosecond VECSELs [16]; the observation of self-mode-locking in VECSELs [17]; and the introduction of colliding pulse mode-locking [18]; and these will be described in depth elsewhere in this volume. In particular, the effort to generate shorter and more intense optical pulses from semiconductor lasers is profoundly informed by new insights gained from rigorous microscopic modeling of semiconductor gain and absorber dynamics, also described in this book. Advances in the computational techniques used to apply these models to ML-VECSELs are helping to identify ultimate performance levels for these lasers [19].

Vertical External Cavity Surface Emitting Lasers: VECSEL Technology and Applications, First Edition.
Edited by Michael Jetter and Peter Michler.
© 2022 WILEY-VCH GmbH. Published 2022 by WILEY-VCH GmbH.

8.1.1 Ultrafast Lasers

The ultrashort pulsed laser has become an indispensable tool of contemporary science and engineering, and there exist mature commercial technologies based on bulk and fiber gain media that can reliably deliver sub-picosecond light pulses for materials processing, medical and biomedical imaging, precision time and frequency measurement, and much else [20]. One invention in particular made these lasers stable and useable: the semiconductor saturable absorber mirror (SESAM) has been used to establish self-starting passive mode-locking in an immense range of diverse laser systems [21].

Passive mode-locking is the name given to the state of a spectrally broadband continuous-wave laser that spontaneously generates a repetitive train of precisely equidistant ultrashort pulses. Nonlinear coupling drives the oscillating longitudinal modes of the laser cavity into fixed-phase relationships, localizing the intracavity light within a time window of reduced loss: a single optical pulse circulates in the cavity and couples to the externally emitted pulse train on each reflection off the output coupler mirror. The bleachable loss that localizes the pulse is very often the saturable absorption of a SESAM; it may also be attenuation at a hard or soft aperture of low-intensity background intracavity light, too weak to induce a Kerr lens. Other interactions that shape the emerging pulse include gain saturation, dispersion, and nonlinear phase modulation. The speed of response of these nonlinearities greatly exceeds what can be achieved by, for example, acousto-optic intracavity loss modulation: thus the very shortest pulses invariably result from passive rather than active mode-locking.

The optical pulse train emitted by a mode-locked laser has a comb-shaped frequency-domain structure defined by two degrees of freedom; the repetition rate, f_{rep}, which is the inverse of the cavity round-trip group delay time T_R; and the carrier-envelope offset frequency, f_0. Figure 8.1 shows schematically how these physical quantities are related to the time variation of the E-field for an idealized noise-free pulse train. A development of immense historic significance has been the invention and proof of schemes for stabilizing all the degrees of freedom of such a frequency comb [22], recognized with the award of the Nobel Prize for Physics in 2005. The need for immensely complex frequency chains with which to relate near-infrared and visible frequencies to microwave standards was superseded: for the first time it was possible to count optical frequencies with a precision approaching 1 part in 10^{18}. We shall return to the topic of frequency combs later in this chapter; for now we note that a particularly lucid and intuitive description of mode-locked pulse trains is found in [23]. It is a tricky measurement challenge, especially for new researchers, to characterize the shortest optical pulses with confidence: the tutorial by Monmayrant et al. may be found particularly helpful [24].

Commercial femtosecond lasers have become robust instruments that offer turn-key operation for nonspecialist users, who can access reliable pulse trains attaining many kilowatts of peak power. They are also large, power-hungry, non-portable, expensive, and suited primarily to the laboratory environment. They address quite specific combinations of operating parameters, with pulse repetition

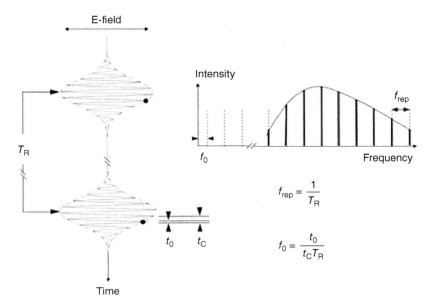

Figure 8.1 Coherent optical pulse train. Left: time-varying E-field. The carrier wave is harmonic with optical period t_c; the pulse envelope (dotted line) is periodic with the round-trip time, T_R. The black dot is a phase marker for a peak in the carrier wave, which shifts relative to the pulse envelope by time t_0 per round trip. The E-field of the pulse train (solid line) repeats after a time $t_c T_R / t_0$. Right: frequency power spectrum of the pulse train, showing a comb of modes, spaced at the pulse repetition rate, f_{rep}, and extrapolating back to the carrier-envelope-offset frequency, f_0, which represents the overall periodicity of the E-field.

rate typically around 100 MHz. For operating wavelengths outside the gain regions of the usual fiber or titanium-doped sapphire systems, nonlinear conversion schemes such as second harmonic generation or optical parametric oscillation are needed, increasing complexity with a cost in overall power conversion efficiency. These limitations have motivated a search for alternative ultrafast systems, with potential for better portability and energy efficiency, as well as wider choice of wavelength and repetition rate. Semiconductor quantum well (QW) and quantum dot (QD) gain media attracted particular interest, as artificial nanofabricated materials, where band-gap engineering techniques might be recruited to enhance gain bandwidth or access desirable wavelengths, perhaps even with current-injection pumping. In the next section, we describe the ML-VECSELs and MIXSELs that exploit these capabilities.

8.1.2 Ultrafast Semiconductor Lasers; Diodes, VECSELs, and MIXSELs

The edge-emitting diode laser has such obvious advantages of tiny size, excellent electrical-to-optical power conversion efficiency, and cheapness of manufacture, that this must be the first direction in which to seek compact alternatives to ultrafast solid-state lasers. A passively mode-locked laser diode is constructed with

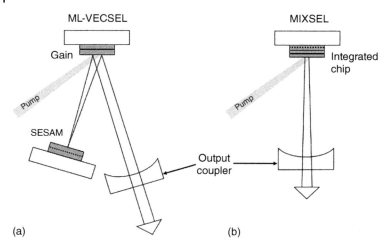

Figure 8.2 Schematic comparison of (a) ML-VECSEL and (b) MIXSEL.

an edge-emitting waveguide fabricated in two sections; a forward-biased section to provide gain, and a reverse-biased saturable absorber section. Devices of this type have been shown to emit pulses of duration ranging from many-picosecond to sub-picosecond, at repetition rates of tens or hundreds of GHz; very promising for application to high-bandwidth optical communications and ultrafast data processing [25]. Their power output, however, is typically tens of mW, with pulse energies limited to the pJ range. Even when fabricated with an integrated amplifier, the peak pulse power of such a device is at most a few W. Mode-locked diodes are therefore unlikely to be suitable for applications such as multiphoton fluorescence spectroscopy, or self-referenced frequency comb generation, where the pulse must be sufficiently intense to drive a nonlinear process. It is passively mode-locked optically pumped semiconductor disk lasers – VECSELs – with the architecture introduced in 1999 by Kuznetsov et al. [26], which offer the most promising route for an all-semiconductor laser system to generate intense pulses. The external cavity of these lasers readily allows the inclusion of a SESAM with which to induce passive mode-locking.

Figure 8.2a shows the layout of a simple SESAM mode-locked VECSEL. The folding mirror of the V-shaped cavity is a gain chip, with light amplification in optically pumped QW or QD layers embedded in a barrier medium just under the surface of the chip, in front of a highly reflecting distributed Bragg reflector (DBR) semiconductor multilayer mirror. The incident pump radiation is absorbed in barrier regions that confer quantum confinement – and correspondingly high intrinsic gain – on the delocalized band-edge states involved in the laser transition. The resulting free carriers must be trapped into the confined states of the QW or QD regions before a population inversion can be established. Both trapping efficiency and gain per carrier are thermally sensitive: the gain is rapidly lost to rising active region temperature, leading to the well-known phenomenon of thermal rollover of the output power characteristic. The creation of a successful gain chip thus draws on three areas of deep expertise; (i) bandgap engineering, and the design of high-gain nanostructures at desired operating wavelengths; (ii) epilayer growth

(MBE or MOCVD), with layer thickness, composition, and crystallinity controlled sufficiently well to realize designed performance; and (iii) post-growth fabrication of the chip, potentially including substrate removal for thermal management, and the application of optical coatings. Much of this is not specific at all to mode-locked VECSELs and has been extensively reviewed [27, 28]. Some aspects of gain chip design specific to ultrafast performance will be discussed in the next section.

For a practical VECSEL source, it is immensely attractive to dispense with the external pump diode and inject carriers electrically into the active region of the gain chip; however, the technical difficulties involved in confining a uniform pump current distribution over the emitting aperture, and combining low electrical impedance with low optical loss, are considerable. For a mode-locked laser in particular, additional cavity loss in the form of free-carrier absorption exacerbates the pulse-lengthening effect of gain filtering and limits the efficiency of the laser. Progress with these electrically pumped EP-VECSELs will be reviewed in depth elsewhere in this book. Work on SESAM mode-locking of these lasers has seen steady improvement of pulse quality [29, 30]. The latest generation of ML-EP-VECSELs has reached pulse durations as short as 2.5 ps, with average output power of more than 50 mW [31].

An important breakthrough for ultrafast semiconductor lasers is represented by the MIXSEL [32], depicted schematically in Figure 8.2b. The functions of gain and saturable absorber are united in one chip, allowing construction of a mode-locked laser with a two-mirror cavity, winning ease of alignment and reduction in cavity size. This must be a promising candidate for a mass-produced ultrafast laser; nevertheless the integrated chip poses formidable design and fabrication challenges. To shape short pulses, the saturable absorber region must be shielded from the light that pumps the gain region by a blocking multilayer; however, this multilayer is inside the mode-locked laser cavity and has the potential to disrupt the pulse formation. Furthermore, the integrated chip must function with the same laser mode area on the gain and on the absorber mirror surfaces. The mode area ratio is an important degree of freedom, used by ML-VECSEL designers to adjust the relative degree of saturation of gain and absorber. In a MIXSEL the mode area ratio is constrained to be unity: relative saturation is controlled by careful distribution of the standing-wave amplitude through precise multilayer design and growth. Perhaps most challenging of all, absorber layer may require a lower growth temperature than the rest of the structure.

This chapter will briefly describe the the physics of pulse formation in VECSELs and MIXSELs and review the most recent advances in power, pulse duration, and repetition rate, as well as some applications of these lasers.

8.2 Ultrafast Pulse Formation in a Surface-Emitting Semiconductor Laser

8.2.1 Surface-Emitting Gain Chip Design

The fundamental task for the designer of an ultrafast semiconductor disk laser chip is to provide a gain bandwidth that is spectrally broad enough to sustain pulses of the desired short duration. The gold standard for ultrafast lasers is set by the green-pumped titanium-doped sapphire crystal, with a gain bandwidth extending over more than 100 THz, able to support few-cycle optical pulse generation. The intrinsic gain bandwidth of a semiconductor quantum well, on the other hand, is closer to 10 THz: the *effective* gain bandwidth of a surface-emitting chip may be significantly less than that, reduced by optical interference effects that control the amplitude of the standing wave in the chip at the position of each quantum well. A chip with a microcavity resonance at the operating wavelength will exhibit sharply peaked gain and strong dispersion of second or higher order, incompatible with ultrashort pulse generation [15].

Recent progress in the generation of sub-200-fs pulses from these lasers has mostly been achieved with the help of dielectric antireflection coatings applied to the gain chip surface, suppressing the microcavity resonance and reducing the dispersion. A simple configuration is shown schematically in Figure 8.3. The active region of the gain chip is a $5\lambda/2$-thick layer of pump-laser-absorbing GaAs, which forms the barrier region for four embedded pairs of InGaAs quantum wells. The laser mode is incident from the left, through a $\lambda/4$ dielectric layer. The squared modulus of the standing-wave E-field inside the chip is plotted in Figure 8.3 for the 1040-nm design wavelength. Fine details of the active region, such as the window and capping layers, and the P-doped strain-balancing layers, have been omitted for clarity.

The spectral variation of reflectivity, group delay dispersion (GDD), reflectivity, and mode overlap may be calculated numerically for a specific multilayer design using matrix methods to represent the multiple boundary constraints: the principles of one possible method for a calculation of this type are described by Tropper and Hoogland [33]. The results for the structure of Figure 8.3 are displayed in Figure 8.4. GDD is plotted as a function of wavelength in Figure 8.4a; over a large part of the stop band of the DBR, it has a near-linear variation passing through zero close to the design wavelength. An expanded view of the curve near to the design wavelength is shown in Figure 8.4b, to which a band of values corresponding to growth thickness

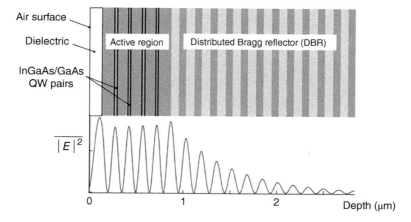

Figure 8.3 Layer design of a broadband gain chip with plot of $\overline{|E|^2}$ as a function of depth for a standing wave at the design wavelength.

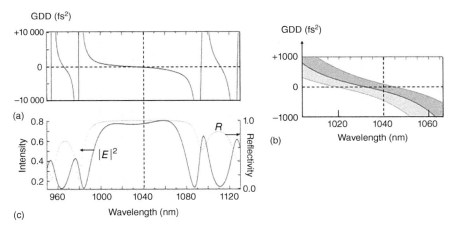

Figure 8.4 Dispersion and reflectivity calculations for the structure of Figure 8.3. (a) GDD as a function of wavelength for a 180-nm region centered on the 1040-nm design wavelength; (b) expanded view of GDD variation near 1040 nm with tolerance corresponding to ±1% growth error; (c) spectrum of reflectivity (dotted) and longitudinal confinement factor (solid).

variations of ±1% have been added, representative of typical minimum machine tolerances. Even exactly on the design wavelength, natural growth variations can evidently introduce GDD ranging from 0 to $-1000\,\text{fs}^2$ per round trip (double pass on gain chip): elsewhere within the stop band the effect is likely to be even bigger. Figure 8.4c shows the reflectivity spectrum of the multilayer, omitting absorption or gain due to the quantum wells. Comparison with Figure 8.4a shows that the outer edges of the stop band are likely to be unsuitable for ultrashort pulse generation due to the high dispersion present there.

In Figure 8.4c the spectral variation of the quantity known as the field enhancement or the longitudinal confinement factor [33] is also shown: this is the average value of $\overline{|E|^2}$ at the positions of the QW in the active region. It is normalized to a peak value of 4 in air, corresponding to the standing wave set up by a 100% reflecting mirror. A traditional resonant periodic gain design, with one well per antinode of an active region many $\lambda/2$ in length, will be restricted in bandwidth by a peak around the design wavelength. The design of Figure 8.3 mitigates this effect by using QW pairs, slightly displaced either side of an antinode, in a shorter active region. For this design, there is a band some 60 nm (17 THz) in width over which the longitudinal confinement factor is essentially flat: for QW emitting within this region, the effective gain profile will be near-intrinsic in shape.

Under operating conditions, the performance of a gain chip may deviate substantially from design values. Pumping heats the active region: the temperature rise is determined by the absorbed pump power, the optical-to-optical conversion efficiency, and the thermal impedance of the chip. Rising active region temperature depresses the gain per carrier of the chip, causing a compensating rise in carrier concentration that changes the gain profile. Optical layer thicknesses in the active region and the DBR change with temperature and carrier concentration, affecting all the optical properties of the chip. Much effort has therefore been devoted to precise measurement of gain chip characteristics under operating conditions. In 2011, Borgentun et al. described a direct measurement of the spectral reflectance of a broadband VECSEL gain chip under pumped and unpumped conditions [34], with an absolute precision of 0.13%. Their chip, which contained six QW pairs in a $13\lambda/2$ active region, finished with a broadband semiconductor multilayer antireflection coating, exhibited a small-signal gain close to 3% over a wavelength range of nearly 35 nm (8 THz).

Mangold et al. [35] describe a comparison of the small-signal gain profile of two similar gain chips, both designed with a $15\lambda/4$ active region containing seven nanostructures and a composite semiconductor/dielectric antireflection multilayer; however, where one contains self-assembled InAs/GaAs QDs, and the other InGaAs/GaAS QW, with intrinsic gain bandwidths likely to be in the region of 90 and 20 nm respectively. They report FWHM gain bandwidth values of 26 nm for the QD chip, and 30 nm for the QW chip. In mode-locked operation, these chips generated pulses of duration 784 and 625 fs respectively.

8.2.2 Gain Filtering

The excited state of a Ti^{3+} ion in a sapphire crystal decays radiatively to the ground state by a first-order-forbidden electric dipole transition, with a lifetime in the region of 2 µs. Electrons and holes in a quantum well recombine radiatively by a dipole-allowed transition, with a lifetime in the range 300 ps–1 ns, depending on temperature, carrier density, and well quality. The dynamical behavior of semiconductor lasers is thus fundamentally different from that of most solid-state (i.e. impurity-doped dielectric) lasers, in which dipole transitions are only weakly allowed between electronic states of the same parity, through small admixtures of electronic configurations of the opposite parity.

The implication for a VECSEL, with a low-loss external cavity, characterized by a photon lifetime that may be of order 100 ns, is that the carrier number in the active region will adjust fast enough to follow the photon number adiabatically. This has been described by the laser dynamics community as a "class A" regime, in which carrier number can be adiabatically eliminated from the dynamical equations. At the onset of lasing, the photon number in the cavity builds up smoothly to its steady-state value without relaxation oscillations. A distinctive feature of the initial transient is the long timescale over which the laser spectrum narrows or "condenses" to its final value. A simple analysis of mode coupling by gain saturation shows that the *inverse* of the spectral bandwidth of the laser as it settles to a steady state should increase in proportion to the time elapsed since the onset of lasing [36]. A value for the gain bandwidth of the laser can therefore be inferred from spectrotemporal measurements of the laser output during the settling period. This measurement is complementary to the precise reflectance measurements of [34], since it tracks the laser dynamics into a strongly saturated regime. It is also different because it measures the spectral curvature (second derivative) of the gain, $g''(\omega)$, around the operating frequency $\omega = \frac{2\pi c}{\lambda}$ of the laser, rather than mapping out the full spectral profile. In a parabolic approximation, the dependence of gain on angular frequency may be written $g(\omega) = g_0 + \frac{g''}{2}(\omega - \omega_0)^2$, for a gain spectrum that assumes a maximum value of g_0 at frequency ω_0. The quantity $-g_0/(g'')^2$ features in the Haus theory of mode-locking as the gain dispersion, D_g: the quantity that controls the amount by which the duration of the intracavity pulse in a mode-locked laser is extended, each time it is filtered by amplification in the gain medium. The Haus theory assumes that the circulating pulse suffers only a small perturbation to its envelope and its phase profile over a single cavity roundtrip. ML-VECSELs are low-gain lasers that fulfill this condition well. In this approximation, the fractional increase in duration of a pulse, initial duration τ, caused by a single pass of the gain medium, is given by $\Delta\tau/\tau = \kappa D_g/\tau^2$, where κ is a constant of proportionality of order unity that depends on the pulse shape and on the measure used for pulse duration.

Barnes et al. [36] used an acousto-optic modulator to sample the output of a cw VECSEL as a function of time delay after an intracavity chopper uncovered the lasing mode, and the VECSEL started to turn on. They verified an inverse relationship between output bandwidth and elapsed time for their laser and inferred an effective gain bandwidth of up to 50 nm for their antiresonant gain chip, based on Findlay–Clay determination of the cavity loss. The time resolution of their acousto-optic modulator did not allow them to define the time window of each spectrum better than about 50 μs. A refinement of their experiment with improved time resolution was reported by Head et al. [37], who used a fast photodiode at the output of a 1-m grating spectrometer, improving the time resolution of the measurement. Head et al. reported a value for $(g'')^{-2}$ of $-36 \pm 6 \, \text{fs}^2$, corresponding to a wavelength bandwidth of about 18 nm, and showed that there was little change in the curvature of the gain spectrum as it saturated toward a steady-state value. A 200-fs pulse might therefore be stretched by about 200 attoseconds per pass over this gain structure. In steady-state mode-locking, it will be the task of the SESAM, or other bleachable loss agent, to counteract this change.

8.2.3 Gain Saturation and Recovery

The energy of an intracavity pulse is multiplied by $\exp g_p$ in one reflection off the pumped gain chip, where the gain exponent g_p depends on the fluence (energy per unit area), F, of the incident pulse. A key dynamical parameter governing the stability of mode-locked operation is the gain saturation fluence, F_{sat}; defined as value of F at which the gain drops to $(1 - e^{-1})$, or about 63%, of its initial unsaturated value. We describe the performance of the gain chip using the gain saturation parameter, $S_g = F/F_{\text{sat}}$, and write $g_p(S) \approx g_0(1 - e^{-S})/S$, where g_0 is the initial value of the gain exponent. The relationship describes incoherent extraction of energy from a two-level system, with no gain recovery during the passage of the pulse: it assumes that the gain is small enough to justify a first-order expansion of the exponential. As the pulse leaves the gain chip, the gain exponent dips to its smallest value of $g_0 e^{-S}$: it is then repumped to the starting value before the pulse returns. Gain saturation can contribute to the formation of a short pulse by attenuating its trailing edge.

Stable mode-locking requires the correct degree of gain saturation; too much, and the laser will be unstable against multipulsing, with an uncontrolled number of intracavity pulses that will not necessarily be equidistant from each other. Too little gain saturation creates a tendency to the Q-switching instability that imposes an upper limit on the repetition rate of a mode-locked laser [38].

The respect in which ultrafast semiconductor lasers least resemble their dielectric counterparts is in the complexity of the underlying physics that governs the interaction between quantum-confined carriers and light. Electron-hole pairs are generated in the barriers; the carriers diffuse through the active region and are captured into QW or QDs on a nanosecond timescale. The k-space distribution of the quantum-confined carriers relaxes through a number of scattering and recombination processes, occurring on a hierarchy of timescales from 10 fs to 10 ns [39]. The heat released in the active region gets transported away by diffusion, on a timescale in the range 100 ns–1 μs. An intracavity light pulse may couple to carriers strongly enough to cause kinetic holeburning; a large transient deviation from the equilibrium Fermi–Dirac distribution [40]. Although experimentally the gain dynamics may superficially resemble those of an ultrafast solid-state gain medium, in practice there are major differences. The effective saturation fluence will get smaller for pulses that are shorter than the timescale for carrier thermalization, promoting complex behavior, including multiple pulsing, and poor optical conversion efficiency [41].

Mangold et al. present experimentally determined values for the gain saturation fluence of 7-QW gain structures of $<100\,\mu\text{J cm}^{-2}$ [35]. They used a pulsed nonlinear reflectivity technique based on the theoretical and experimental framework developed for SESAM characterization [42]. The values were shown to be temperature-dependent, with overlapping ranges for QD and QW samples, and for ps and fs probe pulse durations. For comparison, the gain saturation fluence of a Ti^{3+}-doped sapphire crystal is about $1\,\text{J cm}^{-2}$ [43]: the 10^4-fold difference in magnitude reflects the forbidden nature of the Ti:sapphire laser transition, as does the contrast between its 3-μs upper level lifetime and the 300-ps recombination time typical of a QW laser. Note, however, that the size of F_{sat} in a VECSEL gain chip can be usefully increased by designing for a small longitudinal confinement factor, as occurs in an antiresonant structure.

Baker et al. characterized non-exponential gain recovery in situ in a working ML-VECSEL, interrogating the active region with resonant asynchronous 20-fs probe pulses [44]. Depletion of the gain over the passage of the ML-VECSEL pulse was observed, followed by 80% recovery within 5 ps. Thereafter, the recovery slowed down, dominated by the rate of carrier capture from the QW barriers. These authors describe measurements from an 800-fs and a 300-fs ML-VECSEL; it is notable that the shorter pulse laser exhibited significantly faster gain dynamics.

Alfieri et al. used stretched titanium–sapphire laser pulses to measure the saturation gain fluence of an ultrafast VECSEL chip as a function of probe pulse duration [41]. Their pumped chip exhibited a small-signal gain of about 5% for durations from 1890 to 170 fs; however, the saturation fluence dropped from 70 to $25\,\mu\text{J cm}^{-2}$ over this range. Time-resolved pump probe measurements revealed time constants for the fast and slow temporal recovery processes of 300–400 fs and 130–140 ps respectively.

The sub-100-fs pulses described by Quarterman et al., in a ML-VECSEL with exceptionally low dispersion and broad gain, could not be formed in fundamental mode-locking, but circulated within the cavity in groups of up to 70, equally spaced in time at a separation around 800 fs [45]. The pulse duration could be tuned by adjusting the SESAM temperature: it was observed that as the pulse duration dropped from 135 to 75 fs, the energy of a single pulse in the intracavity group fell from 0.35 to 0.12 nJ. It seems likely that the number of pulses in the cavity was controlled primarily by gain dynamics, adjusting spontaneously to keep a constant level of gain saturation. If this assumption is correct, a twofold reduction in pulse duration over this range corresponded to a threefold reduction in saturation fluence. The shortest pulses generated by this laser (60 fs) hit the gain with a fluence of about $1\,\mu J\,cm^{-2}$.

8.2.4 Saturable Absorbers for ML-VECSELs and MIXSELs

The precise experimental characterization of SESAMs described by Haiml et al. [42] has advanced the performance of many kinds of passively mode-locked laser and established a quantitative foundation for pulse formation simulation. The observed dependence of SESAM reflectivity on pulse fluence follows a curve that can typically be represented using four parameters; the modulation depth ΔR; the non-saturable loss ΔR_{ns}; the saturation fluence F_{sat}; and the F_2 parameter that represents induced loss at high pulse fluence, for example, by a two-photon absorption mechanism. The modulation depth of the SESAM controls the strength of the pulse shaping and also determines the size of the gain that will be required. The non-saturable loss makes no contribution to pulse shortening and increases the pulse stretching effect of gain dispersion by raising the overall cavity loss. The two F-parameters are pulse-duration-dependent. F_{sat} may be smaller for the shortest pulses, of duration less than the recovery time constant of the absorber. F_2 effects are likely to be determined by the instantaneous intensity of the optical field, rather than the integrated power delivered to the SESAM. Pump-probe spectroscopy is used to determine the rate at which the absorption recovers to its unsaturated value.

The first sub-500-fs ML-VECSEL used a specially designed SESAM, with a single QW absorber embedded in a quarter-wave window layer, a few nm from the air surface [46]. Carriers generated in the well tunneled through to the air surface, where the high density of defect states led to rapid recombination. A streak camera measurement of photoluminescence from this structure showed a signal decaying on a timescale of about 20 ps. The reason for developing the surface recombination design was to avoid low temperature growth as a route to fast recovery, since any associated non-saturable loss would lengthen the pulses via added gain dispersion. Small modulation depth tends to be an inherent problem with this design, because the QW sits under a field node at the air surface of the $\lambda/4$ window layer.

Wilcox et al. were able to demonstrate 260-fs pulses using a surface recombination SESAM with a thinner window layer: they moved the node far enough away from the QW to double the modulation depth compared with the earlier designs that had generated pulses of >400 fs duration. The excess GDD incurred by this change had to be carefully offset against GDD of opposite sign in the gain structure at the operating wavelength [47]. The cavity mode in this laser was focused to a 28-μm radius spot on the SESAM, corresponding to a pulse fluence of >300 μJ cm^{-2}, about 18× that on the gain structure. With the SESAM so strongly saturated, experiencing peak pulse intensity in the region of 1 GW cm^{-2}, power broadening of the absorbing resonance also contributed to bleaching of the SESAM loss. The fast response of this optical Stark effect broadening mechanism helped to explain how this laser could form pulses of duration 100× shorter than the carrier recombination time [48].

The surface recombination design was adapted by Klopp et al. with the addition of a dielectric AR coating to suppress dispersion, leading to the demonstration of 107-fs pulses [4]. Pump-probe spectroscopy of their structures showed a fast initial response on a 1-ps timescale followed by a few-ps tail that became slightly slower at higher pump fluences [49].

A surface recombination SESAM was used in an ML-VECSEL that generated pulses with >4 kW peak power [50]. Head et al. present a characterization of the nonlinear reflectivity of the surface recombination SESAM with which they were able to reach >1 kW of peak power in a 193-fs pulse train [9]. In this device, the single InGaAs QW was separated from the last (low index) DBR layer by a 13.5-nm-thick window layer and protected from the air surface by a 2-nm-thick capping layer of GaAs. From the dependence of reflectivity on fluence, measured with 200-fs pulses, the authors were able to extract a saturation fluence value of 1.22 ± 0.08 μJ cm^{-2}. The modulation depth and non-saturable loss were about 2.3 and 0.2% respectively.

Fast absorption recovery can also be ensured using low-temperature MBE growth of the absorbing QW. In a recent demonstration of sub-100-fs pulse generation, the SESAM incorporated a single InGaAs QW grown at 260 °C between AlAs barriers [5]. The authors measured a saturation fluence value of 4.3 μJ cm^{-2} for this structure, and a recovery characteristic with a fast component of 560 fs and a slow component of 5.5 ps. This type of design, incorporating a dielectric coating, allows great freedom of parameter selection [51]. With the addition of a highly reflecting top coating, the structure can withstand mJ pulse energies at kW average power levels without damage [52], allowing the development of high-energy disk laser oscillators.

The integration of gain and saturable absorber into a single chip, in the MIXSEL, is achieved with a multilayer designed in six main sections. The surface antireflection structure and the active region containing multiple QW spaced by barrier regions broadly resemble those found in ML-VECSEL gain chips. The third section is a short DBR that reflects unabsorbed pump radiation back through the active region and protects the absorber from saturation, while transmitting the laser light. The fourth section is an intermediate DBR designed for strong field enhancement on the absorber layer, which forms the fifth section: the sixth and final section is the bottom DBR, which completes the laser cavity [1]. The original MIXSEL concept made use of a single self-assembled InAs QD absorber layer, with saturation fluence inherently lower than that of a QW in the active region. The whole structure could be grown in a single run, at a growth temperature of 600 °C for all the sections except the QD absorber layer, for which it was reduced to 430 °C. An average output power of 6.4 W was recorded from a flip chip MIXSEL of this type; a record for any ultrafast semiconductor laser [8]. The 28-ps pulse duration of this laser is typical of MIXSELs with QD absorber layers, which become partly annealed during the growth of the bottom DBR, slowing down their recovery time. The first demonstration of a sub-picosecond MIXSEL, emitting 620-fs pulses, used as the absorber structure a single low-temperature-grown InGaAs QW embedded between AlAs spacers [53]. The fast characteristics of this absorber survived annealing well; a pump probe measurement revealed a bitemporal response, with a fast component of 380 fs and a slow component of 4.13 ps, contrasted with the 188-ps time constant previously reported for the annealed QD structure [54]. More recently, the generation of pulses shorter than 300 fs has been reported from a QW absorber MIXSEL, designed with an exceptionally low and flat dispersion spectrum [55].

The unique nonlinear-optical properties of carbon nanostructures have recently attracted much interest as saturable absorbers for ultrafast lasers [56, 57]. Carbon nanotubes have been proposed as a generic laser mode-locking technology for the whole of the near-infrared region; highly nonlinear, and with a fast response [58]. Solutions of single-walled carbon nanotubes (SWCNTs) can readily be deposited on a wide range of substrates by spray or spin coating, offering an inexpensive and versatile saturable absorber technology, which lends itself to use in reflection or transmission, unlike its semiconductor counterpart. High non-saturable loss levels, however, are introduced by residual populations of multiple-walled tubes that persist even after purification; and this potentially makes these saturable absorbers unsuitable for use in low gain VECSELs. Seger et al. were nevertheless able to generate 1.23-ps pulses at 1074 nm from an InGaAs/GaAs VECSEL using a transmission-type SWCNT absorber spin-coated onto an intracavity Brewster-angle quartz etalon placed at a beam waist [59]. Their absorber had a modulation depth of 0.25% and a saturation fluence of 11.4 μJ cm^{-2}: its single-pass non-saturable loss was 0.74%. The authors report bitemporal recovery time constants of 150 fs and 1.1 ps. Single-layer graphene (SLG) is an even broader and faster material: the challenge that it presents for VECSEL mode-locking is an optical absorption of 2.3% per pass, more than can be tolerated within a VECSEL cavity. Zaugg et al. describe a design strategy for controlling the modulation depth of an SLG saturable absorber: they transferred CVD-grown SLG films to silica-coated GaAs/AlAs DBRs. By adjusting the strength of the E-field on the SLG, they were able to tailor the modulation depth of their "GSAMs" from 0.2 to 2%; an intermediate device had modulation depth of 5% and non-saturable loss of 5.1%, with a saturation fluence of 100 μJ cm^{-2}. The GSAMs were observed to damage for incident pulse fluences in the 100, or few-100 μJ cm^{-2} range. The shortest pulses that these authors report from a GSAM-ML-VECSEL were 466 fs in duration, with average power of 12.5 mW; the GSAM would mode-lock VECSELs over a 46-nm wavelength range [60]. Husaini and Bedford also reported evidence of GSAM mode-locking in a high-power 1045-nm VECSEL, emitting an average of 10 W [61].

An interesting recent development has been the publication of reports of SESAM-free mode-locking, or self-mode-locking, in VECSELs [17, 62–65], prompting some controversy [66, 67], as well as several experimental investigations of nonlinear lensing in VECSEL gain chips [68–70], aimed at quantifying a physical mechanism that could establish mode-locking of the Kerr lens type. This intriguing story will be expounded in detail elsewhere in this book. If generally applicable principles for the design of self-mode-locked lasers can be established, it will greatly enhance the prospects for commercializing this technology.

8.3 Performance of Passively Mode-Locked Semiconductor Lasers

8.3.1 Pulse Duration

The testbed for ultrashort pulse generation in an optically-pumped VECSEL has been the InGaAs/GaAsP/AlAs/GaAs laser, grown on a GaAs substrate, and operating at a wavelength near 1 μm. The lasers whose properties are summarized in Table 8.1, presented roughly in chronological order, use gain and absorber structures incorporating compressively strained InGaAs/GaAs quantum wells, with or without GaAsP strain-balancing layers; or alternatively multiple self-assembled InAs quantum dot layers. The wide range of nanostructure designs used underpins the large wavelength range spanned by the mode-locking results, from about 960 to 1040 nm. Table 8.1 is kept to a manageable length by recording only sub-picosecond lasers: many notable picosecond results, however, will feature elsewhere in this chapter.

The earliest demonstrations of sub-300 fs mode-locking of VECSELs used SESAMs of the surface-recombination design introduced by Garnache et al. [12]. Following their 290-fs result [3], Klopp et al. were able to report the generation of first 190-fs pulses [49] and then 105-fs pulses [4]. Although they made use of a thermally managed gain structure, the average power of their device was limited to 3 mW; with increasing pump power, the laser adopted a harmonically mode-locked state, up to a pulse repetition frequency of 90 GHz, with 30 pulses circulating in the cavity. The gain chip used for this laser was of the resonant type with a dielectric antireflection coating, extremely well controlled as to dispersion, but allowing full penetration of the laser mode into the active region, which therefore saturated at low output power levels.

With no thermal management of the gain chip, the average power in mode-locked operation is typically limited to about 100 mW. Wilcox et al. described a laser of this type that reached a peak power of 315 W in 335-fs pulses at a repetition frequency of 1 GHz [72]. Further advances in peak power were realized using thermally managed gain chips that could be operated with average output powers of a few W. In 2012 Scheller et al. described mode-locking of a thermally managed gain chip with 5.1-W average power, setting a new record for a femtosecond VECSEL [7]. With a pulse duration of 685 fs at a repetition frequency of 1.7 GHz, the peak power of these output pulses reached 3.7 kW. The following year, Wilcox et al. combined a similar gain

Table 8.1 Overview of femtosecond mode-locking results in InGaAs-based VECSELs and MIXSELs.

Gain structure	Laser wavelength (λ_0)	Pulse duration (τ_p)	Average output power (P_{av})	Repetition rate (f_{rep})	Reference
6 QW	1035 nm	260 fs	25 mW	1.008 GHz	[2]
4 QW	1036 nm	290 fs	10 mW	3.013 GHz	[3]
6 QW	1028 nm	870 fs	45 mW	0.895 GHz	[71]
4 QW	1044 nm	190 fs	5 mW	2.998 GHz	[49]
6 QW	1025 nm	60 fs	35 mW[a]	1 GHz	[45]
6 QW	999 nm	335 fs	120 mW	1.005 GHz	[72]
3 QW	1030 nm	107 fs	3 mW	5.136 GHz	[4]
7×9 QD	960 nm	784 fs	1 W	5.435 GHz	[73]
7×9 QD	961 nm	416 fs	143 mW	4.471 GHz	[73]
6 QW	995 nm	450 fs	56 mW	1.042–1.126 GHz	[74]
6 QW	991 nm	290–2500 fs	40 mW	2.78–7.87 GHz	[75]
7 QW	964 nm	625 fs	169 mW	6.5–11.3 GHz	[76]
6 QW	1030 nm	400 fs	300 mW	175 GHz	[77]
10 QW	1030 nm	682 fs	5.1 W	1.71 GHz	[7]
2×9 QD	950 nm	466 fs	5 mW	2.5 GHz	[60]
10 QW	1013 nm	400 fs	3.3 W	1.67 GHz	[6]
12 QW	1026 nm	482 fs	900 mW	1 GHz	[78]
MIXSEL	968 nm	620 fs	101 mW	4.8 GHz	[53]
10 QW	1014 nm	860 fs	460 mW	504 MHz	[65]
5×7 QD	1040 nm	830 fs	500 mW	1.545 GHz	[17]
MIXSEL	964 nm	790 fs	201 mW	60 GHz	[79]
MIXSEL	964 nm	570 fs	127 mW	101.2 GHz	[79]
10 QW	1038 nm	231 fs	100 mW	1.75 GHz	[80]
7 QW	1050 nm	830 fs	900 mW	193 GHz	[81]
MIXSEL	1045 nm	253 fs	235 mW	3.350 GHz	[82]
MIXSEL	1043 nm	279 fs	310 mW	10.02 GHz	[82]
MIXSEL	1045 nm	<400 fs	>225 mW	2.9–3.4 GHz	[82]
10 QW	1034 nm	128 fs	80 mW	1.81 GHz	[5]
10 QW	1034 nm	107/96 fs[b]	100 mW	1.63 GHz	[5]
8 QW	1040 nm	193 fs	400 mW	1.6 GHz	[83]
10 QW	1035 nm	230–388 fs	18 mW	0.88–1.88 GHz	[84]
8 QW	992 nm	250 fs	300 mW	2.2 GHz	[18]
8 QW	992 nm	195 fs	225 mW	2.2 GHz	[18]
QW	990 nm	130 fs[b]	850 mW	200 MHz	[10]
MIXSEL	1048 nm	184 fs	115 mW	4.33 GHz	[41]

a) 70 intracavity pulses.
b) With external pulse compression.

chip with a surface-recombination-type SESAM to yield a 3.1-W average power train of 400-fs pulses [50]. The peak power of these pulses was 4.1 kW, which at time of writing appears still to represent a record for semiconductor lasers.

In 2016, Waldburger et al. reported an ML-VECSEL that revisited the 100-fs pulse duration of the 2011 result of Klopp et al., but with a 30-fold improvement in average power [5]. These authors were able to generate a 128-fs pulse train at a prf of 1.81 GHz, average power 80 mW. Reduction of the prf to 1.63 GHz, with an increase of average power to 100 mW generated pulses that could be externally compressed to a duration of 96 fs, corresponding to a peak power of 560 W. These results were achieved using an incident 808-nm pump power of 21 W. A better overall optical efficiency was reported by Head et al., although with the relatively long pulse duration of 193 fs [9]. With an incident pump power of 30 W the laser emitted an average power of 400 mW at a repetition rate of 1.6 GHz, corresponding to a peak power of >1 kW.

In 2010, Rudin et al. reported the first thermally managed flip-chip MIXSEL structure and set a new record of 6.4 W for the greatest average power emitted by any mode-locked semiconductor laser [8]. The integrated chip combined a QW gain region with a QD saturable absorber and emitted the long pulses (15 ps) typical of these structures. More recently, Mangold et al. introduced a novel MIXSEL structure in which the absorbing region contained a single low-temperature-grown quantum well, with fast recovery characteristics [53]. This was the first femtosecond MIXEL, emitting 620-fs pulses at a 4.8 GHz prf at 101 mW average power. Further refinements in the growth settings led to the demonstration of a new short pulse record from a MIXSEL at 253 fs, this time at a 1030-nm wavelength compatible with Yb-doped fiber amplifiers [82]. Most recently, Alfieri et al. were able to generate 184-fs pulses from this same MIXSEL chip, using precise control of the thickness of the PECVD-deposited fused silica coating, adapted to partially compensate small growth-error deviations from design of the MIXSEL multilayer [41]. The sub-200-fs MIXSEL operated at a repetition rate of 4.33 GHz with average power of 115 mW.

8.3.2 Pulse Repetition Rate

A physical property of the ML-VECSEL that is strongly contrasted with that of existing commercial solid-state ultrafast lasers is the potential for stable mode-locked operation at pulse repetition rates in the GHz regime, owing to the intrinsically small gain saturation fluence discussed earlier. This property is potentially interesting to speed up data acquisition rates or to generate frequency combs with large tooth spacing and high power per mode. Given current limits to the average power that ML-VECSELs can reach, however, and the fundamental physical constraints that limit optical conversion efficiency for the shortest pulses, it becomes increasingly difficult to hit peak powers in the kW regime as the repetition rate increases.

The highest repetition rates attained by VECSELs have involved harmonically mode-locked lasers. Saarinen et al. [85] describe a 1040-nm VECSEL in which up to 4 pulses, of 10–20 ps duration, could circulate in the cavity, with a linear correlation between pump power and number of pulses. With multiple pulses in the cavity, the fundamental 350 MHz repetition rate was suppressed by 50 dB in the rf spectrum, implying that the intracavity pulses adjusted themselves to be precisely equidistant, with identical pulse envelopes. Griebner et al., as noted earlier, observed 18th-order mode-locking at 92 GHz; with no intracavity element to stabilize the interval between pulses, the number of pulses in the cavity was liable to fluctuate, although modulation of the optical spectrum confirmed a degree of mutual coherence between adjacent pulses [4]. Quarterman et al. used an intracavity etalon to stabilize operation at 147 GHz [86]. Wilcox et al. took advantage of an intracavity heatspreader in contact with the gain chip to demonstrate a 175-GHz prf train of 400-fs pulses [77]: Saarinen et al. also used an intracavity heatspreader and generated a 193-GHz train of 930-fs pulses [87].

The short gain lifetime of the mode-locked VECSEL sets a lower limit on the pulse repetition rate at which it will be practical to operate these lasers: with too long an interval between pulse transits of the gain, pump energy will be wasted, and the pulse risks being destabilized by amplified spontaneous emission. This is challenging for applications such as biomedical imaging, where a repetition rate of about 200 MHz is preferred, allowing fluorescence decay to be tracked between pulses. It also prevents scaling up of the pulse energy by extending cavity round trip time at a given average power, as has been used to raise the pulse energy of mode-locked thin disk solid-state lasers to 10 µJ and beyond. A way round this limitation has been described by Zaugg et al. [88], who constructed a 60-cm long cavity with four gain passes per round trip. Their laser emitted a train of 11-ps pulses at a 253-MHz repetition frequency, with nanosecond intervals between transits of the gain chip. The alignment of such a cavity is, however, nontrivial. Butkus et al. investigated the mode-locking of an 85.7-MHz VECSEL, with two passes of the gain chip per round trip [89].

Since the ML-VECSEL, unlike the KLM mode-locked Ti:sapphire laser, does not rely on a cavity poised near the stability limit, it can readily be configured as a tunable-repetition-rate device, in which the interval between adjacent pulses in the train can be continuously varied by mechanical translation of one cavity mirror, typically the output coupler. This unusual functionality has potential for applications; the Optical Sampling by Cavity Tuning (OSCAT) technique described by Wilk et al. [90] offers a compact alternative to conventional time-resolved pump and probe measurements. Figure 8.5 summarizes the repetition rate tuning ranges that have been reported to date, with the corresponding pulse durations indicated. The dark gray ranges on the left correspond to experiments using ML-VECSELs. The first report of a prf-tunable laser described a 1-GHz device that could be tuned over 8% of its center repetition rate [91]. Sieber et al. described a VECSEL tuned continuously from 6.3 to 11.3 GHz with no replacement of cavity optics [76]: the pulse duration and center operating wavelength remained constant to within 2% over the entire range, representing a 1.8-fold change in cavity length. An even more extreme range was reported by Wilcox et al., with a 2.8-fold length change from 2.78 to 7.87 GHz [75]; these researchers, however, noted a rapid switch in operating pulse characteristics from the femtosecond to the picosecond regime toward the high-prf end of the range. Chen Sverre et al. described a laser with a 2.1-fold tuning range ratio, in which the pulse duration was <400 fs over the entire tuning range, and <300 fs for the longer cavities [84]. Tunable ML-VECSELs use specially

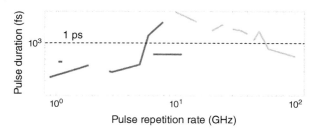

Figure 8.5 Repetition rate tuning ranges (GHz) for ML-VECSELs (dark gray [74, 75, 84]) and MIXSELs (light gray [79]) indicating approximate pulse duration variation (fs).

designed four-mirror Z-cavities, in which the laser mode is highly collimated in the region of the output coupler, which can therefore be translated with minimal change to the focusing of the cavity, and the spot sizes on gain and SESAM chips. The pulse may nevertheless have a tendency to get longer as the cavity length is reduced because the pulse energy varies inversely with the repetition rate, and the SESAM is driven less strongly. The light gray ranges on the right of Figure 8.5 correspond to an experiment reported by Mangold et al. [79] using an MIXSEL. A single integrated gain/absorber structure was used in a two-mirror cavity. The four tuning range curves in the figure were measured using output couplers with four different curvature values. Interestingly, the pulse durations of these lasers were found to decrease with decreasing pulse energy.

8.3.3 Mode-Locked VECSELs: Visible to Mid-Infrared

Although by far the greatest body of research on ML-VECSELs makes use of InGaAs/GaAs quantum well gain structures emitting in the region of 1 μm, there are a number of instances where the design principles emerging from this research have been successfully translated into other semiconductor systems. The first femtosecond VECSEL to emit more than 1 W of average power used a gain chip with 63 layers of self-assembled InAs quantum dots, positioned in the active region to achieve optimal flattening of the gain profile [73]. The laser also used a QD SESAM, emitting 785-fs pulses at a wavelength of 960 nm and prf of 5.4 GHz.

The chart in Figure 8.6 indicates the wavelength ranges explored to date. The first mode-locked VECSEL in the visible part of the system was reported by Ranta et al. [92], who used GaInP/AlGaInP/GaAs gain and SESAM chips to realize a train of 5-ps pulses with 45 mW at 973 MHz repetition rate. The design of AlGaInP/GaInP mode-locked VECSELs for operation is in the red, and the demonstration of a sub-250-fs pulse train at 664 nm from such a laser is described in detail elsewhere in this book [16]. The same group subsequently reported an all-quantum dot VECSEL at 655 nm [93].

The mode-locked red InP QD VECSEL has been made to emit ultrashort pulses in the ultraviolet part of the spectrum, at 325 nm, by intracavity second harmonic generation in a nonlinear beta barium borate crystal [94]. The intracavity nonlinear conversion techniques that have been applied to narrow-band VECSELs to offer custom wavelengths in the visible and UV are not easily adapted for broadband lasers: nevertheless these authors measured a pulse duration of only 1.22 ps for the fundamental radiation. An earlier paper by Casel et al. [95] reported blue-green 489-nm

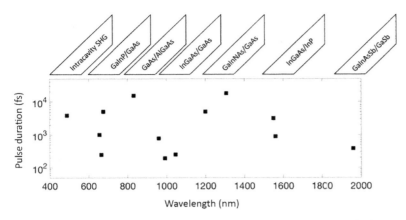

Figure 8.6 Pulse durations of selected mode-locked semiconductor disk lasers of different material systems, as a function of operating wavelength.

pulses from intracavity doubling of the fundamental radiation in a mode-locked InGaAs/GaAs VECSEL using a lithium triborate crystal; in this case the pulses were 5.8 ps long.

The shortest near-infrared wavelength addressed to date has been 830 nm, using a GaAs/AlGaAs quantum well structure, optimized for narrow-band operation, and emitting correspondingly long 15-ps pulses [96].

The dilute nitride system, GaInNAs, has allowed several reports of VECSEL mode-locking in the 1200–1300 nm wavelength band [97–99]. Rautiainen et al. demonstrated a 1.2-µm VECSEL emitting a train of 5-ps pulses at a repetition rate of 840 Hz with 275 mW of output power [98].

ML-VECSELs based on lattice-matched InGaAs/InP quantum wells in the telecoms C-band have also been reported [100, 101]: a challenge for this material system is to compensate for the low index contrast of available DBR materials. Zhao et al. were able to demonstrate sub-ps pulses at 1560 nm [102] using a SESAM in which an absorbing InGaAsNSb QW was sandwiched between layers of GaAsN; fast recombination of carriers that tunneled out of the absorber into these layers gave this SESAM a recovery time of about 13 ps.

A GaSb-based material system in the mid-infrared has allowed the demonstration of 384-fs pulses at 1960 nm from a VECSEL with a fast SESAM [103].

8.3.4 Simulation and Modeling

An immensely important development has been the application of fully microscopic many-body modeling to semiconductor lasers; for VECSELs this has aided, for example, the accurate design of gain structures for "difficult" wavelengths. Most recently, this approach has been developed to study mode-locking dynamics, showing how competition between different regimes of mode-locking may be related to k-space holeburning and holefilling processes [39, 40, 104–107]. A detailed account of these theoretical developments, and the insights that they yield into VECSEL performance, can be found elsewhere in this volume. My aim here is to review the much simpler models that have, nevertheless, provided useful "back of envelope" insights into the performance of mode-locked VECSELs and needed only relatively modest computer power (and operator expertise).

An early example of a simulation that included the effect of coherent material polarization was published by Wilcox et al., who used a two-level-atom model to simulate the operation of a SESAM in a ML-VECSEL emitting 260-fs pulses; the peak intensity of the focused laser mode on the surface of the SESAM approached 1 GW cm^{-2} [2]. They showed that, even in the absence of any pulse shaping by carrier bleaching, power broadening of the two-level absorber resonance could act like a fast saturable absorber, strong enough to form pulses of the observed duration [108, 109].

The numerical pulse propagation models that have been most extensively used in ML-VECSEL research do not involve either many-body calculations or the solution of Maxwell–Bloch equations; they are based on two-level rate equtions and assume that the net gain can be described by a parabolic spectrum. Coherence of the material polarization induced in the gain medium is assumed to decay in a time much less than the pulse duration. The dominant interactions that modify the pulse envelope and phase structure as it propagates from one round trip to the next are characterized by a small number of empirical parameters. Effective values for some of these parameters can be determined rather precisely by experimental measurement, notably those that govern small signal reflectivity and nonlinear reflectivity. Others are more problematic; theoretical calculations of the linewidth enhancement factor that relates refractive index to carrier number show that this quantity varies strongly with wavelength in the vicinity of the band edge. The overall GDD of the cavity, dominated by the semiconductor multilayers, can readily be calculated; however, the tolerance on this number, when standard growth thickness errors and temperature excursions of the active regions are taken into account, may be large.

Paschotta et al. presented the earliest model of this type [110] and argued that the shortest pulses should occur when a small amount of net positive GDD in the cavity was present to offset the nonlinear phase shifts created by carrier number excursions in the wake of the pulse, through the linewidth enhancement factor. A beautiful study by Hoffmann et al., using a range of Gires–Tournois mirrors to introduce controlled amounts of GDD into the cavity, validated this model, at least for pulse durations from 1 ps upward [73]. More recently, Sieber et al. showed that a model of this type could replicate the near transform-limited sub-440-fs pulses experimentally observed in a ML-VECSEL [111]. Their numerical simulation included absorber recovery with two time constants, as observed in a pump-probe characterization of their SESAM. They were able to set their simulation off from a 10 nW noise floor and observe the emergence of a pulse that settled into a stable state after about 10^4 round trips. Not all combinations of laser parameters resulted in stable pulse formation; however, they were able to show that in a two-dimensional phase space defined by the linewidth enhancement factors of SESAM and gain respectively (being the least precisely known of the simulation parameters), there was an extended region where pulse duration and energy were fairly constant. Alfieri et al. extended the rate equation model to represent the pulse-duration-dependent gain saturation effect described earlier [41]. Three coupled carrier populations are considered; a reservoir in the barriers; a reservoir in higher-lying quantum-confined states, at recombination energies outside the laser bandwidth; and a population in quantum-confined band-edge states resonant with the laser field. While this scheme is immensely simplified compared with a rigorous microscopic model, it can represent the observed gain recovery dynamics, with distinct fast and slow components, in a physically plausible way. These authors were able to account for the spectroscopic and laser properties of their gain chips in an internally consistent way.

8.3.5 Noise

The pulse train emerging from a real mode-locked laser differs from that depicted in Figure 8.1 by the admixture of noise. In a ML-VECSEL the round-trip group and phase delay times of the external cavity are modulated by mechanical vibration. Intensity fluctuations of the pump laser modulate carrier population and temperature in the active region, adding noise via gain and optical path length. Quantum noise is injected by spontaneous emission and by vacuum noise that enters the cavity through the output coupler mirror and other lossy elements. The impact of these noise processes on the emitted pulse train can be quantified by measuring RMS values for fractional power fluctuation, and for timing jitter, or fluctuations in the pulse arrival time. These RMS measurements only have meaning in conjunction with specified lower and upper limits for the frequency band over which the fluctuation spectrum is integrated. In addition, the carrier-envelope offset frequency has noise characteristics that will critically influence the ease with which this degree of freedom can be stabilized. Overall, the laser output will have a characteristic coherence length, l_c, which imparts a finite frequency width c/l_c to the longitudinal modes. Paschotta et al. present a rigorous framework for the description and numerical simulation of noise in mode-locked lasers, including optical phase noise, and carrier-envelope offset frequency noise [112].

Where ultrafast pulse trains are used for their temporal precision, as in optical sampling, frequency metrology, or high-bandwidth communication, pulse timing jitter is liable to limit the performance that can be achieved. Wilcox et al. demonstrated a nearly threefold reduction in the timing jitter of a 2.3-ps, 897-MHz ML-VECSEL pulse train when the repetition rate was locked to an electronic oscillator. The stabilized laser exhibited a timing jitter of 160 fs over the bandwidth between 1 kHz and 15 MHz [113]. Quarterman et al. locked the 1-GHz repetition rate of a 470-fs ML-VECSEL to a 10-MHz oscillator using a programmable phase-locked loop frequency synthesizer. This is a low-cost route to the stabilization of GHz lasers, useful because, although frequency division by a factor N makes the low frequency oscillator equivalent to a high-frequency oscillator with $20\log(N)$-fold greater noise, the low-frequency oscillator is considerably cheaper and has better noise characteristics. The jitter of the stabilized laser was 190 fs between 300 and 1.5 MHz [114]. Baili et al. were able to demonstrate a 16-fold reduction in the timing jitter of a passively mode-locked VECSEL when the reflectivity of their SESAM was modulated by an external optical beam at a frequency close to the pulse repetition rate: they report a value of 423 fs between 100 Hz and 10 MHz [115].

State-of-the-art values for the timing jitter of a free-running ML-VECSEL were reported by Wittwer et al., who constructed a low-vibration housing for their 4.6-ps, 2-GHz laser, which reached an average output power of 40 mW using only passive air-cooling. Over the bandwidth from 100 Hz to 10 MHz, they measured a timing jitter of 212 fs for this laser [116]. With the repetition rate locked to the 2-GHz output from a synthesized signal generator, the timing jitter over the same frequency band was reduced to 43 fs [117]. When a 14-ps, 2 GHz MIXSEL was operated in this type of stable housing, jitter values of 129 fs in free-running operation, and 31 fs with repetition rate stabilization were measured over the 100 Hz to 10 MHz band [118]. This MIXSEL operated, moreover, with an average power of 700 mW, and used a water-cooled Peltier to stabilize the gain chip.

The first experimental investigation of carrier-envelope offset frequency noise in an ML-VECSEL was reported by Brochard et al., who used a frequency heterodyning scheme mediated by an auxiliary narrow-band laser [119]. They estimated a linewidth of 1.5 MHz over a 1-s observation time for the free-running carrier-envelope offset beat and measured a 3-dB bandwidth of 300 kHz for modulation transfer from the pump current to the carrier-envelope offset frequency, consistent with the demands of a self-referenced stabilization scheme. An earlier demonstration used optical amplification of ML-VECSEL seed pulses in order to generate the octave-spanning continuum required for f-to-2f interferometry, and the observed beat note presumably acquired some noise characteristics from the fiber power amplifier [80].

8.4 Applications

8.4.1 Biological Imaging

A powerful suite of imaging tools for biological and biomedical scientists is based on femtosecond laser pulse trains. The 3-D imaging capability of a confocal microscope can be enhanced using an infrared ultrafast laser to excite fluorescence via multiphoton absorption. Nonlinear excitation confers better spatial resolution, while sensitive tissue tolerates infrared light better than visible light. As a crude rule of thumb, a desirable source for this application might emit a train of 200-fs pulses with average power of 200 mW at a repetition rate of 200 MHz.

The first demonstration of multiphoton-excited fluorescence microscopy using an ultrafast VECSEL was reported by Aviles-Espinosa et al. [120], who presented in vivo images of neuronal processes and cell bodies of the Caenorhabditis elegans organism. Their VECSEL was mode-locked at a repetition rate of 500 MHz using a quantum dot SESAM and emitted 1.5-ps pulses at an average power of 287 mW. These researchers estimated that at the microscope focus the sample was exposed to only 34 mW average power and 40 W peak power, so that, even at this relatively high pulse repetition rate, sample viability was preserved well. An advantageous feature of the VECSEL source was its 965-nm operating wavelength, which aligned with the maximum in the two-photon action cross section spectrum of the fluorescent marker that was used (GFP). It is of interest to consider other key wavelengths where an ML-VECSEL might constitute a simple and cheap alternative to an OPO-based source for biomedical imaging. For example, a recent article showed images of spontaneous activity of neurons deep in the hippocampus of an intact mouse brain, with single-cell resolution. The measurements were made possible by three-photon excitation with 1.3- and 1.7-µm light.

8.4.2 Quantum Optics

Intense current interest in quantum-enhanced photonic technologies [121] is stimulating much work on the development of robust and simple photon pair sources, aiming to combine a high rate of successful single pair generation with turn-key operation. A particularly promising approach involves launching ultrashort optical pulses into the core of a photonic crystal fiber (PCF), where strong optical confinement of the light favors photon pair generation by four-wave mixing. A suitable pump laser must emit transform limited pulses with peak power around 100 W in a near-diffraction-limited beam that can be coupled efficiently into PCF with tiny mode size. The brightness of the resulting photon pair source cannot be scaled with pump power, which must typically be set at a level where the pair generation probability is <0.01 per pulse, so that the output is not contaminated with two-pair generation events.

The GHz-repetition rate typical of a ML-VECSEL offers an attractive way of raising the pair generation rate without degrading the quality of the source. Morris et al. describe photon-pair generation in PCF driven by a 1030-nm VECSEL that emitted 4.5-ps pulses at a 1.5-GHz repetition rate [122]. The thermally managed chip was able to emit up to 1 W in the mode-locked output beam. Avalanche photodiodes monitoring the signal and idler photons emerging from the fiber recorded coincidence count rates up to 80 times the rate of accidental background events, confirming the good quality of the pair source. It was possible to tune the laser wavelength through a range of ±4.5 nm by adjustment of an intracavity etalon, a very practical feature for this application, since small fiber fabrication errors are liable to shift the phasematching wavelength by offsets of this order.

8.4.3 Supercontinuum Generation and Frequency Combs

Since the gain bandwidth of double-clad ytterbium-doped power amplifiers (YDFAs) is well matched to the 1-µm emission wavelength of InGaAs/GaAS QW ML-VECSELs, a number of groups have created high repetition rate pulse sources with high average power, using an ML-VECSEL as a source of seed pulses for an YDFA. Some key results appear in Table 8.2.

The 4.5-W result from Kerttula et al. [125] actually used not an YDFA, but an erbium-doped fiber amplifier, together with a 1500-nm ML-VECSEL. The amplified pulses were launched into germania-silica fiber to generate a supercontinuum extending from 1320 to 2000 nm (20 dB level). Head et al. used PCF to generate a supercontinuum with components between 750 and 1300 nm [126]: for pulses of this length (>100 fs) the supercontinuum was undoubtedly incoherent. Zaugg et al.,

Table 8.2 Overview of fiber-amplified ML-VECSEL results.

Average power	Repetition rate	Pulse duration	Reference
200 W	1 GHz	5.8 ps	[123]
53 W	1 GHz	110 fs[a]	[123]
1.5 W	6 GHz	545 fs	[124]
4.5 W	1.6 GHz	15.5 ps	[125]
40 W	3 GHz	400 fs[a]	[126]
5.5 W	1.75 GHz	85 fs[a]	[127]
53 W	193 GHz	930 fs	[128]

a) With external pulse compression.

on the other hand, were able to recompress their amplified pulses to 85 fs [127], with which they were able to generate a coherent octave-spanning continuum from PCF. They detected a beatnote at the CEO frequency in an f-to-$2f$ interferometer, the first time that this has been achieved for an ML-VECSEL. At 17 dB above noise floor, the signal-to-noise ratio of the beat note was not strong enough to stabilize the laser.

A key objective for ML-VECSELs is to generate a stabilized self-referenced frequency comb directly from the laser without a power amplifier; however, these sources do not yet reach high enough energy in a short enough pulse to develop the broad coherent supercontinuum needed for standard self-referencing schemes [22]. A different kind of frequency comb metrology, however, has recently been shown to lie fully within the existing capability of the MIXSEL, even with pulse durations in the few-picosecond regime. Dual-comb spectroscopy can achieve excellent frequency resolution, accuracy, sensitivity, and speed of data acquisition without the need for a spectrometer and with the potential for a compact system, since its performance is not fundamentally determined by optical path length [129]. A drawback can be the requirement for two stabilized frequency combs with slightly different tooth spacings. Recently, Link et al. have demonstrated dual-comb spectroscopy of water vapor with a single free-running MIXSEL [130]. They use the technique of De et al., who measured and modeled intensity noise correlation between the two orthogonally polarized modes of a cw VECSEL with an intracavity birefringent crystal [131]. The dual-comb MIXSEL emits colinear, cross-polarized GHz mode-locked pulse trains, with pulse repetition rates mutually offset by 5 MHz, corresponding to the polarization-dependent path length in the crystal [132, 133]. This approach promises an exceptionally compact and rugged implementation of a high-performance measurement.

8.4.4 Terahertz Imaging and Spectroscopy

Quantum cascade lasers are the dominant sources of narrow-band continuous THz radiation; but single-cycle electromagnetic pulses of ps duration for THz spectroscopy are generated by rectification of ultrafast optical pulses, using either photoconductive antennas or difference frequency generation in nonlinear crystals. For these applications, it is desirable to have an optical source that is compact, with a high repetition rate pulse train to reduce the (typically rather long) data acquisition time. A number of publications explore the possibility of using ML-VECSELs for this application.

The earliest report of THz technology involving an ultrafast VECSEL source concerned an imaging experiment, in which a train of 450-fs pulses from a 1.04-μm VECSEL was used to drive a photoconductive antenna, and a bolometer was used to detect transmitted radiation [134]. Mihoubi et al. subsequently described an all-semiconductor time domain THz spectrometer, in which both the transmitting and the receiving antennas were ML-VECSEL-driven [135].

Chen Sverre et al. pursued the possibility of adapting an ML-VECSEL for intracavity generation of THz pulses, in order to shrink the footprint of a time-domain THz spectrometer operating at high repetition rate. They were able to pattern a gold strip-line antenna onto the face of a surface recombination SESAM without impairing its ability to generate 300-fs pulses, even with bias voltage applied [136]. The antenna was shown to emit coherently detectable levels of THz radiation when irradiated with optical pulses of wavelength and energy fluence comparable with those incident on the SESAM-antenna during ML-VECSEL operation. The repetition rate tunability of this laser will allow the delay-line-free pump-probe technique of OSCAT (optical sampling by cavity tuning) [90] to be adopted. An alternative approach to intracavity generation of THz pulses from a ML-VECSEL is reported by Scheller et al. [137], who use an intracavity etalon to enforce two-color oscillation of their SESAM-mode-locked VECSEL, at wavelengths about 4 nm (1 THz) apart.

8.5 Summary and Outlook

Over the past seven years, the field of ultrafast semiconductor disk lasers has undoubtedly matured; however, it does not seem as if the rate of advance is slowing. This period has seen the introduction of ML-VECSELs that generate 100-fs pulse trains in fundamental mode-locking. It has been proved for the first time that a semiconductor QW gain medium can sustain laser pulses with several kW of peak optical power. Rigorous microscopic modeling suggests that there may be further performance gains to extract from these lasers. Femtosecond mode-locking has now been demonstrated in other material systems, including the red-emitting GaInP/GaAs composition, in a wavelength region where no simple compact pulse sources are available yet, despite the existence of important potential applications.

The MIXSELs described by Sudmeyer in 2010 emitted trains of 30-ps pulses with almost 200 mW of average power. Since then, improvements in layer design and the control of dispersion and intensity enhancement, as well as improvements in fabrication enhancing the speed of saturable absorbers, have yielded a 160-fold reduction in pulse duration and a 30-fold increase in power output. The MIXSEL offers ultrafast pulses from a small and simple device and accesses multi-GHz pulse repetition rates that are not otherwise available.

Quite new concepts have emerged; notably self-mode-locked operation, with the potential for exceptionally simple device designs. It will be of great interest to see whether VECSEL equivalents of Kerr lens mode-locked solid-state lasers are able to push performance into new regimes.

A very recent and surprising development is the introduction of semiconductor gain media in membrane form, with a surface-emitting active region that has been lifted off its parent wafer and bonded to a dielectric supporting and heat-spreading structure, apparently with minimal crystallographic damage, retaining excellent power conversion efficiency in optically pumped operation [138, 139]. The technique is described in detail elsewhere in this volume: the implications for mode-locking and ultrashort pulse generation will be profound. The membrane external cavity surface-emitting laser, or MECSEL, dispenses with the semiconductor DBR and uses a cavity formed only by dielectric multilayer mirrors, with wide stop bands and low dispersion. The membrane can be used in either reflection or transmission: many of the traditional ML-VECSEL constraints are removed, and it will be exceptionally interesting to see how this technology advances both the mode-locked performance of semiconductor lasers and the wavelength regions in which they can be realized.

A major question for this field is whether it will eventually be possible to generate 100-fs pulses with sufficient pulse energy to drive a coherent GHz-repetition-rate octave-spanning continuum. At present the best experimental pulse energy reported for a sub-200-fs pulse is 250 pJ [83]: for a 100-fs pulse the value is 60 pJ [5]. Kilen et al. use microscopic modeling to identify an optimized ML-VECSEL design, for which they simulate sub-20-fs pulse generation, with a peak intensity above $1\,\mathrm{mW\,cm^{-2}}$ [107], potentially consistent with an *intracavity* pulse energy of 10 pJ. For reference, we note that it has been possible to stabilize a GHz frequency comb generated in a silicon nitride waveguide, using 64-fs pulses at 1055 nm with coupled (incident) pulse energy of 36 (230) pJ [140]. If it is possible to build a portable GHz frequency comb around a ML-VECSEL, it will be transformative; not least for the practical application of quantum metrology and sensing. Meanwhile, progress in two-photon microscopy and dual-comb spectroscopy indicates the growing range of application areas now addressed by ultrafast semiconductor disk lasers.

References

1 Südmeyer, T., Maas, D.J.H.C., and Keller, U. (2010). Mode-locked semiconductor disk lasers. In: *Semiconductor Disk Lasers*, pp. 213–261. Weinheim, Germany: Wiley-VCH Verlag GmbH & Co. KGaA. doi: 10.1002/9783527630394.ch6.

2 Wilcox, K., Mihoubi, Z., Daniell, G. et al. (2008). Ultrafast optical Stark mode-locked semiconductor laser. *Optics Letters* 33 (23). doi: 10.1364/OL.33.002797.

3 Klopp, P., Saas, F., Zorn, M. et al. (2008). 290-fs pulses from a semiconductor disk laser. *Optics Express* 16 (8): 5770. doi: 10.1364/OE.16.005770.

4 Klopp, P., Griebner, U., Zorn, M., and Weyers, M. (2011). Pulse repetition rate up to 92 GHz or pulse duration shorter than 110 fs from a mode-locked

semiconductor disk laser. *Applied Physics Letters* 98 (7): 071 103. doi: 10.1063/1.3554751.

5 Waldburger, D., Link, S.M., Mangold, M. et al. (2016). High-power 100 fs semiconductor disk lasers. *Optica* 3 (8): 844. doi: 10.1364/OPTICA.3.000844.

6 Wilcox, K.G., Tropper, A.C., Beere, H.E. et al. (2013). 4.35 kW peak power femtosecond pulse mode-locked VECSEL for supercontinuum generation. *Optics Express* 21 (2): 1599–1605. doi: 10.1364/OE.21.001599.

7 Scheller, M., Wang, T.L., Kunert, B. et al. (2012). Passively modelocked VECSEL emitting 682 fs pulses with 5.1 W of average output power. *Electronics Letters* 48 (10): 588. doi: 10.1049/el.2012.0749.

8 Rudin, B., Wittwer, V.J., Maas, D.J.H.C. et al. (2010). High-power MIXSEL: an integrated ultrafast semiconductor laser with 64 W average power. *Optics Express* 18 (26): 27 582. doi: 10.1364/OE.18.027582.

9 Head, C., Hein, A., Turnbull, A. et al. (2016). High-order dispersion in sub-200-fs pulsed VECSELs. In: *Proceedings of SPIE – The International Society for Optical Engineering*, vol. 9734. doi: 10.1117/12.2212690.

10 Lubeigt, W., Bialkowski, B., Lin, J. et al. (2017). Commercial mode-locked vertical external cavity surface emitting lasers, p. 100870D. International Society for Optics and Photonics. doi: 10.1117/12.2254830.

11 Hoogland, S., Dhanjal, S., Tropper, A. et al. (2000). Passively mode-locked diode-pumped surface-emitting semiconductor laser. *IEEE Photonics Technology Letters* 12 (9). doi: 10.1109/68.874213.

12 Garnache, A., Hoogland, S., Tropper, A. et al. (2002). Sub-500-fs soliton-like pulse in a passively mode-locked broadband surface-emitting laser with 100 mW average power. *Applied Physics Letters* 80 (21). doi: 10.1063/1.1482143.

13 Tropper, A.C., Foreman, H.D., Garnache, A. et al. (2004). Vertical-external-cavity semiconductor lasers. *Journal of Physics D: Applied Physics* 37: 75–85. doi: 10.1088/0022-3727/37/9/R01.

14 Keller, U. and Tropper, A.C. (2006). Passively modelocked surface-emitting semiconductor lasers. *Physics Reports* 429 (2): 67–120. doi: 10.1016/j.physrep.2006.03.004.

15 Tropper, A.C., Quarterman, A.H., and Wilcox, K.G. (2012). *Ultrafast Vertical-External-Cavity Surface-Emitting Semiconductor Lasers*, vol. 86, 1e. Elsevier Inc. doi: 10.1016/B978-0-12-391066-0.00007-1.

16 Bek, R., Kahle, H., Schwarzbäck, T. et al. (2013). Mode-locked red-emitting semiconductor disk laser with sub-250 fs pulses. *Applied Physics Letters* 103 (24): 242 101. doi: 10.1063/1.4835855.

17 Gaafar, M., Nakdali, D.A., Möller, C. et al. (2014). Self-mode-locked quantum-dot vertical-external-cavity surface-emitting laser. *Optics Letters* 39 (15): 4623. doi: 10.1364/OL.39.004623.

18 Laurain, A., Marah, D., Rockmore, R. et al. (2017). High power sub-200fs pulse generation from a colliding pulse modelocked VECSEL, p. 100870E. International Society for Optics and Photonics. doi: 10.1117/12.2252525.

19 Kilen, I., Koch, S.W., Hader, J., and Moloney, J.V. (2016). Fully microscopic modeling of mode locking in microcavity lasers. *Journal of the Optical Society of America B* 33 (1): 75. doi: 10.1364/JOSAB.33.000075.

20 Keller, U. (2010). Ultrafast solid-state laser oscillators: a success story for the last 20 years with no end in sight. *Applied Physics B* 100 (1): 15–28. doi: 10.1007/s00340-010-4045-3.

21 Keller, U., Weingarten, K., Kartner, F. et al. (1996). Semiconductor saturable absorber mirrors (SESAM's) for femtosecond to nanosecond pulse generation in solid-state lasers. *IEEE Journal of Selected Topics in Quantum Electronics* 2 (3): 435–453. doi: 10.1109/2944.571743.

22 Telle, H., Steinmeyer, G., Dunlop, A. et al. (1999). Carrier-envelope offset phase control: a novel concept for absolute optical frequency measurement and ultrashort pulse generation. *Applied Physics B* 69 (4): 327–332. doi: 10.1007/s003400050813.

23 Arissian, L. and Diels, J.C. (2009). Investigation of carrier to envelope phase and repetition rate: fingerprints of mode-locked laser cavities. *Journal of Physics B: Atomic, Molecular and Optical Physics* 42 (18): 183 001. doi: 10.1088/0953-4075/42/18/183001.

24 Monmayrant, A., Weber, S., and Chatel, B. (2010). A newcomer's guide to ultrashort pulse shaping and characterization. *Journal of Physics B: Atomic, Molecular and Optical Physics* 43 (10): 103 001. doi: 10.1088/0953-4075/43/10/103001.

25 Marsh, J.H. and Hou, L. (2017). Mode-locked laser diodes and their monolithic integration. *IEEE Journal of Selected Topics in Quantum Electronics* pp. 1–1. doi: 10.1109/JSTQE.2017.2693020.

26 Kuznetsov, M., Hakimi, F., Sprague, R., and Mooradian, A. (1999). Design and characteristics of high-power (<0.5-W CW) diode-pumped vertical-external-cavity surface-emitting semiconductor lasers with circular TEM/sub 00/ beams. *IEEE Journal of Selected Topics in Quantum Electronics* 5 (3): 561–573. doi: 10.1109/2944.788419.

27 Kuznetsov, M. (2010). VECSEL semiconductor lasers: a path to high-power, quality beam and UV to IR wavelength by design. In: *Semiconductor Disk Lasers*, pp. 1–71. Weinheim, Germany: Wiley-VCH Verlag GmbH & Co. KGaA. doi: 10.1002/9783527630394.ch1.

28 Calvez, S., Hastie, J.E., Kemp, A.J. et al. (2010). Thermal management, structure design, and integration considerations for VECSELs. In: *Semiconductor Disk Lasers*, pp. 73–117. Weinheim, Germany: Wiley-VCH Verlag GmbH & Co. KGaA. doi: 10.1002/9783527630394.ch2.

29 Pallmann, W., Zaugg, C., Mangold, M. et al. (2012). Gain characterization and passive modelocking of electrically pumped VECSELs. *Optics Express* 20 (22): 24 791. doi: 10.1364/OE.20.024791.

30 Pallmann, W.P., Zaugg, C.A., Mangold, M. et al. (2013). Ultrafast electrically pumped VECSELs. *IEEE Photonics Journal* 5 (4): 1501 207–1501 207. doi: 10.1109/JPHOT.2013.2274773.

31 Zaugg, C.A., Gronenborn, S., Moench, H. et al. (2014). Absorber and gain chip optimization to improve performance from a passively modelocked electrically pumped vertical external cavity surface emitting laser. *Applied Physics Letters* 104 (12). doi: 10.1063/1.4870048.

32 Maas, D., Bellancourt, A.R., Rudin, B. et al. (2007). Vertical integration of ultrafast semiconductor lasers. *Applied Physics B* 88 (4): 493–497. doi: 10.1007/s00340-007-2760-1.

33 Tropper, A. and Hoogland, S. (2006). Extended cavity surface-emitting semiconductor lasers. *Progress in Quantum Electronics* 30 (1). doi: 10.1016/j.pquantelec.2005.10.002.

34 Borgentun, C., Bengtsson, J., and Larsson, A. (2011). Direct measurement of the spectral reflectance of OP-SDL gain elements under optical pumping. *Optics Express* 19 (18): 16 890. doi: 10.1364/OE.19.016890.

35 Mangold, M., Wittwer, V.J., Sieber, O.D. et al. (2012). VECSEL gain characterization. *Optics Express* 20 (4): 4136. doi: 10.1364/OE.20.004136.

36 Barnes, M.E., Mihoubi, Z., Wilcox, K.G. et al. (2010). Gain bandwidth characterization of surface-emitting quantum well laser gain structures for femtosecond operation. *Optics Express* 18 (20): 21 330–21 341. doi: 10.1364/OE.18.021330.

37 Head, C.R., Wilcox, K.G., Turnbull, A.P. et al. (2014). Saturated gain spectrum of VECSELs determined by transient measurement of lasing onset. *Optics express* 22 (6): 6919–6924. doi: 10.1364/OE.22.006919.

38 Hönninger, C., Paschotta, R., Morier-Genoud, F. et al. (1999). Q-switching stability limits of continuous-wave passive mode locking. *Journal of the Optical Society of America B* 16 (1): 46. doi: 10.1364/JOSAB.16.000046.

39 Hader, J., Scheller, M., Laurain, A. et al. (2017). Ultrafast non-equilibrium carrier dynamics in semiconductor laser mode-locking. *Semiconductor Science and Technology* 32 (1): 013 002. doi: 10.1088/0268-1242/32/1/013002.

40 Kilen, I., Hader, J., Moloney, J.V., and Koch, S.W. (2014). Ultrafast nonequilibrium carrier dynamics in semiconductor laser mode locking. *Optica* 1 (4): 192. doi: 10.1364/OPTICA.1.000192.

41 Alfieri, C.G.E., Waldburger, D., Link, S.M. et al. (2017). Optical efficiency and gain dynamics of modelocked semiconductor disk lasers. *Optics Express* 25 (6): 6402. doi: 10.1364/OE.25.006402.

42 Haiml, M., Grange, R., and Keller, U. (2004). Optical characterization of semiconductor saturable absorbers. *Applied Physics B* 79 (3): 331–339. doi: 10.1007/s00340-004-1535-1.

43 Ertel, K., Hooker, C., Hawkes, S.J. et al. (2008). ASE suppression in a high energy titanium sapphire amplifier. *Optics Express* 16 (11): 8039–8049. doi: 10.1364/OE.16.008039.

44 Baker, C., Scheller, M., Koch, S.W. et al. (2015). In situ probing of mode-locked vertical-external-cavity-surface-emitting lasers. *Optics Letters* 40 (23): 5459. doi: 10.1364/OL.40.005459.

45 Quarterman, A.H., Wilcox, K.G., Apostolopoulos, V. et al. (2009). A passively mode-locked external-cavity semiconductor laser emitting 60-fs pulses. *Nature Photonics* 3 (November): 729–731. doi: 10.1038/NPHOTON.2009.216.

46 Garnache, A., Hoogland, S., Tropper, A.C. et al. (2002). Sub-500-fs soliton-like pulse in a passively mode-locked broadband surface-emitting laser with

100 mW average power. *Applied Physics Letters* 80 (21): 3892–3894. doi: 10.1063/1.1482143.

47 Wilcox, K.G., Mihoubi, Z., Daniell, G.J. et al. (2008). Ultrafast optical Stark mode-locked semiconductor laser. *Optics Letters* 33 (23): 2797–2799. doi: 10.1364/OL.33.002797.

48 Quarterman, A., Daniell, G., Carswell, S. et al. (2011). Numerical modelling of optical Stark effect saturable absorbers in mode-locked femtosecond VECSELs. In: *Proceedings of SPIE – The International Society for Optical Engineering*, vol. 7919. doi: 10.1117/12.874649.

49 Klopp, P., Griebner, U., Zorn, M. et al. (2009). Mode-locked InGaAs-AlGaAs disk laser generating sub-200-fs pulses, pulse picking and amplification by a tapered diode amplifier. *Optics Express* 17 (13): 10 820. doi: 10.1364/OE.17.010820.

50 Wilcox, K., Tropper, A., Beere, H. et al. (2013). 4.35 kW peak power femtosecond pulse mode-locked VECSEL for supercontinuum generation. *Optics Express* 21 (2). doi: 10.1364/OE.21.001599.

51 Spühler, G.J., Weingarten, K.J., Grange, R. et al. (2005). Semiconductor saturable absorber mirror structures with low saturation fluence. *Applied Physics B* 81 (1): 27–32. doi: 10.1007/s00340-005-1879-1.

52 Alfieri, C.G.E., Diebold, A., Emaury, F. et al. (2016). Improved SESAMs for femtosecond pulse generation approaching the kW average power regime. *Optics Express* 24 (24): 27 587. doi: 10.1364/OE.24.027587.

53 Mangold, M., Wittwer, V.J., Zaugg, C.A. et al. (2013). Femtosecond pulses from a modelocked integrated external-cavity surface emitting laser (MIXSEL). *Optics Express* 21 (21): 24 904. doi: 10.1364/OE.21.024904.

54 Wittwer, V., Mangold, M., Hoffmann, M. et al. (2012). High-power integrated ultrafast semiconductor disk laser: multi-Watt 10 GHz pulse generation. *Electronics Letters* 48 (18): 1144–1145.

55 Alfieri, C.G.E., Waldburger, D., Link, S.M. et al. (2016). Recent progress in high power ultrafast MIXSELs. International Society for Optics and Photonics, p. 973407. doi: 10.1117/12.2212165.

56 Scardaci, V., Sun, Z., Wang, F. et al. (2008). Carbon nanotube polycarbonate composites for ultrafast lasers. *Advanced Materials* 20 (21): 4040–4043. doi: 10.1002/adma.200800935.

57 Sun, Z., Hasan, T., Torrisi, F. et al. (2010). Graphene mode-locked ultrafast laser. *ACS Nano* 4 (2): 803–810. doi: 10.1021/nn901703e.

58 Rotermund, F., Cho, W.B., Choi, S.Y. et al. (2012). Mode-locking of solid-state lasers by single-walled carbon-nanotube based saturable absorbers. *Quantum Electronics* 42 (8): 663–670. doi: 10.1070/QE2012v042n08ABEH014775.

59 Seger, K., Meiser, N., Choi, S.Y. et al. (2013). Carbon nanotube mode-locked optically-pumped semiconductor disk laser. *Optics Express* 21 (15): 17 806. doi: 10.1364/OE.21.017806.

60 Zaugg, C.A., Sun, Z., Wittwer, V.J. et al. (2013). Ultrafast and widely tuneable vertical-external-cavity surface-emitting laser, mode-locked by a

graphene-integrated distributed Bragg reflector. *Optics Express* 21 (25): 31 548. doi: 10.1364/OE.21.031548.

61 Husaini, S. and Bedford, R.G. (2014). Graphene saturable absorber for high power semiconductor disk laser mode-locking. *Applied Physics Letters* 104 (16): 161 107. doi: 10.1063/1.4872258.

62 Kornaszewski, L., Maker, G., Malcolm, G. et al. (2012). SESAM-free mode-locked semiconductor disk laser. *Laser & Photonics Reviews* 6 (6): L20–L23. doi: 10.1002/lpor.201200047.

63 Albrecht, A.R., Wang, Y., Ghasemkhani, M. et al. (2013). Exploring ultrafast negative Kerr effect for mode-locking vertical external-cavity surface-emitting lasers. *Optics Express* 21 (23): 28 801. doi: 10.1364/OE.21.028801.

64 Gaafar, M., Möller, C., Wichmann, M., and Heinen, B. (2014). Harmonic self-mode-locking of optically pumped semiconductor disc laser. *Electronics Letters* 50 (7). doi: 10.1049/el.2014.0157.

65 Gaafar, M., Richter, P., Keskin, H. et al. (2014). Self-mode-locking semiconductor disk laser. *Optics Express* 22 (23): 28 390. doi: 10.1364/OE.22.028390.

66 Wilcox, K. and Tropper, A. (2013). Comment on SESAM-free mode-locked semiconductor disk laser. *Laser and Photonics Reviews* 7 (3). doi: 10.1002/lpor.201200110.

67 Kornaszewski, L., Maker, G., Malcolm, G. et al. (2013) Reply to comment on SESAM-free mode-locked semiconductor disk laser. *Laser & Photonics Reviews* 7 (4): 555–556. doi: 10.1002/lpor.201300008.

68 Quarterman, A.H., Tyrk, M.A., and Wilcox, K.G. (2015). Z-scan measurements of the nonlinear refractive index of a pumped semiconductor disk laser gain medium. *Applied Physics Letters* 106 (1): 011 105. doi: 10.1063/1.4905346.

69 Quarterman, A.H., Mirkhanov, S., Smyth, C.J.C., and Wilcox, K.G. (2016). Measurements of nonlinear lensing in a semiconductor disk laser gain sample under optical pumping and using a resonant femtosecond probe laser. *Applied Physics Letters* 109 (12): 121 113. doi: 10.1063/1.4963352.

70 Shaw, E.A., Quarterman, A.H., Turnbull, A.P. et al. (2016). Nonlinear lensing in an unpumped antiresonant semiconductor disk laser gain structure. *IEEE Photonics Technology Letters* 28 (13): 1395–1398. doi: 10.1109/LPT.2016.2543302.

71 Wilcox, K., Butkus, M., Farrer, I. et al. (2009). Subpicosecond quantum dot saturable absorber mode-locked semiconductor disk laser. *Applied Physics Letters* 94 (25). doi: 10.1063/1.3158960.

72 Wilcox, K., Quarterman, A., Beere, H. et al. (2010). High peak power femtosecond pulse passively mode-locked vertical-external-cavity surface-emitting laser. *IEEE Photonics Technology Letters* 22 (14). doi: 10.1109/LPT.2010.2049015.

73 Hoffmann, M., Sieber, O.D., Wittwer, V.J. et al. (2011). Femtosecond high-power quantum dot vertical external cavity surface emitting laser. *Optics Express* 19 (9): 8108. doi: 10.1364/OE.19.008108.

74 Wilcox, K.G., Quarterman, A.H., Beere, H.E. et al. (2011). Variable repetition frequency femtosecond-pulse surface emitting semiconductor laser. *Applied Physics Letters* 99 (13): 34–37. doi: 10.1063/1.3644162.

75 Wilcox, K., Quarterman, A., Beere, H. et al. (2011). Repetition-frequency-tunable mode-locked surface emitting semiconductor laser between 2.78 and 7.87 GHz. *Optics Express* 19 (23). doi: 10.1364/OE.19.023453.

76 Sieber, O.D., Wittwer, V.J., Mangold, M. et al. (2011). Femtosecond VECSEL with tunable multi-gigahertz repetition rate. *Optics Express* 19 (23): 23 538. doi: 10.1364/OE.19.023538.

77 Wilcox, K., Quarterman, A., Apostolopoulos, V. et al. (2012). 175 GHz, 400-fs-pulse harmonically mode-locked surface emitting semiconductor laser. *Optics Express* 20 (7). doi: 10.1364/OE.20.007040.

78 Albrecht, A.R., Wang, Y., Ghasemkhani, M. et al. (2013). Exploring ultrafast negative Kerr effect for mode-locking vertical external-cavity surface-emitting lasers. *Optics Express* 21 (23). doi: 10.1364/OE.21.028801.

79 Mangold, M., Zaugg, C.A., Link, S.M. et al. (2014). Pulse repetition rate scaling from 5 to 100 GHz with a high-power semiconductor disk laser. *Optics Express* 22 (5): 6099. doi: 10.1364/OE.22.006099.

80 Zaugg, C.A., Klenner, A., Mangold, M. et al. (2014). Gigahertz self-referenceable frequency comb from a semiconductor disk laser. *Optics Express* 22 (13): 16 445. doi: 10.1364/OE.22.016445.

81 Saarinen, E., Rantamaki, A., Chamorovskiy, A., and Okhotnikov, O. (2012). 200 GHz 1 W semiconductor disc laser emitting 800 fs pulses. *Electronics Letters* 48 (21): 1355. doi: 10.1049/el.2012.2443.

82 Mangold, M., Golling, M., Gini, E. et al. (2015). Sub-300-femtosecond operation from a MIXSEL. *Optics Express* 23 (17): 22 043. doi: 10.1364/OE.23.022043.

83 Head, C.R., Hein, A., Turnbull, A.P. et al. (2016). High-order dispersion in sub-200-fs pulsed VECSELs. *Proceedings of SPIE* 9734: 973 408. doi: 10.1117/12.2212690.

84 Chen Sverre, T., Head, C.R., Turnbull, A.P. et al. (2016). Tunable repetition rate VECSEL for resonant acoustic-excitation of nanostructures. *Proceedings of SPIE* 9734: 97 340Z. doi: 10.1117/12.2212748.

85 Saarinen, E.J., Härkönen, A., Herda, R. et al. (2007). Harmonically mode-locked VECSELs for multi-GHz pulse train generation. *Optics Express* 15 (3): 955. doi: 10.1364/OE.15.000955.

86 Quarterman, A., Perevedentsev, A., Wilcox, K. et al. (2010). Passively harmonically mode-locked vertical-external-cavity surface-emitting laser emitting 1.1 ps pulses at 147 GHz repetition rate. *Applied Physics Letters* 97 (25). doi: 10.1063/1.3527973.

87 Saarinen, E.J., Filippov, V., Chamorovskiy, Y. et al. (2015). 193-GHz 53-W sub-picosecond pulse source. *IEEE Photonics Technology Letters* 27 (7): 778–781. doi: 10.1109/LPT.2015.2392155.

88 Zaugg, C.A., Hoffmann, M., Pallmann, W.P. et al. (2012). Low repetition rate SESAM modelocked VECSEL using an extendable active multipass-cavity approach. *Optics Express* 20 (25): 27 915. doi: 10.1364/OE.20.027915.

89 Butkus, M., Viktorov, E.A., Erneux, T. et al. (2013). 857 MHz repetition rate mode-locked semiconductor disk laser: fundamental and soliton bound states. *Optics Express* 21 (21): 25 526. doi: 10.1364/OE.21.025526.

90 Wilk, R., Hochrein, T., Koch, M. et al. (2011). OSCAT: novel technique for time-resolved experiments without moveable optical delay lines. *Journal of Infrared, Millimeter, and Terahertz Waves* 32 (5): 596–602. doi: 10.1007/s10762-010-9670-8.

91 Wilcox, K., Quarterman, A., Beere, H. et al. (2011). Variable repetition frequency femtosecond-pulse surface emitting semiconductor laser. *Applied Physics Letters* 99 (13). doi: 10.1063/1.3644162.

92 Ranta, S., Härkönen, A., Leinonen, T. et al. (2013). Mode-locked VECSEL emitting 5 ps pulses at 675 nm. *Optics Letters* 38 (13): 2289. doi: 10.1364/OL.38.002289.

93 Bek, R., Kersteen, G., Kahle, H. et al. (2014). All quantum dot mode-locked semiconductor disk laser emitting at 655 nm. *Applied Physics Letters* 105 (8): 082107. doi: 10.1063/1.4894182.

94 Bek, R., Baumgärtner, S., Sauter, F. et al. (2015). Intra-cavity frequency-doubled mode-locked semiconductor disk laser at 325 nm. *Optics Express* 23 (15): 19947. doi: 10.1364/OE.23.019947.

95 Casel, O., Woll, D., Tremont, M.A. et al. (2005). Blue 489-nm picosecond pulses generated by intracavity frequency doubling in a passively mode-locked optically pumped semiconductor disk laser. *Applied Physics B* 81 (4): 443–446. doi: 10.1007/s00340-005-1931-1.

96 Wilcox, K.G., Mihoubi, Z., Elsmere, S.P. et al. (2008). Passively mode-locked 832-nm vertical-external-cavity surface-emitting semiconductor laser producing 15.3-ps pulses at 1.9-GHz repetition rate. In: *2008 Conference on Lasers and Electro-Optics*, pp. 1–2. IEEE. doi: 10.1109/CLEO.2008.4552021.

97 Rutz, A., Liverini, V., Maas, D. et al. (2006). Passively modelocked GaInNAs VECSEL at centre wavelength around 1300 nm. *Electronics Letters* 42 (16): 926. doi: 10.1049/el:20061793.

98 Rautiainen, J., Korpijärvi, V.M., Puustinen, J. et al. (2008). Passively mode-locked GaInNAs disk laser operating at 1220 nm. *Optics Express* 16 (20): 15964. doi: 10.1364/OE.16.015964.

99 Rautiainen, J., Lyytikainen, J., Toikkanen, L. et al. (2010). 1300 nm mode-locked disk laser with wafer fused gain and SESAM structures. *IEEE Photonics Technology Letters* 22 (11): 748–750. doi: 10.1109/LPT.2010.2045494.

100 Hoogland, S., Garnache, A., Sagnes, I. et al. (2003). Picosecond pulse generation with 1.5 µm passively modelocked surface-emitting semiconductor laser. doi: 10.1049/el:20030576.

101 Lindberg, H., Sadeghi, M., Westlund, M. et al. (2005). Mode locking a 1550 nm semiconductor disk laser by using a GaInNAs saturable absorber. *Optics Letters* 30 (20): 2793. doi: 10.1364/OL.30.002793.

102 Zhao, Z., Bouchoule, S., Song, J. et al. (2011). Subpicosecond pulse generation from a 1.56 µm mode-locked VECSEL. *Optics Letters* 36 (22): 4377. doi: 10.1364/OL.36.004377.

103 Harkonen, A., Grebing, C., Paajaste, J. et al. (2011). Modelocked GaSb disk laser producing 384 fs pulses at 2 [micro sign]m wavelength. *Electronics Letters* 47 (7): 454. doi: 10.1049/el.2011.0253.

104 Kolesik, M. and Moloney, J. (2007). Time-domain vertical-external-cavity semiconductor laser simulation. *IEEE Journal of Quantum Electronics* 43 (7): 588–596. doi: 10.1109/JQE.2007.898833.

105 Love, D., Kolesik, M., and Moloney, J.V. (2009). Optimization of ultrashort pulse generation in passively mode-locked vertical external-cavity semiconductor lasers. *IEEE Journal of Quantum Electronics* 45 (5): 439–445. doi: 10.1109/JQE.2009.2013729.

106 Moloney, J., Kilen, I., Bäumner, A. et al. (2014). Nonequilibrium and thermal effects in mode-locked VECSELs. *Optics Express* 22 (6): 6422. doi: 10.1364/OE.22.006422.

107 Kilen, I., Koch, S.W., Hader, J., and Moloney, J.V. (2017). Non-equilibrium ultrashort pulse generation strategies in VECSELs. *Optica* 4 (4): 412. doi: 10.1364/OPTICA.4.000412.

108 Quarterman, A., Carswell, S., Daniell, G. et al. (2011). Numerical simulation of optical Stark effect saturable absorbers in mode-locked femtosecond VECSELs using a modified two-level atom model. *Optics Express* 19 (27). doi: 10.1364/OE.19.026783.

109 Tropper, A., Quarterman, A., and Wilcox, K. (2012). Ultrafast vertical-external-cavity surface-emitting semiconductor lasers. In: *Semiconductors and Semimetals*, vol. 86 (J.J. Coleman, A.C. Bryce, and C. Jagadish, eds.). Elsevier. doi: 10.1016/B978-0-12-391066-0.00007-1.

110 Paschotta, R., Häring, R., Garnache, A. et al. (2002). Soliton-like pulse-shaping mechanism in passively mode-locked surface-emitting semiconductor lasers. *Applied Physics B: Lasers and Optics* 75 (4–5). doi: 10.1007/s00340-002-1014-5.

111 Sieber, O.D., Hoffmann, M., Wittwer, V.J. et al. (2013). Experimentally verified pulse formation model for high-power femtosecond VECSELs. *Applied Physics B* 113 (1): 133–145. doi: 10.1007/s00340-013-5449-7.

112 Paschotta, R., Schlatter, A., Zeller, S. et al. (2006). Optical phase noise and carrier-envelope offset noise of mode-locked lasers. *Applied Physics B* 82 (2): 265–273. doi: 10.1007/s00340-005-2041-9.

113 Wilcox, K., Foreman, H., Roberts, J., and Tropper, A. (2006). Timing jitter of 897 MHz optical pulse train from actively stabilised passively mode-locked surface-emitting semiconductor laser. *Electronics Letters* 42 (3). doi: 10.1049/el:20063844.

114 Quarterman, A., Wilcox, K., Elsmere, S. et al. (2008). Active stabilisation and timing jitter characterisation of sub-500 fs pulse passively modelocked VECSEL. *Electronics Letters* 44 (19): 1135. doi: 10.1049/el:20081452.

115 Baili, G., Alouini, M., Morvan, L. et al. (2010). Timing jitter reduction of a mode-locked VECSEL using an optically triggered SESAM. *IEEE Photonics Technology Letters* 22 (19): 1434–1436. doi: 10.1109/LPT.2010.2058796.

116 Wittwer, V.J., Zaugg, C.A., Pallmann, W.P. et al. (2011). Timing jitter characterization of a free-running SESAM node-locked VECSEL. *IEEE Photonics Journal* 3 (4): 658–664. doi: 10.1109/JPHOT.2011.2160050.

117 Wittwer, V.J., van der Linden, R., Tilma, B.W. et al. (2013). Sub-60-fs timing jitter of a SESAM modelocked VECSEL. *IEEE Photonics Journal* 5 (1): 1400107–1400107. doi: 10.1109/JPHOT.2012.2236546.

118 Mangold, M., Link, S.M., Klenner, A. et al. (2014). Amplitude noise and timing jitter characterization of a high-power mode-locked integrated external-cavity surface emitting laser. *IEEE Photonics Journal* 6 (1). doi: 10.1109/JPHOT.2013.2295464.

119 Brochard, P., Jornod, N., Schilt, S. et al. (2016). First investigation of the noise and modulation properties of the carrier-envelope offset in a modelocked semiconductor laser. *Optics Letters* 41 (14): 3165. doi: 10.1364/OL.41.003165.

120 Aviles-Espinosa, R., Filippidis, G., Hamilton, C. et al. (2011). Compact ultrafast semiconductor disk laser: targeting GFP based nonlinear applications in living organisms. *Biomedical Optics Express* 2 (4): 739. doi: 10.1364/BOE.2.000739.

121 Pan, J.W., Chen, Z.B., Lu, C.Y. et al. (2012). Multiphoton entanglement and interferometry. *Reviews of Modern Physics* 84 (2): 777–838. doi: 10.1103/RevModPhys.84.777.

122 Morris, O.J., Francis-Jones, R.J.A., Wilcox, K.G. et al. (2014). Photon-pair generation in photonic crystal fibre with a 1.5 GHz modelocked VECSEL. *Optics Communications* 327: 39–44. doi: 10.1016/j.optcom.2014.02.003.

123 Dupriez, P., Finot, C., Malinowski, A. et al. (2006). High-power, high repetition rate picosecond and femtosecond sources based on Yb-doped fiber amplification of VECSELs. *Optics Express* 14 (21). doi: 10.1364/OE.14.009611.

124 Elsmere, S.P., Mihoubi, Z., Quarterman, A.H. et al. (2008). High-repetition-rate subpicosecond source of fiber-amplified vertical-external-cavity surface-emitting semiconductor laser pulses. *IEEE Photonics Technology Letters* 20 (8): 623–625. doi: 10.1109/LPT.2008.919461.

125 Kerttula, J., Chamorovskiy, A., Okhotnikov, O., and Rautiainen, J. (2012). Supercontinuum generation with amplified 1.57 μm picosecond semiconductor disk laser. *Electronics Letters* 48 (16): 1010–1012. doi: 10.1049/el.2012.1359.

126 Head, C.R., Chan, H.Y., Feehan, J.S. et al. (2013). Supercontinuum generation with GHz repetition. *IEEE Photonics Technology Letters* 25 (5): 464–467.

127 Zaugg, C.A., Klenner, A., Mangold, M. et al. (2014). Gigahertz self-referenceable frequency comb from a semiconductor disk laser. *Optics Express* 22 (13): 16445. doi: 10.1364/OE.22.016445.

128 Saarinen, E.J., Filippov, V., Chamorovskiy, Y. et al. (2015). 193-GHz 53-W subpicosecond pulse source. *IEEE Photonics Technology Letters* 27 (7): 778–781. doi: 10.1109/LPT.2015.2392155.

129 Coddington, I., Newbury, N., and Swann, W. (2016). Dual-comb spectroscopy. *Optica* 3 (4): 414. doi: 10.1364/OPTICA.3.000414.

130 Link, S.M., Maas, D.J.H.C., Waldburger, D., and Keller, U. (2017). Dual-comb spectroscopy of water vapor with a free-running semiconductor disk laser. *Science* 356 (6343): 1164–1168.

131 De, S., Pal, V., El Amili, A. et al. (2013). Intensity noise correlations in a two-frequency VECSEL. *Optics Express* 21 (3): 2538. doi: 10.1364/OE.21.002538.

132 Link, S.M., Klenner, A., Mangold, M. et al. (2015). Dual-comb modelocked laser. *Optics Express* 23 (5): 5521. doi: 10.1364/OE.23.005521.
133 Link, S.M., Klenner, A., and Keller, U. (2016). Dual-comb modelocked lasers: semiconductor saturable absorber mirror decouples noise stabilization. *Optics Express* 24 (3): 1889. doi: 10.1364/OE.24.001889.
134 Wilcox, K., Rutz, F., Wilk, R. et al. (2006). Terahertz imaging system based on LT-GaAsSb antenna driven by all-semiconductor femtosecond source. *Electronics Letters* 42 (20). doi: 10.1049/el:20061825.
135 Mihoubi, Z., Wilcox, K.G., Elsmere, S. et al. (2008). All-semiconductor room-temperature terahertz time domain spectrometer. *Optics Letters* 33 (18): 2125–2127. doi: 10.1364/OL.33.002125.
136 Chen Sverre, T., Turnbull, A.P., Gow, P.C. et al. (2017). Mode-locked VECSEL SESAM with intracavity antenna for terahertz emission, p. 100870O. International Society for Optics and Photonics. doi: 10.1117/12.2252346.
137 Scheller, M., Baker, C.W., Koch, S.W., and Moloney, J.V. (2016). Dual-wavelength passively mode-locked semiconductor disk laser. *IEEE Photonics Technology Letters* 28 (12): 1325–1327. doi: 10.1109/LPT.2016.2541626.
138 Yang, Z., Albrecht, A.R., Cederberg, J.G., and Sheik-Bahae, M. (2016). 80 nm tunable DBR-free semiconductor disk laser. *Applied Physics Letters* 109 (2): 022 101. doi: 10.1063/1.4958164.
139 Kahle, H., Mateo, C.M.N., Brauch, U. et al. (2016). Semiconductor membrane external-cavity surface-emitting laser (MECSEL). *Optica* 3 (12): 1506. doi: 10.1364/OPTICA.3.001506.
140 Klenner, A., Mayer, A.S., Johnson, A.R. et al. (2016). Gigahertz frequency comb offset stabilization based on supercontinuum generation in silicon nitride waveguides. *Optics Express* 24 (10): 11 043. doi: 10.1364/OE.24.011043.

9

Ultrafast Nonequilibrium Carrier Dynamics in Semiconductor Laser Mode-Locking

I. Kilen[1], J. Hader[2,3], S.W. Koch[2,4], and J.V. Moloney[1,2,3,5]

[1] Program in Applied Mathematics, University of Arizona, Tucson, AZ, USA
[2] College of Optical Sciences, University of Arizona, Tucson, AZ, USA
[3] Nonlinear Control Strategies Inc., Tucson, AZ, USA
[4] Department of Physics and Material Sciences Center, Philipps-Universität Marburg, Marburg, Germany
[5] Department of Mathematics, University of Arizona, Tucson, AZ, USA

9.1 Introduction

Vertical external-cavity surface-emitting lasers (VECSELs) as seen in Figure 9.1, operating in continuous-wave (CW) mode or as mode-locked pulse sources, have emerged as an active area of research into novel low-noise, wavelength-versatile sources covering a broad swath of potential applications [1–16] (S. Husaini and R.A. Bedford (2013). Antiresonant graphene saturable absorber mirror for mode-locking VECSELs, personal communication). Record CW performance so far includes 23 W single frequency operation [13] and 106 W multimode operation [6, 7]. Stable mode-locked pulses have been generated using diverse cavity components such as: semiconductor external saturable absorber mirrors (SESAMs) [1, 2, 4, 8, 9], graphene, and carbon nanotube saturable absorbers (GSAM) [10, 11] (S. Husaini and R.A. Bedford (2013). Antiresonant graphene saturable absorber mirror for mode-locking VECSELs, personal communication), as well as integrated quantum well (QW) and quantum dot saturable absorbers [10]. Currently, stable mode-locked pulses have reached pulse widths around 100 fs [3, 12, 15]. These include a train of 60 fs pulses inside a picosecond pulse molecule envelope [3], and a single 107 fs mode-locked pulse that was then externally compressed down to 96 fs [15].

The seminal work by H. Haus describes master equations for mode locking with a saturable absorber [17]. These equations relate the pulse shape to the saturable gain and absorption in a simple resonator cavity. By analyzing the differential equations and their solutions, he provided access to a deeper understanding for the dynamics

---

This material is based upon work supported by the Air Force Office of Scientific Research under award numbers FA9550-14-1-0062 and FA9550-17-1-0246.

Vertical External Cavity Surface Emitting Lasers: VECSEL Technology and Applications, First Edition.
Edited by Michael Jetter and Peter Michler.
© 2022 WILEY-VCH GmbH. Published 2022 by WILEY-VCH GmbH.

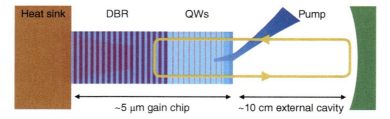

Figure 9.1 A schematic of a linear VECSEL cavity with a reflecting mirror on the right. From the left: A copper heat sink, a distributed Bragg reflector (DBR), a region with optically active quantum wells (QWs), an optical pump, and the propagating cavity field.

of mode locking. Sieber et al. would later build on similar ideas of the saturable gain and absorption dynamics in order to model mode locking in VECSELs [18]. Here each part of the round trip propagation is treated by applying the appropriate operator to the pulse, resulting in a circuit diagram description of the VECSEL. These models can generally be referred to as a *lumped model*, which are simple to understand and have seen use in many diverse fields including, but not limited to climate science, biology, electrical engineering, and heat transfer [19–21]. In the model of Sieber et al., the microscopic QW polarization has been adiabatically eliminated, producing differential equations that describe the gain dynamics. When these rate models are applied to mode locking of VECSELs, there are dozens of parameters that must be fitted to a given experiment. These parameters are integrated quantities that are meant to describe the microscopic QW dynamics, e.g. the gain shape is assumed to be a parabola, which is then fitted to an experimentally measured equilibrium gain. The gain shape is close to a parabola only near the peak gain, and the gain shape is nearly stationary only for weak interactions with the field. This approach leads to fast numerical simulations, but these approximations are harder to justify for mode-locked pulses, which tend to be fairly strong.

This chapter addresses the question of how ultrafast microscopic dynamics of the carriers within their respective bands (valence and conduction) influences mode-locking stability, stability, and robustness of the final pulse that emerges, controls the buildup of a pulse from noise, induces a transient nonlinear refractive index, allows for a mode locked pulse spectrum that is wider than the net gain region, and sets limits on the shortest possible achievable pulses.

In order to produce a quantitative theory for mode-locking VECSELs with a saturable absorber, we will model the microscopic dynamics using the Maxwell-semiconductor Bloch equations [22]. Here the electric field is propagated inside the cavity using Maxwell's equations, and the light matter interaction is examined by simulating the QW carrier dynamics using the semiconductor Bloch equations. These equations are computationally demanding to solve in the context of mode locking, where the microscopic dynamics has to be simulated for thousands of round trips in the cavity. Higher-order correlation effects such as polarization dephasing, carrier scattering, and carrier relaxation are critically important and need to be computed at the level of the second Born–Markov approximation. Fortunately, for pulses on the order of a few hundred femtoseconds and shorter, effective scattering rates for carrier–carrier and carrier–phonon relaxation can be extracted from a one-off solution of this model. This is possible because the mode-locked pulse does not depend too strongly on these rates as long as they are consistently calculated using the microscopic theory. Ironically, such microscopic many-body effects prove to be relevant for longer duration pulses and even CW operation [23, 24]. This is because the mechanism of kinetic hole filling leads to a significant boost of the gain and a shift of the gain peak relative to the simple gain model.

In this chapter, we give an overview of the theoretical modeling of mode locking in VECSELs with a SESAM, as well as microscopic modeling of a GSAM. The authors published these results over the course of several years, and they are summarized here along with some additional background [25–34]. First an overview of the pulse propagation and microscopic theory is given with additional background information in Section 9.2; then the domain modeling, gain region modeling, and some additional background are given in Section 9.3; next the numerical results for mode-locking VECSELs, with an analysis of SESAM and GSAM properties, are presented in Section 9.4; and finally an outlook can be found in Section 9.5.

9.2 Background Theory

9.2.1 Pulse Propagation

In a VECSEL, the electromagnetic field propagates through all materials, including the air cavity, and interacts with the optically active QWs in the gain chip and SESAM. The type of nonequilibrium solution depends on the cavity setup: In a mode-locked VECSEL with a SESAM, the cavity field will form a pulse, and without the SESAM the cavity field will be CW with one or more modes. In order to systematically study the behavior of a VECSEL cavity, we need a model for the propagating field and the optically active QWs. The propagation of the electromagnetic field is naturally modeled by Maxwell's equations, and the QWs will be modeled by using the semiconductor Bloch equations. In order to simplify the problem, first we reduce the 3D Maxwell's equations for the full cavity down to a one-dimensional wave equation for field propagation perpendicular to the QW planes, this will make the model numerically tractable while at the same time also capture the essential aspects of mode locking. The cavity field, $E(z, t)$, is thus modeled by

$$\left[\frac{\partial^2}{\partial z^2} - \frac{n^2}{c_0^2}\frac{\partial^2}{\partial t^2}\right] E(z, t) = \mu_0 \frac{\partial^2}{\partial t^2} P(z, t), \tag{9.1}$$

where c_0 is the speed of light in vacuum, n is the background index of refraction in a material, and μ_0 is the vacuum permeability. The macroscopic polarization $P(z, t)$ couples the equations to the microscopic dynamics of the QWs. This describes the propagation inside a homogeneous material, and a VECSEL cavity naturally consists of many material layers. At any interface of two layers with different refractive indices, there will be reflection and transmission of the cavity field. In order to couple the solution in two different materials, we use the natural boundary conditions from Maxwell's equations. Numerically Maxwell's equations are solved in the time domain with the optically active QW. The spacial simulation domain consists of material layers that represent designs in real experimental devices.

Mode locking occurs in a laser cavity when the output pulse reaches a stage where the output pulse shape after each round trip is constant. However, the internal dynamics in the cavity exhibit strong pulse shape changes undergoing recurring amplification and absorption as it undergoes nonlinear interactions with the active VECSEL and passive SESAM and deforms through linearly dispersive components.

As a pulse propagates through a medium, the pulse shape will be influenced by any material losses or amplification, as well as any effects that change the phase, i.e. the dispersive effects [5, 15, 18, 35–37] A VECSEL consists of many material layers where every interface will add to the pulse dispersion, and this becomes a problem when one realizes that all orders of dispersion above the first will alter the shape of a given pulse. Figure 9.2 shows an example of how a 100 fs pulse shape can be changed under the influence of different orders of dispersion. The amount of dispersion added is frequency-independent and added to the spectral phase of the original pulse while leaving the pulse energy constant. When we add an even order of dispersion, $(\omega - \omega_0)^{2n}$ for $n \geq 1$, to the spectral pulse phase, the time domain pulse becomes wider and a small symmetric *tail* appears in both directions. When adding odd orders, $(\omega - \omega_0)^{2n+1}$ for $n \geq 1$, the pulse also becomes longer and a *tail* is created in only one direction. Changing the sign of the odd-order dispersion will change the direction of the pulse tail. In all cases the presence of nonzero dispersion of any order above the first serves to elongate the pulse and should be offset if the goal is to produce short pulses. The amount of dispersion is important in a real cavity, and one should strive to offset the most detrimental dispersion first.

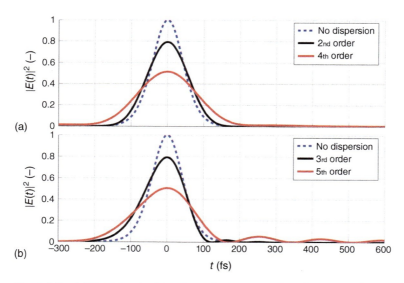

Figure 9.2 A figure showing the effect of adding constant dispersion of (a) even and (b) odd orders to a 100 fs sech^2 pulse.

The optical path length of the pulse will change when the electric field propagates into a material layer at an angle (angle of incidence), i.e. the path that the field takes through the material layer is now longer than it would be at a right angle to the surface. The added propagation length results in a change of the dispersion experienced by the field, and this change is well studied in the literature [38]. The details of the change in dispersion are in general complicated, but for small angles the most noticeable effect is a translation of the dispersion to higher wavelengths. This is something that is taken into account when designing devices for a VECSEL cavity. It allows the user to design a device for a predetermined angle and then fine-tune the dispersion when setting up the cavity.

A distributed Bragg reflector (DBR) is a stack of dielectric materials with an arrangement that causes incident light in a specific wavelength range around a central wavelength to be reflected. This range is referred to as the stopband. The DBR is built up as a stack of base units, where each unit of the DBR is a pair of material layers with a high and a low refractive index, and their individual lengths are a quarter of the central wavelength in that material. The stopband reflectivity will grow closer to 100% when the number of stacked base units increases. The theoretical width of the stopband, when the number of base units goes to infinity, can be determined from the refractive indices and the central wavelength [38]. Figure 9.3 shows the DBR stopband and calculated dispersion for a typical gain chip consisting of AlAs/AlGaAs DBR centered at 980 nm.

This model allows us numerically simulate a VECSEL in the time domain, which consists of an arbitrary combination of material layers. A few different domain models are summarized in Section 9.3 where we talk about the cavity devices and

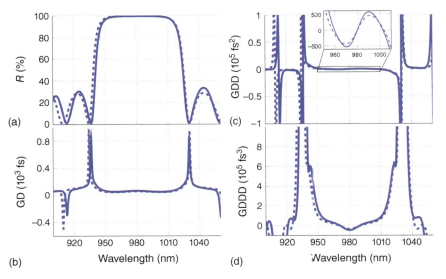

Figure 9.3 A comparison of the reflection (a), and three orders of dispersion (b–d) of the passive material in a gain chip computed with constant refractive indices (dashed) and frequency-dependent refractive indices (solid). The structure does not have dispersion compensating coating in order to highlight curves in the dispersion.

their impact on mode locking. The goal of this project is to model the dynamics of mode-locked pulses and how to make them shorter. In order to do this, we need the model to give a realistic representation of the pulse propagation. The limitations of the above propagation model are in what types of materials can be modeled. The LHS of Eq. (9.1) is linear, and thus one cannot readily model nonlinear field interactions with materials, i.e. the index of refraction is independent of the electric field. The obvious exception comes from the nonlinear interaction with the optically active QWs, which has been separated from the background and appears on the RHS. In real materials the index of refraction is in general frequency-dependent $n(\omega)$. For our wavelength range the refractive indices are nearly constant, and thus this influence is minimal. For example, in Figure 9.3 we see a comparison of the simulated reflection, and the first-, second-, and third orders of dispersion from a typical resonant periodic gain (RPG) structure without the optically active QWs or dispersion compensating layers. The dashed lines represent the gain structure computed with constant refractive indices, while the solid lines are the same layer thicknesses computed with $n(\omega)$. Above we computed a few orders of dispersion for comparison, but in general the model includes all dispersive effects that result from the propagation through material layers. The influence of $n(\omega)$ introduces a change in the material dispersion that, if left uncompensated for, would elongate a pulse. We can estimate how much a single pass over this structure would change the phase by applying the dispersion to a sech^2 pulse of full width half maximum (FWHM) 50 fs and 100 fs. The dispersion from $n(\omega)$ will elongate the pulses by 0.3 fs and 0.06 fs respectively more than with the dispersion from the constant refractive index. Including these effects could improve the accuracy of the predicted mode-locked pulses; however, they would also increase the simulation time considerably. Another limitation would be that in a one dimensional model, one cannot directly compute transverse effects that come from more transverse dimensions or effects that come from using a different cavity mode. Finally, if one wants to make the domain even closer to real experiments, then one should also include transition regions in the material interfaces. These regions will alter the reflection and dispersion from a structure, but we will simply assume perfect transitions in the following.

The cavity in a VECSEL experiment is usually configured in such a way as to facilitate the efficient study of some desired property. In Figure 9.4 we illustrate a few cavity configurations that are all capable of producing mode-locked pulses in order to illustrate the diversity of published work. Figure 9.4a) illustrates the V-cavity configurations where the pulse propagates into one of the elements twice in a round trip [9, 39]. If the gain chip is at the center of the cavity, the pulse will enter into the chip at an angle. This will change the optical path length through the device and thus the dispersion that the pulse will experience. The round trip pulse amplification will increase with the number of times the pulse is folded over the gain chip. However, this also increases the accumulated dispersion, and thus it makes balancing the round trip GDD more challenging. In addition, the pulse will interact with two different states of the gain chip when the two arm lengths are unequal, i.e. the inverted QW carriers, which recover at a constant rate, will recover to two different levels each time the pulse arrives. These carrier dynamics have been observed in experiments

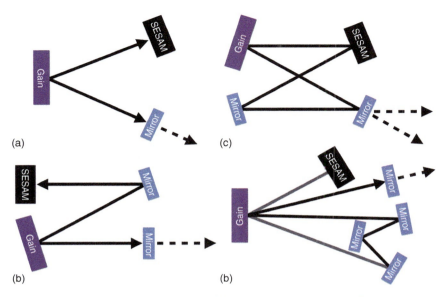

Figure 9.4 A schematic representation of four cavities: (a) V-cavity, (b) Z-cavity, (c) ring cavity, and (d) F-cavity. The solid black lines represent the pulse propagating through the cavity, and the dashed lines are the output. In the F-cavity the first pass over the gain chip is highlighted in gray.

and simulations [32, 40]. Figure 9.4b) illustrates the Z-cavity, which is a perturbation of the V-cavity where there is an additional mirror to reflect the beam [1]. This design allows for good control of the focusing on the SESAM and VECSEL. In Figure 9.4c) we can see the colliding pulse ring cavity where there are two counter-propagating pulses in the cavity. The two pulses are synchronized to simultaneously interact with the SESAM. This cavity has been found to be favorable for stable mode locking of pulses, where the stability might be due to spacial interference effects on the chips [41]. In Figure 9.4d) we can see the folding cavity (F-cavity) where the propagating pulse will take another pass over the gain chip during a round trip [42]. These are just a few example cavities, and it is also possible to join the absorbing QW with the gain chip into a mode-locked integrated external-cavity surface-emitting lasers (MIXEL) structure [28, 43, 44]. The cavity becomes linear when the MIXSEL is used with an output coupling mirror; however, one can use the MIXEL structure in any of the above cavities.

9.2.2 Microscopic Theory

The microscopic dynamics of the QWs will be modeled by the multiband semiconductor Bloch equations (SBE). In this framework the electric field is treated classically using Maxwell's equations and the dynamical equations for the quantum well carriers are derived from the Hamiltonian [22]. The microscopic polarization, $p_{\lambda,\nu,\mathbf{k}}$, couples back into Maxwell's equations through the macroscopic polarization $P(z,t) = \sum_{\lambda,\nu,\mathbf{k}} d_{\mathbf{k}}^{\lambda,\nu} p_{\lambda,\nu,\mathbf{k}}$. The SBE describes the time dynamics of the microscopic polarizations and the carrier occupation numbers, $n_{\lambda(\nu),\mathbf{k}}^{e(h)}$, through

$$\frac{\partial}{\partial t}p_{\lambda,\nu,\mathbf{k}} = -\frac{i}{\hbar}\sum_{\lambda_1,\nu_1}\left(e_{\lambda,\lambda_1,\mathbf{k}}^{e}\delta_{\nu,\nu_1} + e_{\nu,\nu_1,\mathbf{k}}^{h}\delta_{\lambda,\lambda_1}\right)p_{\lambda_1,\nu_1,\mathbf{k}}$$
$$-i\left(n_{\lambda,\mathbf{k}}^{e} + n_{\nu,\mathbf{k}}^{h} - 1\right)\Omega_{\lambda,\nu,\mathbf{k}} + \Gamma_{\lambda,\nu,\text{deph}} + \Lambda_{\lambda,\nu,\text{spont}}^{p}, \qquad (9.2)$$

$$\frac{\partial}{\partial t}n_{\lambda(\nu),\mathbf{k}}^{e(h)} = -2\,\text{Im}\left(\Omega_{\lambda,\nu,\mathbf{k}}(p_{\lambda,\nu,\mathbf{k}})^{*}\right) + \Gamma_{\lambda(\nu),\text{scatt}}^{e(h)} + \Gamma_{\lambda(\nu),\text{fill}}^{e(h)} + \Lambda_{\lambda,\nu,\text{spont}}^{n}.$$

Where $\mathbf{k}$ is the crystal momentum in the QW plane, and $\lambda\,(\nu)$ indicates the electrons (holes) in the conduction (valence) band. For the Coulomb potential, $V_{|\mathbf{k}-\mathbf{q}|}^{\lambda_1,\lambda_2,\nu_1,\nu_2}$, we get that renormalization of the electric field and carrier energy at the Hartree–Fock level gives

$$e_{\lambda,\lambda_1,\mathbf{k}}^{e} = \epsilon_{\lambda,\mathbf{k}}^{e}\delta_{\lambda,\lambda_1} - \sum_{\lambda_2,\mathbf{q}}V_{|\mathbf{k}-\mathbf{q}|}^{\lambda,\lambda_2,\lambda_1,\lambda_2}n_{\lambda_2,\mathbf{q}}^{e},$$
$$e_{\nu,\nu_1,\mathbf{k}}^{h} = \epsilon_{\nu,\mathbf{k}}^{h}\delta_{\nu,\nu_1} - \sum_{\nu_2,\mathbf{q}}V_{|\mathbf{k}-\mathbf{q}|}^{\nu,\nu_2,\nu_1,\nu_2}n_{\nu_2,\mathbf{q}}^{h}, \qquad (9.3)$$

and the effective Rabi frequency is given by

$$\hbar\Omega_{\lambda,\nu,\mathbf{k}} = d_{\mathbf{k}}^{\lambda,\nu}E(z,t) + \sum_{\lambda_1,\nu_1,\mathbf{q}\neq\mathbf{k}}V_{|\mathbf{k}-\mathbf{q}|}^{\lambda,\nu_1,\nu,\lambda_1}p_{\lambda_1,\nu_1,\mathbf{q}}\;. \qquad (9.4)$$

Where $d_{\mathbf{k}}^{\lambda,\nu}$ is the dipole matrix element. A simplification of the microscopic model happens when one assumes strong confinement of the electrons and holes. In this case the band structure is approximated using parabolic bands with effective electron (hole) masses $m_{e(h)}$, which will give the transition energy $\hbar\omega_{\mathbf{k}} = E_g + \frac{\hbar^2\mathbf{k}^2}{2m_e} + \frac{\hbar^2\mathbf{k}^2}{2m_h}$ where E_g is the band gap. The higher-order correlation contributions in Eq. (9.2), which can be approximated with effective rates, are: the Coulomb screening, the dephasing of the polarization ($\Gamma_{\lambda,\nu,\text{deph}}$), the carrier scattering ($\Gamma_{\lambda(\nu),\text{scatt}}^{e(h)}$), and the kinetic hole-filling ($\Gamma_{\lambda(\nu),\text{fill}}^{e(h)}$). The spontaneous emissions are modeled with $\Lambda_{\lambda,\nu,\text{spont}}^{p/n}$.

The higher-order correlation contributions are calculated when trying to solve the fully microscopic many-body problem on the level of second-order Born–Markov approximation [45]. This procedure requires one to evaluate the very time-consuming multiple dimensional scattering integrals at every time step, and in a simulation of mode locking a VECSEL, these equations have to be integrated on the timescale of thousands of cavity roundtrips. However, under experimental conditions the gain chip QWs are inverted at high carrier densities, which do not change much during mode locking. In this situation one can make an effective relaxation approximation of the correlation contributions. This is accomplished by extracting the effective rates from full microscopic calculations before attempting any mode locking simulations. With this method the simulation time used to attain mode locking is reduced from months to days. The mode-locked pulse that results from using effective rates is similar to the one that would come from the full microscopic dynamics as seen later for carrier scattering in Figure 9.6. There are some visible deviations in the pulse that arrises from the complex many body dynamics, and it is worth emphasizing that many-body effects would have a greater influence on pulses that are longer than the scattering timescale.

We are ready to write the approximations of the higher-order correlation terms once the appropriate characteristic times have been determined. In this framework the injection pumping is approximated by $\Gamma^{e(h)}_{\lambda(\nu),\text{scatt}} = -(n^{e(h)}_{\lambda(\nu),\mathbf{k}} - f^{e(h)}_{\lambda(\nu),\mathbf{k}})/\tau_{\text{scatt}}$, where τ_{scatt} is the relaxation time and $f^{e(h)}_{\lambda(\nu),\mathbf{k}}$ is a Fermi distribution at the lattice temperature with the appropriate background carrier density. The polarization dephasing is approximated by $\Gamma_{\lambda,\nu,\text{deph}} = -(1/\tau_{\text{deph}})p_{\lambda,\nu,\mathbf{k}}$, where τ_{deph} is the characteristic dephasing time. The kinetic hole filling is modeled using $\Gamma^{e(h)}_{\lambda(\nu),\text{fill}} = -(n^{e(h)}_{\lambda(\nu),\mathbf{k}} - F^{e(h)}_{\lambda(\nu),\mathbf{k}})/\tau_{\text{fill}}$, where τ_{fill} is the kinetic hole filling time and $F^{e(h)}_{\lambda(\nu),\mathbf{k}}$ is a dynamic quasi-Fermi equilibrium distribution with the same temperature and density as the carrier distributions $n^{e(h)}_{\lambda(\nu),\mathbf{k}}$. The term $\Gamma^{e(h)}_{\lambda(\nu),\text{fill}}$ approximates the behavior of the carrier scattering while in the presence of the pulse. When the carriers are distorted from their equilibrium distribution by an incoming pulse, this term will cause the carriers to relax back to a quasi-Fermi distribution of the same density and temperature. This process is much faster then the pump injection, and thus the general behavior of the carriers is to first relax toward $F^{e(h)}_{\lambda(\nu),\mathbf{k}}$, and then relax back toward $f^{e(h)}_{\lambda(\nu),\mathbf{k}}$ on a slower timescale.

The inversion, $n^e_\mathbf{k} + n^h_\mathbf{k} - 1$, of a QW has been found to be an important quantity when determining mode-locked pulse properties. This allows one to relate the cavity pulse to the microscopic carrier dynamics and study the inverted carriers, which are found where the inversion is positive. If the QWs are in an equilibrium state, then one can use the gain spectrum to study the influence of the carriers on a low energy pulse. However, the gain is not enough to explain most features of the microscopic dynamics during mode locking. In order to compute the gain (or absorption) spectrum, one has to integrate Eq. (9.2) under the influence of a probe field. The gain spectrum can then be found as the ratio of the macroscopic polarization to the electric field [22]. Unlike the usual textbook linear gain spectrum, this includes the influence of many-body Coulomb interactions. Figure 9.5 shows the gain spectra computed from a gain chip where the QW carriers are either in an equilibrium distribution (dashed), or perturbed (solid). The carriers in equilibrium have a density $1.9 \cdot 10^{16}$ m^{-2} and both of the perturbed carriers have a density of about $2.0 \cdot 10^{16}$ m^{-2}. As seen above, very similar gain curves can come from very different carrier distributions. Thus it is hard to infer anything about the carrier dynamics from the gain unless one assumes the carriers are in equilibrium, and, even then, the gain shape and magnitude can be different – even for CW gain. In order to model mode locking of VECSELs, we

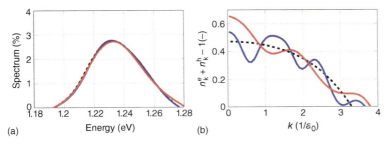

Figure 9.5 The solid (dashed) line in (a) shows an example where the gain is very similar when computed with the solid (dashed) inversion seen in (b) with the corresponding color.

will solve Eq. (9.1) coupled to Eq. (9.2) where the gain can be computed if the need arises.

In the buildup phase of mode locking, the growing cavity pulse will extract more and more inverted carriers from the gain chip QWs. The inversion is *bleached* once the pulse takes out enough carriers to reduce the inversion to zero at some momentum value(s), and at this point the QW does not provide any amplification from the bleached momentum values. If the pulse extracts even more carriers after the inversion is bleached, the negative inversion will make the QW absorbing.

The shape of the kinetic hole in the carrier distributions will be similar to the spectrum of the interacting pulse. A wider pulse spectrum will make a wider kinetic hole, while a narrow pulse spectrum will make a correspondingly narrow kinetic hole. Simultaneously as the pulse extracts carriers, the carrier scattering will start to equilibrate carriers into a quasi-Fermi Dirac distribution. The carriers surrounding the kinetic hole will begin to scatter down into the hole, and higher momentum carrier will scatter down into the available lower momentums states. This process is referred to as *kinetic hole filling* [28, 30, 34].

The source of photons in the cavity is the spontaneous emission from the inverted carriers in the gain chip. In this model the cavity field is treated classically, giving rise to both gain and absorption, but this also means that we need to add spontaneous emissions into our model. Much work has gone into modeling of this effect and several results are very similar, where the final expression can differ by only a prefactor [46–48]. In the following we use the same model as Baumner et al. where $\Lambda^{n}_{\lambda,\nu,\text{spont}} = -\Lambda^{\text{spont}}_{\mathbf{k}} n^{e}_{\lambda,\mathbf{k}} n^{h}_{\nu,\mathbf{k}}$ and $\Lambda^{p}_{\lambda,\nu,\text{spont}} = \beta \Lambda^{\text{spont}}_{\mathbf{k}} n^{e}_{\lambda,\mathbf{k}} n^{h}_{\nu,\mathbf{k}}$. Here $\Lambda^{\text{spont}}_{\mathbf{k}} = \frac{n^{3}_{\text{bgr}}}{\pi^2 \epsilon_0 \hbar^4 c_0^3} |d^{\lambda,\nu}_{\mathbf{k}}|^2 \left(E_g + \frac{\hbar^2 \mathbf{k}^2}{2m_r}\right)^3$, and β is a complex number with a random phase factor whose modulus determines the coupling of noise into the cavity.

In order to speed up calculations, one can use a rate approximation for the kinetic hole filling. A characteristic timescale can be found by tracking the momentum-dependent carrier occupation numbers as they relax back to the equilibrium distribution through carrier scattering. The carriers are first distorted by a pulse of a given amplitude and width centered at some momentum value corresponding to the central pulse frequency. Then one can track the recovery of the carriers as they scatter back toward a quasi-Fermi distribution. The timescale of this process can be found by fitting a decaying exponential to each momentum value. Finally the momentum-dependent characteristic timescale is found by averaging over multiple different distortions of the occupation numbers. The calculated characteristic timescale is a function of the crystal momentum and can be seen in Figure 9.6a. Figure 9.6b shows an example of a mode-locked pulse that is found when using either the rate approximation, the full carrier scattering, or no carrier relaxation at all [26]. In the calculations using the rate approximation, the simulations were completed using a constant timescale of 100 fs for both the electrons and holes. The rate approximation to the kinetic hole filling appears to be good near the peak of the pulse, but deviations appear in the surrounding region. When compared with simulations with no hole filling, one can see that the mode-locked pulses have more energy. This is expected as carrier scattering results

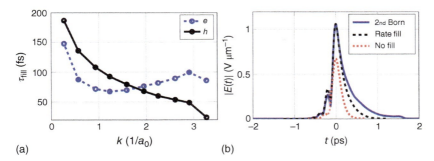

Figure 9.6 (a) The calculated characteristic carrier scattering rate from a QW with background carrier density $1.9 \cdot 10^{16}$ m^{-2}. (b) The mode-locked pulse that results from either using the full second Born, a constant kinetic hole filling rate of 100 fs or without carrier scattering. Source: Reprinted with permission from Ref. [26], (OSA).

in a transport of carriers into the spectral region where the pulse is located. The rate approximation approach leads to faster simulations, but one would have to run the full carrier scattering in order to verify the validity of the approximation.

9.3 Domain Setup/Modeling

9.3.1 The VECSEL Cavity

In order to gain insight into the role of the microscopic dynamics in mode locking, we will pick the simplest cavity that includes all the important components. A minimal VECSEL cavity should contain at least the gain chip, the SESAM, and an output coupling. The linear cavity includes all the important components while also reducing the computational time and complexity of the problem to a manageable amount. Here, the gain chip is placed directly across from the SESAM with an output coupler, and the pulse propagates between the elements. A limitation of a one-dimensional model is that it will be unable to model spatial effects such as the interference patterns from multiple colliding pulses; however, because the pulse propagates essentially along a 1D path through the cavity, it is possible to extend a one-dimensional model to cover more than the linear cavity [32].

The gain chip shown in Figure 9.7a consists of GaAs/AlGaAs distributed Bragg reflector (DBR), a gain region with optically active QWs arranged as RPG, a cap layer, and a dispersion-compensating coating. All these components are centered on the peak gain frequency. This design resembles the design used in real gain chips [6, 7]. The role of the gain chip is to provide spectral amplification to the cavity field, where the amount of amplification depends on the density of carriers in the inverted QWs. As we will show below, the specific QW arrangement can result in widely different mode locking dynamics, and if the goal is to provide ultrashort pulses, there are preferred design strategies.

The dispersion-compensating layers found at the top of the gain chip are used to improve performance of pulsed lasers. These layers are also sometimes referred to as

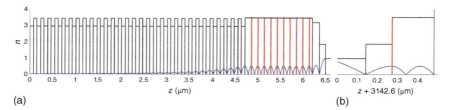

Figure 9.7 A schematic of the material layers in a linear VECSEL cavity with devices: (a) An RPG structure with 10 QWs placed at the antinodes of a standing wave at 980 nm shown in blue. The red lines represent the location of the QWs. (b) A SESAM with a single QW on top of an output coupling mirror on the right. Source: Adapted with permission from Ref. [26], (OSA).

antireflection coating (AR), since they can be used to reduce surface reflection. For the purpose of producing stable mode-locked pulses, the specific configuration, i.e. the number of layers, the specific materials used, and the layer thicknesses become an essential part of the cavity. The coating reduces the amount of dispersion seen by the pulse as it transitions from air into the gain chip and thus helps keep the pulse shape stable. The cavity pulse will accumulate dispersion from all surfaces it propagates through. Thus, when the desire is to achieve ultrashort mode-locked pulses, one has to control the total cavity dispersion. In practice, constructing these devices with coatings can be difficult, as simulations show that a deviation of as little as 1% in the layer thickness can sometimes result in doubling of the pulse length [15].

The role of the SESAM is to provide a saturable loss in order to facilitate pulse generation in the VECSEL cavity. One design of a SESAM is shown in Figure 9.7b) where the absorbing QW is placed on top of an output coupling mirror. It is more practical for the other cavity configurations, to place the absorbing QW on top of a DBR mirror. The main difference between these two SESAM designs is that the DBR will have a limited stopband with dispersion on the edges, whereas the output-coupling mirror has no stopband and zero dispersion. The DBR of the SESAM can only be a limiting factor for short pulse generation and will complicate the total cavity dispersion. In this paper we will only consider the simple SESAM with an AR coating on top of an output-coupling mirror in order to keep the model simple and transparent.

In an experiment the intensity of the cavity field is usually focused onto the SESAM. This increases the cavity field intensity on the SESAM compared with other components, and this will make the SESAM saturate at lower cavity field intensities. A higher level of focus will in general help produce shorter pulses.

9.3.2 The Gain Region

The gain region contains all the optically active QWs. It is responsible for the pulse amplification as well as the pump absorption. For the purpose of attaining high modal gain in a structure, one would want to arrange the QWs in an RPG pattern. When a short pulse is desired, one can vary the QW placement and get a multiple quantum well (MQW) arrangement.

A subcavity can be constructed inside the gain region in order to amplify the available gain, where a gain region can be either resonant or antiresonant as characterized by Tropper et al. [5]. In a resonant cavity the standing wave at a frequency has an antinode on the edge of the gain region. This results in amplification of the cavity field and the gain, but also leads to additional dispersion. In an antiresonant cavity, such as seen in Figure 9.7, one tries to reduce any subcavity effects by ensuring that the standing wave has a node on the edge of the gain region. The resonant cavity has more gain and dispersion, while the antiresonant cavity will have reduced gain and dispersion [9]. In order to generate ultrashort pulses, one needs to reduce the dispersion in the cavity, and thus an antiresonant design is preferred. However, using a multilayer dispersion-compensating coating to "remove" the interface between air and the gain chip will remove the subcavity from both designs. In that case the gain will be comparable for both designs and the dispersion will be removed. Thus for short pulse generation, one will include one or more layers of dispersion compensation.

In a gain chip design that emphasizes high modal gain, or single mode CW output, one should place the QWs on the antinodes of the standing wave corresponding to the frequency of the peak gain i.e. a RPG design. In this design the arrangement of the QWs ensures that the total modal gain from all the QWs preferentially amplifies a single frequency and the design has been used to produce longer pulses with high power or high-power CW output [6, 9]. However, this design does not offer any advantage for producing ultra short mode-locked pulses. The amplification of a single frequency above all others allows the pulse to easily bleach the inversion at this frequency. Thus it is difficult for the pulse spectrum to widen without causing the central frequency to start absorbing. This structure is easy to use as a test bed for studying mode locking dynamics and determining general behavior.

The MQW arrangement of the optically active QWs in the gain region refers to any arrangement that is not an RPG. In recent experimental works MQW structures have been found to be the most successful at producing short mode-locked pulses [12, 15]. A broader range of frequencies will be amplified as the QWs are moved away from the antinodes of the standing wave at the frequency of peak gain. As an example considers the 10 QW MQW structure shown in Figure 9.8a, where the QWs

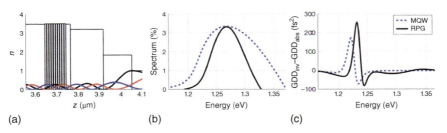

Figure 9.8 (a) A schematic of an MQW structure with 10 QWs placed inside a single antinode of the central wavelength at 980 nm. Standing waves from the DBR at wavelengths 930 nm (blue), 980 nm (black) and 1030 nm (red) are scaled for convenience. (b) The gain from the MQW structure (dashed) compared with the RPG structure (solid). (c) The change in GDD for the MQW and RPG structure. Source: Adapted with permission from Ref. [25], (OSA).

have all been placed inside a single antinode [25]. The gain from the MQW structure is reduced when compared with an RPG structure, because the QWs are no longer placed on the antinodes. In order to compare the gain from the two structures, one can increase the QW carrier density in the MQW structure such as to attain the same peak gain. As is seen in Figure 9.8b, the gain of the MQW structure becomes inhomogeneously broadened, when compared with the gain from an RPG structure. The background carrier density in the QWs of the MQW structure is $3.25 \cdot 10^{16}$ m^{-2} while it is $2.22 \cdot 10^{16}$ m^{-2} for the RPG. Another effect of moving the QWs away from the antinodes of the standing wave is that the resulting dispersion from these QWs is reduced. This reduction will help in the characterization and short pulse generation of the device. In Figure 9.8c) one can see the dispersion from the MQW structure compared with the RPG structure at the densities used to compare the gain in the previous figure. The dispersion is reduced because the QWs are now no longer resonant with the standing wave at the peak gain frequency. Indeed, as is seen from Figure 9.8a), the QWs that are placed off resonance are now in resonance with standing waves at different frequencies from the peak gain. This means that frequencies outside of peak gain will now be amplified at a higher rate than in an RPG structure. The MQW structure above is chosen as an example, because it is a big contrast to the RPG, and it is capable of producing very short pulses. However, if one wanted to build such a structure, the barrier pumping of the QWs and the material strain would be problematic. Therefore spreading the MQW structure over one or two more antinodes can achieve short pulses while also increasing the pump absorption in the barriers and the material.

9.3.3 The Relaxation Rates and the Round Trip Time

The length of the cavity determines the repetition rate of the laser. In terms of comparing the dynamics of the QWs, one is more interested in comparing the round trip time of the intracavity pulse to the relaxation times of the SESAM and inverted QWs. First for stable mode locking, the SESAM should relax fast enough such that the pulse will always see the same SESAM state each round trip. This is not an absolute requirement, but anything else will interfere with a stable mode-locked pulse. Second, if the propagating pulse were to see the same QW gain each round trip, then the pulse will amplify ad infinitum. Thus once the pulse becomes strong enough to bleach the QW inversion, the single strong pulse will start to break up into multiple pulses in order to better take advantage of all the available inversion left over in the QWs. Thus the round trip time of the pulse cannot be too long when compared with the inverted QW relaxation time. In the lower limit, in order to be physical the inverted QW relaxation times cannot be on the same scale as the carrier scattering. For a given chip: the relaxation time of the inverted carriers and the SESAM relaxation time can be determined experimentally [28]. However, as long as the above conditions are met, one will be in an ideal situation to produce mode-locked pulses. Thus, in order to expedite simulation times for mode locking, one can use a shorter cavity with a round trip time on the order of tens of picoseconds and comparably short carrier relaxation times.

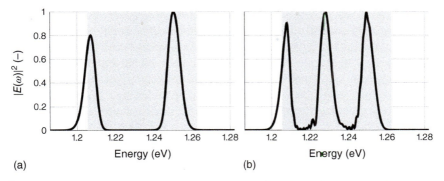

Figure 9.9 An example that shows the emergence of a pulse train in the cavity due to a slow SESAM. (a) A stable mode-locked pulse with two spectral peaks. (b) A third spectral peak appears once the relaxation time of the SESAM is decreased. Source: Adapted with permission from Ref. [27], (OSA).

Another problem could be that the SESAM is not recovered during each round trip, which could happen if the relaxation time of the SESAM is too long compared with the round trip time. This compromises the ability of the SESAM to absorb and can lead to the development of secondary pulses in the cavity. Figure 9.9b shows an example where the slow SESAM relaxation time has led to a third spectral peak developing between the two spectral peaks of a previously mode-locked pulse [27]. The original pulse had two spectral peaks as seen in Figure 9.9a. As the above figure shows, the SESAM was unable to clean up the excess carriers in the system once the relaxation time was made considerably slower. This leads to a third spectral peak appearing, and in the time domain this spectral peak appears as a second pulse in the cavity, which then leads to destabilization of mode locking.

9.3.4 Noise Buildup to Pulse

The spontaneous emission of photons into the VECSEL cavity will start once the gain chip QWs are pumped. Figure 9.10 shows how the cavity field goes from noise to a mode-locked pulse. The uncorrelated low-amplitude noise that is propagating

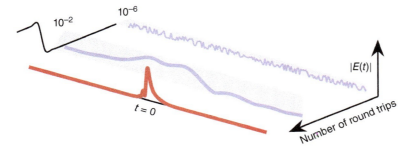

Figure 9.10 A diagram showing the development from noise to a mode-locked pulse. Source: Pulse data from Ref. [28, 32].

in the cavity has a wide range of frequencies. As the noise makes its way through a full round trip of the cavity, the only frequencies that will be amplified are the ones located inside the net gain region, i.e. the spectral range where there is more amplification than loss. Eventually the noise will become correlated and have enough energy to start stimulated emissions in the inverted QWs of the gain chip. The energy will keep growing until the inversion in the gain chip is bleached. It is the transient bleaching of the QW inversion that leads to the final mode-locked pulse. As soon as the inversion is bleached, the field that follows will be absorbed. The resulting pulse shaping leads to the formation of a peak in the cavity field. Over some round trips this process will naturally widen the spectrum of the cavity field, leading to the formation of a pulse. The final mode-locked pulse in a VECSEL strikes a balance between extracting carriers from the gain chip and loss in the SESAM and output coupler.

9.4 Numerical Results

9.4.1 Single-Pass Investigation of QWs and SAMs on the Order of Second Born–Markov Approximation

Before investigating actual mode locking, we examine the response of single quantum wells in the active region and that of saturable absorber mirrors to a single pass of optical pulses. This reduces the observation time from many nanoseconds to just a few picoseconds and allows one to test the full theoretical model. It also enables us to test the viability of potential simplifications like the reduction to only one parabolic electron and hole band and the replacement of detailed carrier scattering processes by effective rates.

Here we solve the full microscopic many-body equations, where an eight-band **k** · **p** model is used and all subbands in the relevant spectral range are included. The electron–electron and electron–phonon scatterings that lead to the dephasing of the polarization and the equilibration of the carrier system are calculated explicitly in the second Born–Markov approximation. For more details of the fully microscopic model, see [28, 49, 50].

First, the model is used to study the response of quantum wells that have carrier occupations typical for lasing operation when hit by pulses with lengths and intensities similar to ultrashort mode-locked pulses. Here we examine the creation of kinetic (spectral) holes in the carrier distributions and the influence of hole filling during pulse amplification. Then we use the model to examine the response of saturable absorber mirrors to similar pulses. We examine semiconductor saturable absorbers (SESAMs) made from a single unpumped quantum well as well as graphene saturable absorber mirrors (GSAMs) made of a single layer of Graphene on a dielectric substrate. For both systems, the absorption bleaching due to carrier buildup is studied to extract recovery times and saturation fluencies and their dependence on pulse widths and spectral positions.

9.4.1.1 Inverted Quantum Well

For weak light pulses the absorption or gain is a good measure for the wavelength-dependent weakening or amplification of the pulse. In this so-called

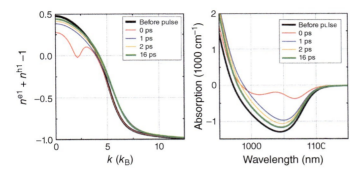

Figure 9.11 Inversion of the lowest electron and hole subband (left) and linear absorption (right) at various times before and after an intense 100 fs pulse passes through the inverted quantum well [28]. The pulse maximum is centered in the well at 0 ps.

linear field regime, the inversion, $n^e_k + n^h_k - 1$, does not change significantly during the presence of the pulse. In the two-band case and apart from Coulomb effects, absorption/gain will mirror the inversion. The occupations remain in thermal equilibrium where they can be described by Fermi distributions. While electron–electron and electron–hole scatterings do take place, they do not change the distributions.

On the other hand, intense pulses, as they are typical for mode locking, change the carrier occupations significantly while passing through the quantum well. Scatterings further modify the nonequilibrium distributions on timescales similar to typical mode-locked pulses of 100 fs or more. Here, the (linear) absorption/gain no longer sreflect the actual response of the system. This is demonstrated in Figure 9.11 for the example of an inverted GaAs/InGaAs well that is hit by a Gaussian pulse, $E(t) = E_0 \exp(-t^2/(2\Delta^2)) \cdot \exp(-i\omega_0 t)$, centered at the frequency ω_0 of the gain maximum and with a width $\Delta = 100$ fs. The pulse intensity, $E_0 = 15$ V μ m^{-1} is strong enough to bleach the inversion completely to transparency (zero inversion) at the time when the pulse maximum hits the well (0 ps). Here, the second half of the pulse will slightly further reduce the occupations and lead to slightly negative inversion at the central frequency of the pulse. This so-called Rabi-flopping beyond transparency is possible since the pulse length is comparable to the dephasing time of the microscopic polarizations. Here, the system remains in a partly coherent state rather then instantly creating free carriers.

While the inversion is completely bleached by the pulse, the absorption spectrum still shows gain at the central lasing frequency. Here, the absorption was calculated by testing the instantaneous system and occupations using a weak and ultrashort, few femtoseconds test pulse.

The spectrally sharp kinetic hole is quickly filled by intraband scattering on a one-hundred femtosecond timescale. After that, the occupations are refilled by the relaxation of pump-injected carriers from higher subbands and the barriers. This is mediated by interband scatterings that take place on a multi-picosecond timescale. Relaxation from high momentum states toward the bandgap is slowed down due to the fact that the initial states, n_i, for these processes have small occupations while the

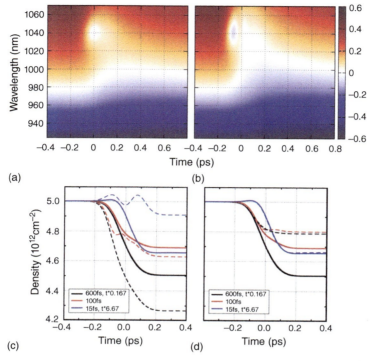

Figure 9.12 (a, b) Time evolution of the inversion in a QW that is excited with a 100 fs [19]. (a) For a pulse with $E_0 = 15$ V μ m^{-1}. (b) For a pulse with $E_0 = 30$ V μ m^{-1}. The pulse maximum is centered in the well at 0 ps. (c, d) Time evolution of the total carrier density in a quantum well that is excited with a 15 fs/100 fs/600 fs pulse (blue/red/black) [19]. Solid (dashed) lines in (c): for a pulse with amplitude E_0 (2 E_0). Solid/dashed lines in (d): when including/neglecting kinetic hole filling. The timescales for different pulse lengths have been adjusted by Δ/100 fs.

final states, n_f are occupied to a high degree. Thus, the terms $n_i(1 - n_f)$ in the scattering equations are reduced by phase-space filling. This leads to a hotter distribution than the initial one before the pulse that persists on a ten picosecond scale.

Figure 9.12a and b shows the time-resolved inversion for the same well as in Figure 9.11 for excitations with 100 fs pulses of two different intensities. The case of the lower intensity is the same as in Figure 9.11. For the lower intensity pulse the Rabi-flopping only leads to a single overshoot beyond transparency and is barely visible on this scale. For a pulse with twice the field strength strong Rabi-flopping is visible. Here, already the leading edge of the pulse is strong enough to completely bleach the gain, and oscillations remain until the pulse has completely passed.

The influence of Rabi-flopping and kinetic hole filling on the pulse amplification is demonstrated in Figure 9.12c and d for pulses of various lengths and two different strengths. The pulse amplitudes have been scaled to result in the same pulse energy for all pulse widths. For very short pulses, the Rabi-flopping becomes more and more important and limits the amount of carriers the pulse can extract and, thus, the pulse

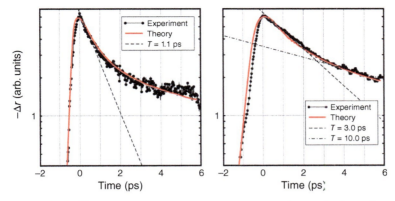

Figure 9.13 Time-dependent (negative) index change of mode-locked VECSELs [28]. Left: VECSEL operating near 980 nm producing 300 fs pulses. Right: VECSEL operating near 1040 nm producing 800 fs pulses. Dots: experiment. Solid line: theory. Dashed lines: exponential fits.

can be amplified. Clearly, this is undesired for lasing operation. For pulses on the timescale of the polarization dephasing (50-100 fs) or longer, the flopping becomes less significant. The kinetic hole filling re-replenishes carriers at the lasing frequency while the pulse is present. This allows the pulse to extract more carriers. Pulses that are shorter than the intraband scattering times cannot take advantage of this effect.

To test the theoretical model, we compare in Figure 9.13 calculated refractive index changes for two mode-locked VECSELs to experimentally measured data. Details of the experimental procedure can be found in Ref. [28]. The two VECSELs shown here produce rather long pulses of several hundred femtoseconds. Thus, the kinetic hole is not very pronounced in these cases and the initial fast recovery of it is not visible. However, the data shows two different timescales in the recovery of the system. The initial faster recovery is due to the return of the system to distributions that resemble hot Fermi distributions. The following slower recovery is due to the cooling of the system and subsequent refilling of the wells with pump-injected carriers.

In both cases we find very good agreement between theory and experiment demonstrating the high quantitative accuracy of the model and its potential for device design and optimization.

9.4.1.2 Saturable Absorber

Figure 9.14 shows the buildup of carriers due to absorption in two saturable absorber mirrors. One is a SESAM for wavelengths around 1000 nm. It consists of a single InGaAs/GaAs quantum well on a GaAs/AlAs DBR. The quantum well is capped by a 4 nm thin GaAs cap-layer to allow for fast carrier recombination on the surface. The other absorber is a GSAM consisting of a single layer of graphene on a SiO_2 substrate. The fundamental model for the GSAM closely follows that of the semiconductors outlined above. The main changes are the different (gapless) dispersion and wavefunctions. Details about the model for Graphene can be found in Ref. [28] and Refs. therein.

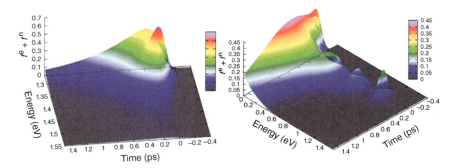

Figure 9.14 Time evolution of the sum of electron and hole occupations in a SESAM for 995 nm (left) and a GSAM (right) after excitation with a 100 fs, 0.1 mJ cm^{-2} pulse centered at 1.32 eV [28].

Both SAMs are excited with a 100 fs pulse with an energy density of 0.1 mJ cm^{-2} and a central frequency of 940 nm. This wavelength is chosen somewhat further above the bandgap than in typical uses of the SESAM in mode locking. This was done in order to make the internal dynamics more visible.

The main difference between the SESAM and GSAM is that in the SESAM the carriers can only relax to the bandgap that is energetically rather close to the excitation. On the other hand, due to vanishing gap in graphene, the carriers can relax much further away from the excitation energy. Thus in the SESAM, carriers will remain near the energy of the excitation after the initial relaxation. This leads to a slower recovery time at the excitation energy than in the GSAM.

Due to the strong phonon coupling in Graphene, carriers relax via phonon scattering on a sub-100 fs timescale. Several LO-phonon replicas can be seen in the time trace. The density of states in graphene scales linearly with the energy. This leads to higher occupation probabilities at lower energies than at the excitation energy.

In the SESAM, the initial pile-up of carriers near the central frequency of the pulse is quickly smoothed out through intraband scattering. This happens on a 100 fs timescale similar to the kinetic hole filling in the inverted quantum well case studied above. It is followed by a slower recovery due to interband scatterings and nonradiative carrier recombination.

The featureless dispersion and density of states in graphene lead to saturation characteristics that vary very little with the excitation wavelength. On the other hand, the density of states in the SESAM shows an abrupt step at the bandgap of the well. This, as well as steps at higher subbands lead to a high wavelength sensitivity, which will require careful optimization in the experiment. In Figure 9.15 the wavelength dependence of the saturation fluence was studied for the SESAM and GSAM from Figure 9.14 in Ref. [50]. In the SESAM, the minimal saturation fluence is found for excitation right at the lowest subband exciton. It quickly increases by almost an order of magnitude with a detuning of a few ten nanometers. For some applications it might be desirable to have a fairly wavelength-independent saturation fluence. Then, the excitation should be centered higher within the first

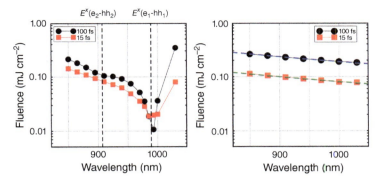

Figure 9.15 Saturation fluence as function of the excitation wavelength for the SESAM (left) and GSAM (right) from Figure 9.14 [50]. Black: using 100 fs pulses. Red: using 15 fs pulses. E^x marks the position of the first and second subband excitonic bandedge. The dashed lines for the GSAM are fits according to $1/\lambda^2$.

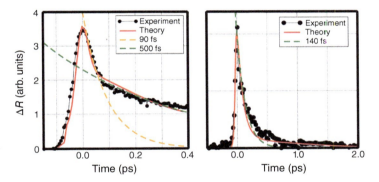

Figure 9.16 Time evolution of the index change in a SESAM (left) and GSAM (right) for the excitation conditions as in Figure 9.14 [28, 50]. Dots: experiment. Solid line: theory. Dashed lines: exponential fits.

subband. The saturation fluence in the GSAM is fairly wavelength-independent with a scaling of about $1/\lambda^2$ due to the linear increase of the density of states.

In Ref. [50] also the dependence of the SESAM recovery time on the wavelength was studied. The fastest recovery times were found for excitation either directly at the lowest subband exciton or just below the onset of the second subband. The recovery time decreases for excitations further away from the bandgap since this allows the carriers to relax away from the excitation energy.

The index change after excitation with pulses was studied experimentally for a SESAM and GSAM using time resolved pump-probe measurements [28]. The results are compared with those of the modeling in Figure 9.16. Here, a SESAM for operation near 1040 nm was examined and the system was excited with a 100 fs pulse at 1040 nm with an energy density of 25 µJ cm^{-2}. The same pulses were used to examine the GSAM.

As for the results for the inverted quantum wells, we find very good quantitative agreement between theory and experiment. Using the theory–experiment comparisons, we deduce recovery times for the SESAM and GSAM of 500 fs and 600 fs, respectively. These times were used for the final data shown in Figure 9.16.

The initial decrease of the reflectivity change in the SESAM is due to an adiabatic following of the pulse envelop (i.e. Rabi-flopping). No clear signature associated with the smearing-out of the carrier pileup near the center frequency can be found here since the excitation is too close to the bandgap. The long-time recovery is due to the nonradiative extraction of the carriers.

In the GSAM a strong initial recovery on a 140 fs timescale can be seen. Since the carriers in the GSAM can relax far away from the excitation energy, the index change quickly recovers to a large degree despite the carriers still being in the system. It should be noted that for some GSAMs a change in sign of the change of the transmission after the initial recovery has been observed [51]. This signature and its dynamics can be reproduced very well by the theory if it is assumed that the graphene is doped [28]. The occurrence of negative differential transmission means that the system is less absorbent than before the excitation. The excitation leads to a heating of the dopant-related carriers that remain after the pulse-created carriers have recombined. The higher carrier temperature leads to higher occupation probabilities near the probe energy, which reduces the absorption there.

9.4.2 Mode-Locked VECSELs

In the following we will examine the dynamics of a mode locked VECSEL cavity as described before. The numerical simulations are run on an SGI UV2000-shared memory machine. Great care has been taken to make sure the simulations run quickly, including parallelizing the numerics when appropriate. Simulating equation Eq. (9.1) coupled to Eq. (9.2) until mode locking is attained still takes between 12 and 48 hours. This might seem to be long, but if the carrier scattering is calculated at the level of the second-order Born–Markov approximation, a simulation can take around 30 days to attain mode locking.

First we go over a few ways that one can use the model to characterize the cavity by measuring the gain, absorption, and dispersion. By measuring this in the same framework as the mode locking, one will get an impression of what the pulse actually experiences during a full simulation. Next the cavity is initialized for longer mode locking simulations where one starts from noise or from a seed pulse. Finally, we discuss how to find even shorter pulses.

9.4.2.1 Gain, Absorption, and Dispersion

The mode locking behavior depends on the energy balance in the cavity, thus it is important to be able to characterize the spectral absorption or amplification from a given gain chip or SESAM in order to control the final mode-locked pulse out of a VECSEL cavity. These variables don't enter into the given model, but can be computed at a given background carrier density. We can find the gain spectrum by computing the spectral change of a weak test pulse that propagates though the

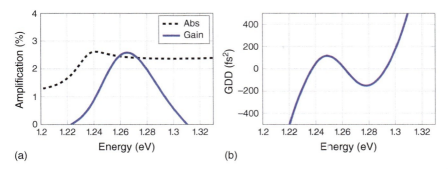

Figure 9.17 The initial loss and gain in the cavity (a) and the GDD in the gain chip (b) [31].

gain chip with inverted QWs. The pulse is propagated through the device and the spectra of the pulse before, $E_0(\omega)$, and after, $E_1(\omega)$, are compared using: $G(\omega) = Log(E_1(\omega)/E_0(\omega))$. Where $G(\omega)$ gives the amplification experienced by the pulse in the device. Figure 9.17a shows the gain from an RPG device with inverted QW density $1.9 \cdot 10^{16}$ m^{-2} (solid line) and the total loss from the SESAM and output coupler computed in a similar way (dashed line). This picture only gives an idea of the mode locking dynamics by indicating if there is enough amplification from the gain chip to initialize a cavity. When comparing the gain spectrum to the total loss in this way, it is possible to see that if the inverted background carrier density is too low, there will be no net round trip amplification, and thus no pulse will build up in the cavity. On the other hand, multiple pulses will appear in the cavity if the gain is too high.

The total dispersion in the VECSEL cavity will influence how the pulse is reshaped during a round trip. The pulse will accumulate dispersion from the SESAM and the gain chip as well as any other devices present in the cavity. A pulse with a wider spectrum, i.e. a shorter duration pulse, will require the dispersion to be controlled over a wider wavelength range. In order to study the influence of the dispersion on the mode-locked pulse, one also has to be able to measure the dispersion in a given structure. This can be accomplished by using the same procedure as described above, in order to characterize the device, but instead of comparing the amplitude change of the pulses we will now compare the change in spectral phase. It is possible to find each order of dispersion by calculating the appropriate Taylor Series expansion of the phase change.

Figure 9.17b shows the second-order dispersion that is found in a gain chip with an RPG gain structure on top of a DBR and dispersion-compensating coating. In order to generate very short pulses in a mode-locked VECSEL, one has to include dispersion-compensating elements into the cavity, and it is common to first reduce the dispersion from each cavity device independently and then reduce the total dispersion wherever possible. Unfortunately, dispersion-compensating coatings cannot remove all dispersion, e.g. it is impossible to remove the dispersion from the DBR stopband edges as seen in Figure 9.3. The simplest configurations utilize a single material layer on the air interface that will reduce the GDD; however, it is possible to use optimization algorithms in order to create combinations of material layers that reduce multiple orders of dispersion [44].

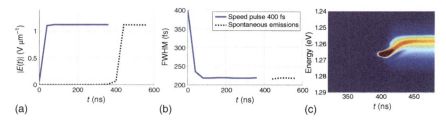

Figure 9.18 Here we see the pulse formation process from spontaneous emission noise (dashed line) as well as from a seed pulse (solid line). (a) The cavity field amplitude. (b) The cavity field pulse FWHM. (c) Snapshots of the cavity pulse at the cusp of evolving into a mode-locked pulse [31, 32].

9.4.2.2 Pulse Buildup and Initial Conditions

Figure 9.18 shows the time-resolved evolution of the pulse amplitude and width from noise [31, 32]. The gain region used in this simulation was an RPG structure, where the QWs had a density of $1.9 \cdot 10^{16}$ m^{-2} resulting in the initial gain seen in Figure 9.17. The simulation was started from noise and eventually reaches the mode-locked pulse around 400 ns. In general, the time that the cavity takes to develop a pulse depends on the round trip time and the net gain in the cavity. The amplitude of the field quickly grows until it bleaches the inversion, at this point the pulse spectrum widens and a mode-locked pulse is formed. Note that the central frequency of the field shifts slightly after the pulse is formed, this is such that the pulse can better take advantage of the available inversion. The formation of a pulse from noise and this behavior has been observed in experiments by Turnbull et al. [52].

Multiple realizations of the same cavity with different random noise initial conditions all give the same mode locked pulse, and another way to start the mode locking of a VECSEL cavity is to inject a seed pulse. This method is used in order to speed up the convergence to a mode-locked pulse, and it does not change the final mode-locked pulse. Note that the FWHM is only defined for a pulse, and thus there are some missing points in the FWHM graph before 400 ns.

9.4.2.3 Self-Phase Modulation from QWs

It is possible to measure the change in refractive index during pulse propagation using the Kramers–Kronig transformation [31]. The procedure is as follows: First a sech^2 pulse with energy and width similar to the mode-locked pulse in Figure 9.18 is used to distort the carriers of the gain chip and SESAM. Then one can probe the distorted carriers with a low-energy short pulse during the interaction of the high-energy pulse with the device. The spectra of each probe are compared, similar to the characterization procedure described earlier, to produce the gain or absorption spectra for the device at the given time. The snapshots of the absorption spectrum are then compared with the absorption spectrum before the pulse distorts the carriers. The Kramers–Kronig transformation finally converts the change in spectrum into a change in the refractive index.

Figure 9.19 shows the change in refractive index as computed with the Kramers–Kronig transformation for the RPG gain chip and SESAM. The pulse used

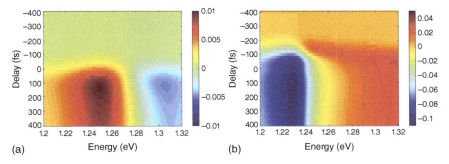

Figure 9.19 The refractive index change as computed with the Kramers–Kronig transformation during interaction with a pulse in the inverted gain chip (a) and the absorbing SESAM (b). Note the different color scales used [31].

for distorting the carriers is a 220 fs FWHM pulse located at 1.265 eV with energy comparable to the mode-locked pulse that came from this cavity. Note that the change in refractive index in the gain chip is much smaller than the change in the SESAM. This is because the high carrier density in the QWs of the gain chip sees less of a change than the low-density QW in the absorbing SESAM. As the pulse enters the SESAM, the change in refractive index becomes abruptly positive and then goes back to zero at the central wavelength once the pulse has bleached the SESAM. The positive refractive index that is seen in the SESAM will make the speed of the front part of the propagating pulse slightly slower, thus compressing the front of the pulse while it is propagating through the SESAM.

9.4.2.4 Mode-Locked Pulse Family

The final mode-locked pulse is a nonequilibrium balance between the energy input in the gain chip and the energy loss through the SESAM and output coupler. Additional constraints on the pulse can be found in the cavity dispersion, which will force the pulse shape to take up certain properties, which might correspond to a stable pulse solution. The pulse is modified nonlinearly in the inverted and absorbing QWs according to the microscopic many-body dynamics. A wide variety of stable mode-locked pulses can be found by varying the pump (inverted QW background density) continuously.

Through systematic numerical studies we have been able to identify the possible mode-locked pulses in a VECSEL. These are divided into the following types: a single stable mode-locked pulse, a single unstable mode-locked pulse, and a "pulse molecule." In addition there can be multiple pulses in a cavity; however, these are usually undesirable solutions that are unstable and consist of samples from the above pulses. The *pulse molecule* should also be considered as semistable, as it is usually oscillating slowly in FWHM and amplitude. Figure 9.20 show samples from all three families, with their respective spectra, autocorrelation traces, and the gain that they result from [26]. These calculations were generated using an RPG gain structure at three different loss/gain settings. Searching through many different QW arrangements has not resulted in any other pulses. The single pulse has a single spectral

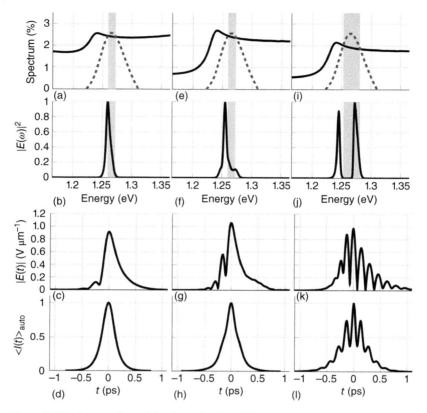

Figure 9.20 An overview of the three families of mode-locked pulses: the single stable pulse (column 1), the single semistable pulse (column 2), and the pulse molecule (column 3). Row 1 shows the initial gain, absorption, and output coupling that produced the pulse, as well as the net gain region shaded in gray. Row 2 shows the normalized spectrum of the pulse. Row 3 shows the pulse amplitude. Row 4 shows the autocorrelation trace of the pulse. Source: Reprinted with permission from Ref. [26], (OSA).

peak that is almost entirely contained inside the net gain region, which is shown in gray. The unstable single pulse has some substructure, and the spectrum of the pulse has a pedestal and is located partially outside the net gain region. Note that the autocorrelation trace of the two single pulses appears very similar for two very different pulse shapes, this emphasizes how difficult it is to recover information about the pulse shape from an autocorrelation trace alone. For the *pulse molecule* there are two spectral peaks that create the observed beating pattern in the envelope. The second spectral peak is located partially outside the region of net linear gain and is dominantly supported by the carrier scattering that is transporting energy into this spectral range. These pulses are selected examples that appear as a result of the chosen gain structure, pumping level, loss, and cavity dispersion. Choosing a different configuration will result in pulses with different properties; however, the final mode-locked pulse will be some variation of these pulses.

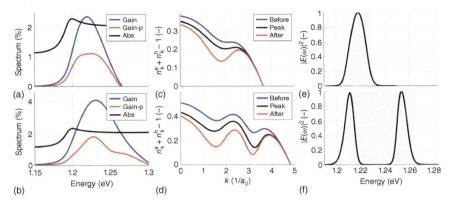

Figure 9.21 An overview of the dynamics for a single stable pulse (a, c, e) and a pulse molecule (b, d, f). Column 1 shows the initial equilibrium gain (blue) and absorption (black) with the nonequilibrium gain (red) calculated during interaction with the pulse. Column 2 shows snapshots of the inversion before (blue), at the peak (black), and after (red) the pulse has passed. Coulumn 3 shows the spectrum of the mode-locked pulses. Note: In these pictures the kinetic hole filling is not included in Eq. (9.2). Source: Adapted with permission from Ref. [27], (OSA).

In order to understand how the different pulse families appear, we will study the dynamics of the inversion. There are three different cases that stem from the different levels of inversion: A low level of inversion that leads to a single pulse, a medium level of inversion that leads to a "pulse molecule," and finally a high level of inversion that leads to multiple pulses in the cavity. At first, it is instructive to observe how these pulses appear in a simpler case where there is no carrier scattering included in the SBE [27].

Figure 9.21 shows, for the cases of a single pulse and pulse molecule: the calculated nonequilibrium gain, snapshots of the inversion during mode locking, and the mode-locked pulse spectra with the net gain region highlighted in gray. The mode-locked pulses are generated using an RPG gain structure and initial background carrier density of $1.75 \cdot 10^{16}$ m^{-2} for the first row, and $2.4 \cdot 10^{16}$ m^{-2} for the second row. In the figure the inversion is shown before the pulse appears, at the peak of the pulse, and after the pulse has passed. In the case of low inversion, the mode-locked pulse will not grow strong enough to bleach the carriers.

From Figure 9.21a and b we can see the nonequilibrium gain that was calculated from the inversion during mode locking. The mode-locked pulse has deformed the QW carriers, which in turn have reshaped the computed gain in such a way that it no longer resembles the initial equilibrium gain. This deformation happens during interaction of the pulse, and the specific gain shape will naturally affect the final mode-locked pulse.

Figure 9.22 shows snapshots of the inversion during the buildup of the pulse molecule at a medium level of inversion. In this case the pulse energy will become high enough to bleach the carriers during the buildup of the pulse. The time dependent inversion is eventually bleached at the central frequency. At this point the pulse cannot extract more carriers at this frequency, and in order to take

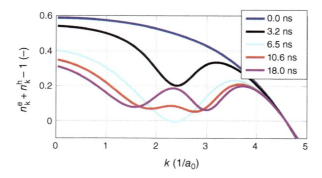

Figure 9.22 Here we see snapshots of the inversion during the buildup of the *pulse molecule*. Note: In this figure the kinetic hole filling is not included in Eq. (9.2). Source: Adapted with permission from Ref. [27], (OSA).

advantage of the carriers in the surrounding spectral region, the pulse spectrum will split into two separate peaks. The two spectral peaks of the final mode-locked pulse both fit into the net gain region and are shown in Figure 9.21f. Contrasting the inversion dynamics from the single mode-locked pulse with the dynamics from the pulse molecule, we can see that the pulse molecule has adapted in order to take out more of the available carriers. If the level of inversion is even higher, there will be too many carriers for a single mode-locked pulse to extract. In this case multiple pulses will appear in the cavity. Note that carrier scattering will result in a transport of carriers from the higher-momentum states into the available lower-momentum states. This causes the carriers to equilibrate into a quasi Fermi distribution between pulse interactions, which increases the available inversion in the lower momentum range. The increased pulse amplification can then lead to inversion bleaching. In this case it is unlikely that the net gain region can be both wide enough to support a *pulse molecule* and still not bleach the inversion. However, the carrier transport enables the pulse molecule to exist at a lower density. The second spectral peak can be supported outside the net gain region, as is the case in Figure 9.20j. In this case the *pulse molecule* is semistable with an oscillating FWHM and amplitude.

9.4.2.5 Influence of Loss on the Mode-Locked Pulse

The nonlinear interplay between dispersion, loss, and amplification complicates the search for the shortest FWHM and/or strongest possible pulse in a mode-locked cavity. Experimental studies by Klopp et al. have found that the mode-locked pulse properties depend on the cavity loss [39]. If the goal were a high-energy output, then one would seek to have a higher-output coupler combined with a high net gain in order to get the most amount of output energy. On the other hand, the amount of saturable absorption both influences the energy loss in the cavity and the pulse shortening capabilities of the absorbing QW. Thus the goal is to see if it is possible to optimize both the energy and the width of the mode-locked pulse, i.e. can we produce a high-energy pulse that is also very short?

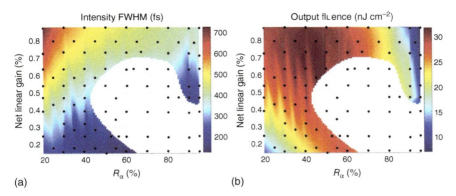

Figure 9.23 Overview of the mode-locked pulse FWHM (a) and output fluence (b) that results from varying R_α and the total absorption. The colored region is an interpolation of the raw data (black dots) with single stable mode-locked pulses. The pulses in the uncolored region are only semistable. Source: Reprinted with permission from Ref. [26], (OSA).

The total loss in the cavity consists of the saturable loss from the absorbing SESAM QW and the nonsaturable loss from the output coupler. In order to study their relationship to the pulse, we define, R_α, the ratio of saturable loss to the total loss at the central frequency. A high R_α means that most of the cavity loss is saturable, and a low R_α means that most of the cavity loss comes from the output coupler. In order to collect this data, we perform a scan over multiple values of R_α and the net gain for a fixed RPG gain chip [26]. For each data point we want to find the FWHM and energy of a mode-locked pulse, thus we iterate the simulation until we are sure that a mode-locked pulse is established. Figure 9.23 shows a collection of the output pulse FWHM and energy for multiple values of R_α and the net gain at the central frequency. The numerical simulations are the black dots, the colored region is an interpolation between the simulation data for the stable mode-locked pulses, and in the region with no color one can find semistable pulses such as the *pulse molecule* and the single unstable pulse from Figure 9.20. Not shown is that for even higher levels of net gain, one ends up with multiple pulses in the cavity. The output pulse energy is highest at high net gain and low R_α, i.e. a high-output coupling compared with the saturable absorption. However, the FWHM of the pulse appears to be the shortest when the net gain is low. These two requirements intersect around $R_\alpha = 50\%$ and at a low net gain, where the output pulse energy is high, and the pulse FWHM is at the lowest point of 140 fs. There is another comparably short pulse at a high R_α, but this solution has a lower energy. These pulses are both located near the region where pulses are no longer clearly mode locked.

The data in Figure 9.23 is for a specific gain chip, and a different setup will produce a different figure. However, it is clear that the influence of the loss on pulse FWHM and energy is not trivial. This analysis is very time-consuming and might be completely unpractical to perform on an experimental setup. Thus we find that, by looking at multiple gain structures, a rule of thumb is that a high R_α is good for producing shorter pulses.

9.4.2.6 Limits on the Shortest Possible Pulse and the Hysteresis Effect

What limits the shortest possible pulse that can be generated in a VECSEL cavity? There are two types of limitations that one encounters when trying to produce a short pulse, the first is that for a given cavity the short pulses are not easy to find in terms of tunable parameters such as the cavity loss, and the pump level of the gain chip. However, the most limiting features for short pulse generation are a result of chip design including: dispersion, the DBR stopband, and the QW arrangement.

In order to find the shortest possible pulse for a given gain chip, one has to scan over the tunable parameters, i.e. the cavity loss and the gain chip background carrier densities (the pump level). First we fix the gain chip design by settling on a specific gain region, DBR, and dispersion compensating coating that will reduce the GDD as far as possible. Next we perform a series of simulations where we try to find a short mode-locked pulse by first reducing the background carrier density in the inverted QWs for a fixed output coupling loss. Then we adjust R_α in order to check if it is possible to further reduce the pulse FWHM. These steps are repeated until one is satisfied with the final mode-locked pulse properties.

The dispersion limitations to short pulse generation are overcome by constructing decent dispersion compensation, which is much harder in an experimental setup than in a simulation. In these numerical simulations a single layer coating is used at a refractive index that makes the GDD as flat as possible. The next limitation on short pulse mode locking is the DBR stopband. A typical AlAs/AlGaAs DBR stopband, with respective refractive indices 2.946 and 3.435, has a theoretical width of about 96 nm for operation at 980 nm [38]. The short pulse cannot fill the entire stopband, because the dispersion from the stopband edges becomes too strong, and it is not possible to remove the dispersion from the edges using dispersion compensation coatings. At most the coating can reduce the dispersion around the center of the stopband. In order to observe the next limit on short pulse generation, one can use a different DBR such as a dielectric DBR composed of Nb_2O_5/SiO_2, with respective refractive indices of 2.26 and 1.45, which has a theoretical width of 280 nm for operation at 980 nm. In the following we are only interested in a wider DBR so for simplicity, the AlAs index of refraction is artificially changed to 1.9 resulting in a DBR with a theoretical stopband width of about 377 nm. The new DBR will also have a different dispersion from the previous structure. Thus, in order to make the GDD of the two structures as similar as possibly, we change the AR coating on the narrow DBR stopband structure. Once the limiting effect of the dispersion and the DBR stopband is reduced, one can observe the influence of the QW arrangement.

Figure 9.24 shows the mode-locked pulse from an MQW10 structure on a wide and narrow DBR stopband and the inversion of the gain chip QWs during mode locking [25]. The dispersion-compensating coating on the narrow DBR structure was changed, in order to reduce the GDD and to provide a fair comparison. This also leads to an amplification of the gain, and thus the background carrier density is also slightly reduced from $3.25 \cdot 10^{16}$ m^{-2} to $2.9 \cdot 10^{16}$ m^{-2} in order for the two simulations to have similar peak gain. The mode-locked pulse from the wide DBR structure has a FWHM of 19 fs and a peak intensity of 5.7 MW/cm^2, while the narrow DBR

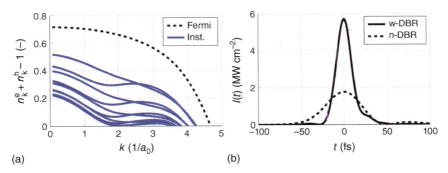

Figure 9.24 (a) A snapshot of the time-dependent inversion in the QWs that results from interaction with the pulse (solid lines) and the initial inversion (dashed). The mode-locked pulse is shown in (b) for the wide (solid) and narrow (dashed) stopband DBR. Source: Reprinted with permission from Ref. [25], (OSA).

produces a 46 fs pulse with peak intensity 1.7 MW/cm^2. Figure 9.24a shows a snapshot of the inversion in the gain chip QWs during mode locking of the 19 fs pulse. The inversion is sampled right after the peak of the pulse, at the time when the total number of carriers in the QWs is the lowest. Notice that the inversion in each QW is slightly different based on their relative placement around the antinode, this is because some of the QWs are in more resonance with the central wavelength of 980 nm while others are deliberately placed off resonance as can be seen in Figure 9.8a). The QW placement in the MQW structure helps with the amplification of a wider range of frequencies than the RPG structure, which in turn allows the pulse to take advantage of more inversion.

The tunable parameters, such as the background carrier density in the gain chip QWs, reveal a surprising effect that complicates the search for the shortest pulse. A memory effect can be observed if one tracks the mode-locked pulse shape while increasing and then gradually decreasing the gain chip background carrier density [25]. In Figure 9.25 one sees the mode-locked pulse intensity and FWHM with increasing (solid black) and decreasing (dashed red) carrier densities, as well as

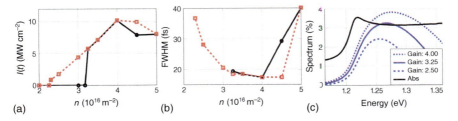

Figure 9.25 Here we see the mode-locked pulse peak intensity (a) and FWHM (b) for various levels of the inverted QW background carrier density. The red curve corresponds to mode-locked pulse shapes found when incrementally decreasing the density from the highest value. In (c) we can see the initial gain and absorption for select inverted QW background carrier densities. Source: Adapted with permission from Ref. [25], (OSA).

9 Ultrafast Nonequilibrium Carrier Dynamics in Semiconductor Laser Mode-Locking

select densities of the linear gain and absorption that was used to initialize the simulations. The measurements start with a density that is below the threshold of the MQW structure. In this density range no mode-locked pulse will appear from noise because there is no net round trip amplification as can be seen in Figure 9.25c. Then, once the density is above the threshold, a mode-locked pulse will appear in the cavity. As the density is increased beyond $4.0 \cdot 10^{16}$ m^{-2} multiple pulses appear in the cavity. At this point one can start to decrease the density, such that a single mode-locked pulse is recovered. The single mode-locked pulses along at the same density of the two branches are nearly the same; however, once the density is decreased below the threshold, the mode-locked pulse unexpectedly persists. If the simulation was started from noise, there would be no pulse, let alone a mode-locked pulse. At the density $3.25 \cdot 10^{16}$ m^{-2} the mode-locked pulse FWHM is 18 fs, slightly shorter than the previous result. Thus the final mode-locked pulse shape has some nonequilibrium memory effects when changing the background carrier density, i.e. the pump of the gain chip. This has been observed experimentally by Waldburger et al. [53].

In order to see how one can use the placement of the QWs in front of a mirror to one's advantage, consider a simpler case where the DBR is replaced by a reflecting surface located at $z_0 = 0$. We place N QWs at locations $z_n > 0$ and consider an electric field $E(t)$ with envelope $E_0(t)$ centered at ω_0, which is propagating into the QWs from the right. The total electric field at the nth QW, $E^n(t)$, is the superposition of the outgoing and incoming field

$$E^n(t) = E^{\text{in}}(t) + E^{\text{out}}(t),$$

which we can relate back to the field incident on the Nth QW with a phase change

$$E^n(t) = E(t - \Delta t_{N \to n}) \, e^{-i\omega_0 \Delta z_{N \to n}/c}$$
$$+ E(t - \Delta t_{0 \to n} - \Delta t_{N \to 0}) \, e^{-i\omega_0 (\Delta z_{0 \to n} + \Delta z_{N \to 0})/c}.$$

Here $\Delta t_{x \to y}$ is the time taken to propagate from z_x to z_y and is related to the length using $\Delta z = c \Delta t$ where $c = c_0/n$. The Fourier transform of this gives

$$E^n(\omega) = E_0(\omega - \omega_0) \, e^{-i\omega \Delta t_{N \to n}}$$
$$+ E_0(\omega - \omega_0) \, e^{-i\omega(\Delta t_{0 \to n} + \Delta t_{N \to 0})}$$
$$= \left(1 + e^{-2i\omega \Delta t_{0 \to n}}\right) E_0(\omega - \omega_0) \, e^{-i\omega \Delta t_{N \to n}}.$$

One can see that the spectrum of the electric field, as seen by the nth QW, is modulated by the factor

$$\Gamma^n(\omega) = \left(1 + e^{-2i\omega \Delta t_{0 \to n}}\right) e^{-i\omega \Delta t_{N \to n}}. \tag{9.5}$$

We will consider two situations for N QWs ordered by $n = 1, 2, \ldots N$: The QWs are arranged as an RPG with $z_n^{\text{RPG}} = n\lambda/2$ or as an MQW with $z_n^{\text{MQW}} = \lambda/2 + \delta(n - (N+1)/2)$ where δ is the spacing between the QWs. In both cases the wavelength is $\lambda = 2\pi c/\omega_0$.

Figure 9.26a and b shows the spectral modulation factor for the 10 QWs of a RPG and MQW structure and how the QW carriers will experience a pulse in momentum space. One can see that the spectrum of a given pulse is modulated by the QW

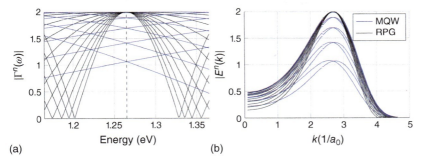

Figure 9.26 (a) The spectral modulation in Eq. (9.5) that comes from placing the QWs in patterns near a mirror for the RPG placement (black) and the MQW placement (blue) with $\delta = 10$ nm. (b) The pulse spectrum, $E^n(k)$, as seen by a 10 QW MQW and RPG structure for a sample envelope.

location relative to the reflecting surface. The field is strongest in the QWs that are resonant, i.e. the QW is placed near the antinode of the standing wave at the central wavelength. For the off-resonant QWs, one can see that the peak of the pulse is shifted to higher or lower momentum values. This has at least two advantages: a single pulse can extract carriers from a wider spectral range than its own spectrum would otherwise allow, and the amplitude modulation lets the pulse extract carriers at different levels from different QWs. The latter could allow for a situation where a few of the resonant QWs are absorbing, while the others are still inverted and thus the pulse could experience net amplification even if the pulse bleaches the carriers in some QWs. This compensation would allow for an ultrashort strong pulse to be more stable with respect to bleaching the inversion. Notice how this pattern is similar to the pattern seen in the inversions of Figure 9.24 during mode locking of the MQW10 on top of a DBR.

For completeness, let us now take a step back and consider the polarizations that are generated by the propagating electric field. If one assumes that the electric field is weak enough such that it does not change the QWs, then at each QW the polarization response is $P(\omega) = \alpha(\omega)E(\omega)$. Since the polarization from each QW will propagate with the electric field, it will also generate a polarization response from the next QW that it interacts with. If one neglects any backward propagating fields, one can track the successive polarization responses as above and finally get that the full expression for the amplification is

$$\Gamma^n(\omega) = \left(1 + (1 + \epsilon_0 \alpha)^{2n-1} e^{-2i\omega \Delta t_{0 \to n}}\right) \left(1 + \epsilon_0 \alpha\right)^{N-n} e^{-i\omega \Delta t_{N \to n}}.$$

Thus we see that the QW feedback will also modulate the above effect; however, the general behavior remains the same.

9.5 Outlook

VECSELs have been widely used for high power, continuous-wave, and short mode-locked pulse generation. They are relatively cheap to build and can be designed for a wide range of frequencies. Currently the challenge for ultrashort pulse generation is overcoming the cavity dispersion and pushing pulse durations well below 100 fs. After this, one would have to build a DBR with even wider stopband and a good dispersion-compensating coating to support even shorter pulses. At the same time, designing the gain structure in such a way as to be advantageous for shorter pulses could make ultrashort mode-locked pulses in VECSELs a reality.

References

1 Garnache, A., Hoogland, S., Tropper, A. et al. (2002). Sub-500-fs soliton-like pulse in a passively mode-locked broadband surface-emitting laser with 100 mW average power. *Applied Physics Letters* 80 (21): 3892–3894.
2 Keller, U. and Tropper, A.C. (2006). Passively modelocked surface-emitting semiconductor lasers. *Physics Reports* 429: 67–120.
3 Quarterman, A.H., Wilcox, K.G., Apostolopoulos, V. et al. (2009). A passively mode-locked external-cavity semiconductor laser emitting 60-fs pulses. *Nature Photonics* 3: 729–731.
4 Hoffmann, M., Sieber, O.D., Wittwer, V.J. et al. (2011). Femtosecond high-power quantum dot vertical external cavity surface emitting laser. *Optics Express* 19: 8108–8116.
5 Tropper, A.C., Quarterman, A.H., and Wilcox, K.G. (2012). Ultrafast vertical-external-cavity surface-emitting semiconductor lasers. *Advances in Semiconductor Lasers* 86: 269–300.
6 Heinen, B., Wang, T.L., Sparenberg, M. et al. (2012). 106 W continuous-wave output power from vertical-external-cavity surface-emitting laser. *Electronics Letters* 48 (9): 516.
7 Wang, T.L., Heinen, B., Hader, J. et al. (2012). Quantum design strategy pushes high-power vertical-external-cavity surface-emitting lasers beyond 100 W. *Laser & Photonics Reviews* 6 (5): L12–L14.
8 Scheller, M., Wang, T.L., Kunert, B. et al. (2012). Passively modelocked vecsel emitting 682 fs pulses with 5.1 W of average output power. *Electronics Letters* 48 (10): 588–589.
9 Wilcox, K.G., Tropper, A.C., Beere, H.E. et al. (2013). 4.35 kW peak power femtosecond pulse mode-locked vecsel for supercontinuum generation. *Optics Express* 21 (2): 1599–1605.
10 Zaugg, C., Sun, Z., Wittwer, V.J. et al. (2013). Ultrafast and widely tuneable vertical-external-cavity surface-emitting laser, mode-locked by a graphene-integrated distributed bragg reflector. *Optics Express* 21: 31 548–31 559.
11 Seger, K., Meiser, N., Choi, S.Y. et al. (2013). Carbon nanotube mode-locked optically-pumped semiconductor disk laser. *Optics Express* 21 (15): 17 806–17 813.
12 Klopp, P., Griebner, U., Zorn, M., and Weyers, M. (2011). Pulse repetition rate up to 92 GHz or pulse duration shorter than 110 fs from a mode-locked semiconductor disk laser. *Applied Physics Letters* 98: 071–103.

13 Zhang, F., Heinen, B., Wichmann, M. et al. (2014). A 23-watt single-frequency vertical-external-cavity surface-emitting laser. *Optics Express* 22 (11): 12817–12822.

14 Laurain, A., Mart, C., Hader, J. et al. (2014). 15 W single frequency optically pumped semiconductor laser with sub-megahertz linewidth. *IEEE Photonics Technology Letters* 26 (2): 131–133.

15 Waldburger, D., Link, S.M., Mangold, M. et al. (2016). High-power 100 fs semiconductor disk lasers. *Optica* 3 (8): 844–852.

16 Alfieri, C., Waldburger, D., Link, S. et al. (2017). Optical efficiency and gain dynamics of modelocked semiconductor disk lasers. *Optics Express* 25 (6): 6402–6420.

17 Haus, H.A. (2000). Mode-locking of lasers. *IEEE Journal of Selected Topics in Quantum Electronics* 6: 1173–1185.

18 Sieber, O.D., Hoffmann, M., Wittwer, V.J. et al. (2013). Experimentally verified pulse formation model for high-power femtosecond vecsels. *Applied Physics B* 113 (1): 133–145.

19 Shah, S.M.S., O'connell, P.E., and Hosking, J.R.M. (1996). Modelling the effects of spatial variability in rainfall on catchment response, 2: experiments with distributed and lumped models. *Journal of Hydrology* 175 (1–4): 89–111.

20 Olufsen, M.S. and Nadim, A. (2004). On deriving lumped models for blood flow and pressure in the systemic arteries. *Mathematical Biosciences and Engineering* 1 (1): 61–80.

21 Ramallo-González, A.P., Eames, M.E., and Coley, D.A. (2013). Lumped parameter models for building thermal modelling: an analytic approach to simplifying complex multi-layered constructions. *Energy and Buildings* 60: 174–184.

22 Haug, H. and Koch, S.W. (2009). *Quantum Theory of the Optical and Electronic Properties of Semiconductors*, 5e. Singapore: World Scientific.

23 Hughes, S., Knorr, A., Koch, S.W. et al. (1996). The influence of electron-hole-scattering on the gain spectra of highly excited semiconductors. *Solid State Communications* 100 (8): 555–559.

24 Hader, J., Moloney, J.V., and Koch, S.W. (1999). Microscopic theory of gain, absorption, and refractive index in semiconductor laser materials-influence of conduction-band nonparabolicity and coulomb-induced intersubband coupling. *IEEE Journal of Quantum Electronics* 35 (12): 1878–1886.

25 Kilen, I., Koch, S.W., Hader, J., and Moloney, J.V. (2017). Non-equilibrium ultrashort pulse generation strategies in vecsels. *Optica* 4 (4): 412–417.

26 Kilen, I., Koch, S.W., Hader, J., and Moloney, J.V. (2016). Fully microscopic modeling of mode locking in microcavity lasers. *JOSA B* 33 (1): 75–80.

27 Kilen, I., Hader, J., Moloney, J.V., and Koch, S.W. (2014). Ultrafast nonequilibrium carrier dynamics in semiconductor laser mode locking. *Optica* 1 (4): 192–197.

28 Hader, J., Scheller, M., Laurain, A. et al. (2016). Ultrafast non-equilibrium carrier dynamics in semiconductor laser mode-locking. *Semiconductor Science and Technology* 32 (1): 013–002.

29 Moloney, J.V., Kilen, I., Bäumner, A. et al. (2014). Nonequilibrium and thermal effects in mode-locked vecsels. *Optics Express* 22 (6): 6422–6427.
30 Böttge, C.N., Hader, J., Kilen, I. et al. (2014). Ultrafast pulse amplification in mode-locked vertical external-cavity surface-emitting lasers. *Applied Physics Letters* 105 (26): 261–105.
31 Hader, J., Kilen, I., Moloney, J.V., and Koch, S.W. (2017). Non-equilibrium effects in vecsels. in *Proceedings of SPIE*, vol. 10087: 1 008 706-1–1 008 706-7.
32 Kilen, I., Koch, S.W., Hader, J., and Moloney, J.V. (2016). Modeling of ultrashort pulse generation In: mode-locked vecsels. In: *SPIE OPTO*, pp. 97 420H-1–97 420H-6. International Society for Optics and Photonics.
33 Kilen, I., Böttge, C.N., Hader, J. et al. (2015). Ultrafast non-equilibrium carrier dynamics in: semiconductor laser mode-locking. In: *SPIE OPTO*, pp. 934 902-1–934 902-8. International Society for Optics and Photonics.
34 Böttge, C.N., Hader, J., Kilen, I. et al. (2015). Influence of non-equilibrium carrier dynamics on pulse amplification in semiconductor gain media. In: *SPIE LASE*, pp. 934 903-1–934 903-12. International Society for Optics and Photonics.
35 Hoffmann, M., Sieber, O.D., Maas, D.J.H.C. et al. (2010). Experimental verification of soliton-like pulse-shaping mechanisms in passively mode-locked vecsels. *Optics Express* 18 (10): 10 143–10 153.
36 Backus, S., Durfee III, C.G., Murnane, M.M., and Kapteyn, H.C. (1998). High power ultrafast lasers. *Review of Scientific Instruments* 69 (3): 1207–1223.
37 Haus, H.A., Moores, J.D., and Nelson, L. (1993). Effect of third-order dispersion on passive mode locking. *Optics Letters* 18 (1): 51–53.
38 Orfanidis, S.J. (2016). Electromagnetic waves and antennas, 2016. Unpublished, available at URL http://www.ece.rutgers.edu/~orfanidi/ewa (accessed 21 April 2021).
39 Klopp, P., Griebner, U., Zorn, M. et al. (2009). Mode-locked ingaas-algaas disk laser generating sub-200-fs pulses, pulse picking and amplification by a tapered diode amplifier. *Optics Express* 17 (13): 10–820.
40 Baker, C., Scheller, M., Koch, S.W. et al. (2015). In situ probing of mode-locked vertical-external-cavity-surface-emitting lasers. *Optics Letters* 40 (23): 5459–5462.
41 Laurain, A., Marah, D., Rockmore, R. et al. (2016). Colliding pulse mode locking of vertical-external-cavity surface-emitting laser. *Optica* 3 (7): 781–784.
42 Baker, C.W., Scheller, M., Laurain, A. et al. (2017). Multi-angle vecsel cavities for dispersion control and peak-power scaling. *IEEE Photonics Technology Letters* 29 (3): 326–329.
43 Rudin, B., Wittwer, V.J., Maas, D.J.H.C. et al. (2010). High-power mixsel: an integrated ultrafast semiconductor laser with 6.4 W average power. *Optics Express* 18 (26): 27 582–27 588.
44 Mangold, M., Zaugg, C.A., Link, S.M. et al. (2014). Pulse repetition rate scaling from 5 to 100 Ghz with a high-power semiconductor disk laser. *Optics Express* 22 (5): 6099–6107.
45 Hader, J., Koch, S.W., and Moloney, J.V. (2003). Microscopic theory of gain and spontaneous emission in GaInNAs laser material. *Solid-State Electron* 47 (3): 513–521.

46 Hess, O. and Kuhn, T. (1996). Spatio-temporal dynamics of semiconductor lasers: theory, modelling and analysis. *Progress in Quantum Electronics* 20 (2): 85–179.

47 Bäumner, A., Koch, S.W., and Moloney, J.V. (2011). Non-equilibrium analysis of the two-color operation in semiconductor quantum-well lasers. *Physica Status Solidi (b)* 248 (4): 843–846.

48 Yokoyama, H. and Ujihara, K. (1995). *Spontaneous emission and laser oscillation in microcavities*, vol. 10. CRC Press.

49 Hader, J., Moloney, J.V., and Koch, S.W. (2014). Microscopic analysis of non-equilibrium dynamics in the semiconductor-laser gain medium. *Applied Physics Letters* 104 (15): 151–111.

50 Hader, J., Yang, H.J., Scheller, M. et al. (2016). Microscopic analysis of saturable absorbers: semiconductor saturable absorber mirrors versus graphene. *Journal of Applied Physics* 119 (5): 053–102.

51 Breusing, M., Kuehn, S., Winzer, T. et al. (2011). Ultrafast nonequilibrium carrier dynamics in a single graphene layer. *Physical Review B* 83 (15): 153–410.

52 Turnbull, A.P., Head, C.R., Shaw, E.A. et al. (2015). Spectrally resolved pulse evolution in a mode-locked vertical-external-cavity surface-emitting laser from lasing onset measurements. In: *SPIE LASE*, pp. 93 490I-1–93 490I-11. International Society for Optics and Photonics.

53 Waldburger, D., Link, S.M., Alfieri, C.G. et al. (2016). High-power 100-fs sesam-modelocked vecsel. In: *Lasers Congress 2016 (ASSL, LSC, LAC)*, p. ATu1A.8. Optical Society of America. doi: 10.1364/ASSL.2016.ATu1A.8. URL http://www.osapublishing.org/abstract.cfm?URI=ASSL-2016-ATu1A.8 (accessed 21 April 2021).

10

Mode-Locked AlGaInP VECSEL for the Red and UV Spectral Range

Roman Bek[1], Michael Jetter[2], and Peter Michler[2]

[1] Twenty-One Semiconductors GmbH, Neckartenzlingen, Germany
[2] Institut für Halbleiteroptik und Funktionelle Grenzflächen - IHFG, University of Stuttgart, Stuttgart, Germany

10.1 Introduction

Ultrashort laser pulses in the red spectral range are of special interest not only for quantum optics research but also for many different applications in the fields of medicine and biology. The high peak power can be used for the fabrication of complicated three-dimensional microstructures by two-photon polymerization [1], for multi-photon fluorescence microscopy [2] and for time-resolved fluorescence lifetime imaging microscopy [3]. In photodynamic therapy, where photosensitizers are activated by light of a specific wavelength, red-emitting laser sources - preferably fiber-coupled - are needed for sensitizers with strong absorbance at 630 nm, but also at wavelengths above 650 nm [4]. With pulsed lasers a higher penetration depth into the tissue and different effects on tumor cells are achieved when compared with CW laser light [5, 6]. For spectroscopy of indium phosphide quantum dots (QDs) with typical charge carrier life times of around 500 ps [7], an excitation source with a repetition rate in the order of 1 GHz is desirable to achieve a high single photon flux [8]. Additionally, by tuning the wavelength, resonant excitation can lead to improved quantum optical properties of the single-photon emission such as higher indistinguishability [9].

A variety of further applications is enabled by second harmonic generation of the laser pulses into the ultraviolet spectral range. Especially time-resolved investigations of gallium nitride and zinc oxide-based materials would benefit from the availability of compact and cost-effective ultrafast laser systems emitting in the UV.

Among the most important requirements for all these applications are short pulse durations with high peak powers, a tunable emission wavelength and a near-diffraction-limited beam profile. State-of-the-art laser sources emitting ultrashort pulses in the red spectral range are complex and expensive optical parametric oscillators and bulky dye lasers. Praseodymium solid-state lasers mode-locked by semiconductor saturable absorber mirrors (SESAMs) represent an alternative, however, with a fixed wavelength of 639.5 nm [10]. Pulses in the ultraviolet range

Vertical External Cavity Surface Emitting Lasers: VECSEL Technology and Applications, First Edition.
Edited by Michael Jetter and Peter Michler.
© 2022 WILEY-VCH GmbH. Published 2022 by WILEY-VCH GmbH.

are typically produced with nitrogen lasers and excimer lasers emitting nanosecond pulses at fixed wavelengths. The latter can be used in complex setups together with cerium-doped oscillators to achieve ultrashort pulses at wavelengths around 330 nm [11]. Femtosecond pulses with tunable emission across 250–355 nm are only available from frequency-doubled optical parametric amplifiers in the visible spectrum [12].

With repetition rates around 1 GHz and pulse durations in the picosecond and femtosecond regime, mode-locked vertical external-cavity surface-emitting lasers (VECSELs) based on the aluminum gallium indium phosphide (AlGaInP) material system are perfectly suited to meet the abovementioned requirements. Due to the semiconductor material in a compact setup, a large range of wavelengths can be accessed, including the UV spectral range by intracavity frequency doubling.

10.2 Epitaxial Layer Design of AlGaInP-SESAM Structures

For passive mode locking of red-emitting VECSELs, semiconductor saturable absorber mirrors have been fabricated both by metal-organic vapor-phase epitaxy (MOVPE) and molecular beam epitaxy (MBE). The SESAM structures are similar to the gain structures, with GaAs substrates and AlGaInP-based active regions grown on top of high-reflective AlAs/Al$_{0.45}$GaAs distributed Bragg reflectors (DBRs). Saturable absorption is provided by one or two compressively strained GaInP quantum wells (QWs) or a layer of InP QDs, embedded in AlGaInP barriers and cladding layers. Different approaches were followed to reduce the recovery time of the absorber and to optimize parameters such as the saturation fluence and the modulation depth. However, due to limited access to suitable ultrafast laser systems in this wavelength range, only few SESAM characterization measurements are available and most design considerations are derived from results obtained in the infrared spectral range.

10.2.1 Quantum Well SESAMs

The first mode-locked operation of an AlGaInP-based VECSEL was reported with an antiresonant SESAM design containing one GaInP QW with 5 nm thickness embedded in Al$_{0.6}$GaInP on top of 40 DBR pairs [13]. The maximum nonlinear reflectivity change deduced from reflectivity measurements of this absorber structure is between 1 and 2%. To enhance the recovery dynamics, the MBE grown absorber structure is treated with ion irradiation. The same technique has already been used for samples operating in the IR with the effect of sub-10 ps recovery times [14].

A different approach was chosen for the design of an MOVPE-grown near-resonant absorber structure with two QWs and 55 DBR pairs [15]. The compressively strained QWs, separated by a 4 nm thick Al$_{0.33}$GaInP barrier, are placed close to the semiconductor surface, as shown in Figure 10.1a. This allows fast surface recombination of the charge carriers by tunneling through the 2 nm

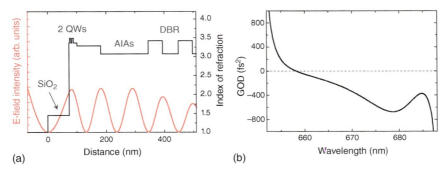

Figure 10.1 (a) Near-resonant SESAM design with two GaInP QWs near the semiconductor surface and an additional fused silica layer. (b) Simulated GDD of the SESAM. Source: Reproduced with permission from [16].

$Al_{0.33}$GaInP cap. The recovery dynamics of this absorber design shows three decay times and is discussed in Section 10.3. On top of the near-resonant semiconductor sample, an additional fused silica layer is deposited. This overall antiresonant design leads to a small negative value for the group delay dispersion (GDD) at the laser wavelength around 664 nm (see Figure 10.1b).

10.2.2 Quantum Dot SESAMs

With indium phosphide (InP) QDs, a large wavelength region from the green to the near infrared spectral range can be covered [7, 17]. Furthermore, QDs as absorber material enable independent control of saturation fluence and modulation depth, since the latter can be adjusted by the quantum dot density [18]. For an emission wavelength of around 650 nm, InP QDs are typically embedded in strain compensating $Al_{0.1}$GaInP barriers and $Al_{0.55}$GaInP cladding layers. For a red-emitting all quantum dot mode-locked laser [19], a near-antiresonant SESAM design was used, similar to the applied antiresonant gain structure [20]. As shown in Figure 10.2a, a fused silica layer can be added, increasing the field enhancement of the absorber by a factor of about 1.8. This reduces the group delay dispersion (see Figure 10.2b) and leads to a higher modulation depth and a lower saturation fluence. With this overall near-resonant design, stable mode-locked operation can be maintained with increased transmission of the outcoupling mirror and therefore higher output power [21].

10.3 Temporal Response of AlGaInP SESAMs

Characterizing the nonlinear reflectivity or the temporal response of a SESAM requires an ultrafast laser source with suitable parameters at the wavelength of interest. Due to the limited availability of these lasers in the red spectral range, there has only been one report about such measurements so far [16]. The dynamics of an absorber structure with two surface-near QWs (see Section 10.2.1) without a fused

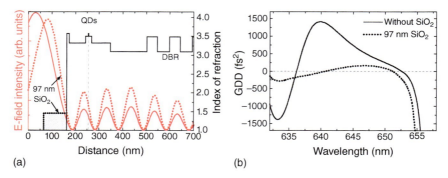

Figure 10.2 (a) Near-antiresonant SESAM design with one InP QD layer. The field enhancement is modified by an additional fused silica layer. (b) Simulated GDD of the SESAM with and without the fused silica layer. Source: Reproduced with permission from [21].

silica layer is investigated and compared to an identically grown structure with only one QW. A conventional degenerate reflective pump-probe setup with balanced lock-in detection and a pulsed laser source tunable from 655 to 668 nm is used. The 200 fs pulses are obtained via second-harmonic generation of an optical parametric oscillator at around 1320 nm, synchronously pumped by a titanium-sapphire laser.

The SESAM with only one QW shows a biexponential temporal response as described for absorber structures in the IR spectral range in several publications [18, 22, 23]. However, the dynamics of the 2 QW SESAM (exemplary curve is plotted in Figure 10.3a) is fitted well with an exponential decay including three characteristic times: The fast time constant of less than 300 fs is dominant with a prefactor of almost 0.5 and can be attributed to intra-band thermalization of charge carriers. The intermediate and slow time constants result from the recombination of charge carriers in two QWs, which are separated by a 4 nm barrier and capped by a layer with only 2 nm thickness. Due to this design, the propability of fast surface

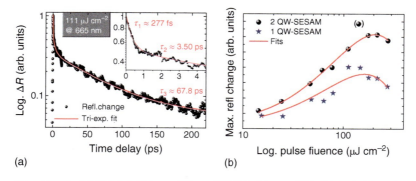

Figure 10.3 (a) Pump-probe result for a SESAM with two QWs close to the surface. In contrast to the typical bi-exponential decay, three time constants are observed. Source: Reproduced with permission from [16]. (b) Maximum reflectivity change fitted after [24] for different pulse fluences of the two QW absorber and an identically grown sample with only one QW.

recombination differs for the charge carriers in the two QWs, leading to similar prefactors, but time constants in the order of a few ps for the QW closer to the surface and about 70 ps for the second QW.

Figure 10.3b shows the maximum reflectivity changes of the absorber structures for varying pulse fluences obtained from the decay curves and corresponding fit curves after [24]. It reveals a rollover for the last third of the measurement range, indicating two-photon absorption. Despite a large uncertainty of the values for the single QW SESAM, a higher modulation depth can clearly be observed for the double QW SESAM and the saturation fluence of the two QW absorber can be estimated to $\sim 90\,\mu\text{J}\,\text{cm}^{-2}$.

10.4 Cavity Designs

Cavity configurations of SESAM mode-locked VECSELs are usually V- or Z-shaped with the SESAM as one of the end mirrors. In both cases, the gain structure can either be used as the second end mirror or as a folding mirror. So far, SESAM mode-locked AlGaInP VECSELs have been realized in two cavity configurations, which are shown in Figure 10.4. In the Z-shaped cavity, the gain chip is pumped under normal incidence and serves as folding mirror and a high-reflective folding mirror as well as an outcoupling mirror complete the resonator. In the V-shaped cavity, the gain chip is positioned as end mirror, and the resonator is folded by the curved outcoupling mirror.

For efficient heat removal from the active region, a heat spreader is placed on top of the gain chip with a good thermal contact obtained either by liquid capillary bonding [25] or simply by mechanically induced pressure. This heat spreader (usually a single-crystal diamond) has different effects on the laser, depending on the cavity geometry and the shape of the heat spreader. If the cavity is folded by the gain chip, an antireflection coating is needed to avoid multiple reflections inside the heat spreader leading to an unstable mode-locked operation (e.g. by higher transverse modes). With the gain chip as end mirror, a plane-parallel heat spreader acts as an etalon introducing Fabry-Pérot fringes in the laser spectrum which leads to side pulses in the time domain (see Section 10.6.1.2). This can be avoided by using a wedged heat spreader, which again needs to be antireflection-coated to minimize the losses due to reflections at the surface.

If the gain and the absorber structure contain the same active material (e.g. the same QWs) and exhibit a similar field enhancement, a smaller spot size on the SESAM is a prerequisite for mode-locked operation, since it leads to saturation of the absorber at lower pulse fluences than needed to saturate the gain. Independently from the cavity design, a large ratio between the mode sizes of about 25 to 30 has shown to be suitable for SESAM mode locking of AlGaInP VECSELs. With the mirror's radii of curvature of 50 mm in the V-shaped and 30 mm (folding mirror) and 50 mm (outcoupling mirror) in the Z-shaped design, the resonator arms are adjusted to give mode diameters of about 100 μm (V-shaped cavity) and 115 μm (Z-shaped cavity) on the gain chip and 20 μm on the SESAM for both cavity configurations.

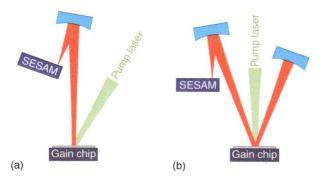

Figure 10.4 Cavity configurations used for mode locking of AlGaInP VECSELs with the absorber structure positioned as end mirror. The gain structure can be used as end mirror in a V-shaped cavity (a) or as folding mirror in a Z-shaped cavity (b). If a plane heat spreader is used on top of the gain chip, it will lead to side pulses due to the etalon effect in the case of a V-shaped cavity.

Assuming the same overall cavity length, a double pass through the gain medium can provide more gain and enable lower repetition rates than with the simpler V-shaped design if these are limited by the charge carrier lifetime in the gain structure. Furthermore, with a high-reflective folding mirror in the Z-shaped cavity, only one beam is emitted through the outcoupling mirror. In the V-shaped cavity, where the folding mirror also serves as outcoupling mirror, two beams are emitted. This reduces the output power per single beam, but it can also be convenient for the simultaneous characterization by two different measurement methods.

10.5 Characterization Methods

In addition to standard measurements performed on laser systems - such as characterizing the output power, the emission wavelength, the beam profile, and the beam propagation - the pulse trains emitted by mode-locked lasers need to be investigated carefully using further methods. The basic characterization typically requires four different measurements yielding the pulse duration, the optical and the radio frequency (RF) spectrum as well as the temporal signal of the pulse train.

Picosecond and femtosecond pulse durations in the red spectral range can be measured by second-harmonic autocorrelation (AC), usually in a noncollinear configuration to obtain background-free intensity AC traces. The optical spectrum can be accessed with standard spectrometers, and RF spectrum analyzers are available with bandwidths of tens of GHz, covering a sufficiently large number of higher harmonics in the RF spectrum for the standard repetition frequencies of around 1 GHz. However, the measurement range is often limited by the photodetector, since for the visible range only few devices are available with the desired multi-gigahertz bandwidth. The same holds for the temporal measurement with photodiode and oscilloscope, where the limited bandwidths of the devices can lead to traces without a clear zero signal between the pulses or produce measurement artefacts such as overshoots

Table 10.1 Overview of reported mode-locked AlGaInP VECSELs

Active material (gain and absorber)	Laser wavelength	Pulse duration (FWHM)	Repetition frequency	Max. average output power	Ref.
GaInP QWs	675 nm	5.1 ps	973 MHz	45 mW	[13]
GaInP QWs	664 nm	220–250 fs	836 MHz	0.5 mW	[15, 16]
InP QDs	665–655 nm	0.7–2.0 ps	836–852 MHz	10 mW	[19, 21, 31]

and ringing. Nowadays, real-time oscilloscopes are available with sufficiently large bandwidths, which can be used to measure the temporal signal over hundreds of µs. These traces contain the same information as RF spectra recorded with the same bandwidth, since time and frequency domain are linked to each other by means of the Fourier transform.

Devices using more comprehensive methods accounting for the phase of the signal, such as frequency-resolved optical gating (FROG), are not commercially available for the red spectral range with picosecond pulse durations and the corresponding narrow spectra. Therefore, the proper characterization of mode-locked VECSELs in the visible spectrum remains rather challenging compared to systems operating in the near-infrared range.

10.6 Mode-Locking Results

Mode-locked semiconductor disk lasers with emission in the red spectral range have been reported since 2013 [13], 11 years after the first demonstration of an AlGaInP VECSEL, while in the IR range, the first SESAM mode-locked VECSEL [26] was realized only 3 years after the first VECSEL in CW operation [27]. One reason is the advanced state of SESAMs in the infrared range at that time, since the development of these absorber structures had been driven by their applications in pulsed solid-state lasers already since 1992 [28–30]. The field of research on SESAMs for mode-locked red-emitting VECSELs, however, is rather new with only few publications to date, covering the wavelength range from 650 to 675 nm. With standard V- or Z-shaped cavities using either QWs or QDs in the gain and absorber structures, picosecond as well as femtosecond pulses can be achieved. An overview of the reported results from SESAM mode-locked VECSELs emitting in the red spectral range is given in Table 10.1.

10.6.1 Quantum Well Mode-Locked AlGaInP VECSELs

10.6.1.1 High Output Power

The highest average output power of 45 mW at 675 nm [13] was achieved with a gain structure containing 20 GaInP QWs used in a Z-shaped cavity with 1% outcoupling as shown in Figure 10.4b. The antiresonant gain chip with an antireflection-coated,

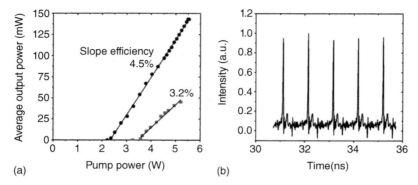

Figure 10.5 High output power results from a QW mode-locked AlGaInP VECSEL. (a) Output power over incident pump power at a heat sink temperature of 10 °C. Gray squares: Mode-locked operation with SESAM. Black dots: CW operation with high-reflective end mirror. (b) Oscilloscope trace of the pulse train. Source: Reprinted with permission from [13]. © The Optical Society.

wedged diamond heat spreader on top is pumped under normal incidence and serves as folding mirror, while the MBE-grown, antiresonant SESAM (see Section 10.2.1) is placed as end mirror.

The average output power of the SESAM mode-locked VECSEL at maximum pump power (~ 5 W) is roughly one-third of the CW output power with a high-reflective end mirror and the slope is reduced from 4.5% to 3.2% (see Figure 10.5a). The typical characterization methods described in Section 10.5 demonstrate stable mode-locked operation when the SESAM is inserted in the cavity and no indication of mode locking with the high-reflective mirror instead.

The autocorrelation trace reveals a 10 ps peak with extended wings, which indicates a pulse form different to the commonly observed sech^2 or Gaussian. By applying the PICASO method [32], an asymmetric pulse shape is retrieved with a FWHM of 5.1 ps. With a spectral width of 0.14 nm this results in a time bandwidth product (TBWP) 3.75 times above the Fourier limit.

10.6.1.2 Femtosecond Operation

Pulse durations below 250 fs [15] can be achieved with an absorber structure containing QWs close to the semiconductor surface (see Section 10.2.1). As shown in Figure 10.4a, the SESAM and a gain structure with 20 QWs are used as end mirrors in a V-shaped cavity with similar mode sizes and repetition frequency as in the Z-shaped cavity above. With nearly the same pump power as used for the picosecond pulses with 45 mW of output power, only 0.5 mW per beam are achieved in this configuration due to the high-reflective outcoupling mirror. However, the pulse duration measured by autocorrelation (see Figure 10.6a) is between 200 and 250 fs, depending on the position of the spot on the SESAM.

A plane diamond heat spreader on top of a gain chip introduces a Fabry-Pérot effect and leads to fringes in the spectrum. In the time domain, this results in side pulses around the main pulse with temporal distances corresponding to the roundtrip time in the subcavity formed by the heat spreader. To calculate the energy

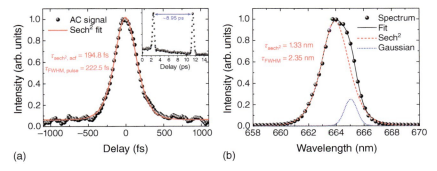

Figure 10.6 Results from a QW mode-locked AlGaInP VECSEL with femtosecond pulses. (a) Intensity autocorrelation fitted by a squared hyperbolic secant (sech^2) with a FWHM pulse duration of 222.5 fs. Inset: Side pulses with temporal delays of 8.95 ps appear for larger scan ranges. (b) Corresponding spectrum indicating a CW contribution. The Fabry-Pérot fringes due to the heat spreader are not resolved. Source: Reproduced from [15], with the permission of AIP Publishing.

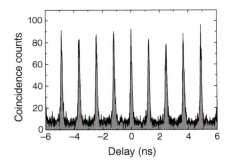

Figure 10.7 Autocorrelation histogram recorded with a HBT type setup using APDs. The constant background supports the "continuum" indicated by the spectrum. Source: Reproduced with permission from [16].

and the peak power of a single pulse, the power distribution among this group of pulses needs to be considered. In principle, this can be obtained by a Fourier transform of the spectrum with sufficiently high resolution. The side pulses also appear in the autocorrelation trace for a sufficiently large scan range or a shifted zero delay of the autocorrelator (see inset of Figure 10.6).

As observed for other mode-locked VECSELs with similar pulse durations [33, 34], the optical spectrum of the femtosecond pulsed laser (see Figure 10.6b) consists if a "soliton-like" part at 664 nm and a "continuum" at 665 nm. The former results in a time bandwidth product close to the Fourier limit, while the latter indicates a CW component in the signal. Autocorrelation measurements with a Hanbury-Brown and Twiss (HBT) type setup [35] using avalanche photodiodes (APD) with 40 ps of temporal resolution support the findings. As shown in Figure 10.7, pulsed operation with a repetition frequency corresponding to the cavity length is observed on top of a background. This background is not measured if laser operation is prevented by tilting the SESAM, demonstrating that it does not origin from dark counts of the APDs or from spontaneous emission, i.e. photoluminescence, in the gain structure.

The characteristic times of this mode-locked laser can be linked to the time constants of the utilized SESAM (see Section 10.3). The dominant fast decay time

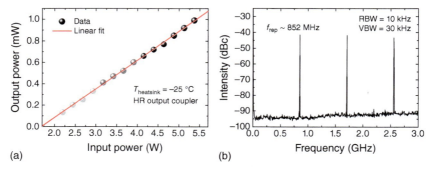

Figure 10.8 (a) Power transfer curve of the QD VECSEL with QD absorber structure. Only for pump powers above ~4 W, stable mode locking is achieved (black dots). For lower pump powers, instabilities are observed (dark gray dots) or the indications of mode-locked operation completely vanish (light gray dots). (b) RF spectrum for mode-locked operation revealing a repetition frequency of 852 MHz. Source: Reproduced from [19], with the permission of AIP Publishing.

observed in the pump-probe measurements is very close to the pulse duration, while the intermediate time constant of a few picoseconds further leads to a roughly 70% recovery of the absorber in the 9 ps delay between two consecutive (side) pulses.

10.6.2 Quantum Dot Mode-Locked AlGaInP VECSELs

InP QDs are not only an alternative to GaInP QWs for emission in the red spectral range. They are the perfect choice as active material in VECSELs in order to reach high output powers and extend the wavelength range. With seven InP QD layers, a CW output power of nearly 1.4 W at around 650 nm could be reached [20], outperforming VECSELs with 20 GaInP QWs. In a V-shaped cavity, as shown in Figure 10.4a, an antiresonant QD gain structure is combined with a near-antiresonant SESAM containing one InP QD layer (see Section 10.2.2) and a high-reflective mirror [19]. Different regimes of operation can be observed, depending on the pump power (see Figure 10.8a). For high pump powers of more than ~4 W, mode locking is indicated by a stable pulse train in the oscilloscope trace, a $sech^2$-shaped pulse in the autocorrelation trace and peaks at the repetition frequency of 852 MHz and its harmonics in the RF spectrum. As plotted in Figure 10.8b, they have nearly equal intensities with a high signal-to-noise ratio. The AC trace shows a FWHM pulse duration of 1 ps and side pulses due to the etalon effect of the diamond heat spreader. The FWHM of the optical spectrum at 655 nm is close to the resolution of the spectrometer, and therefore gives an upper limit for the TBWP of 0.6, which is 1.9 times the Fourier limit.

Instabilities - such as an intensity variation in the time signal, a background appearing in the autocorrelation trace and a decreasing intensity of the higher harmonics in the RF spectrum - are observed in the intermediate regime and no sign of mode locking is visible in any of these measurements for even lower pump powers. For a pump power of 5.4 W, the average output power of 1 mW per beam increases to 6.2 mW in CW operation when the SESAM is replaced by a high-reflective mirror.

As mentioned in Section 10.2.2, the above described near-antiresonant SESAM can be modified by an additional fused silica layer. Depending on the thickness of this layer, the field enhancement is increased by a factor of up to 1.8 for the overall resonant case (see Figure 10.2). This leads to an increased modulation depth, which, according to theory [36], is expected to reduce the pulse duration. In contrast, with an increasing thickness of the fused silica layer, larger pulse durations and narrower spectra are obtained [21]. Besides, the overall resonant design enables stable mode-locking also for an increased outcoupling of the laser. With an outcoupler transmission of 0.3% instead of the high-reflective mirror, the maximum average output power is increased from 1 mW to more than 10 mW per beam. With this cavity configuration, pulsed UV emission can be obtained by inserting a frequency doubling crystal in front of the SESAM as described in the following section.

10.7 Second Harmonic Generation into the UV Spectral Range

As described in Chapter 7.2.4, the open cavity configuration of VECSELs allows for efficient intracavity second harmonic generation (SHG) by including frequency doubling crystals such as beta barium borate (BBO). In CW operation, high UV powers of more than 800 mW can be achieved from VECSELs with fundamental emission in the red spectral range [37].

In a mode-locked laser, the nonlinear effect of frequency doubling counteracts the pulse shaping effect of the SESAM. While the saturable absorber favors short pulses with high intensities over a continuous signal, the second harmonic generation preferably converts these high power signals due to the quadratic dependence on the fundamental power and therefore impedes the pulsed operation. However, this presupposes a degradation of the fundamental wave, which is only the case in the strong conversion regime. For weak conversion, the effect on the fundamental wave is insignificant, and a stable mode-locked laser will remain in mode-locked operation also with intra-cavity second harmonic generation.

The most obvious configurations for such an experiment are the two cavities shown in Figure 10.4 with the frequency doubling crystal placed in front of the SESAM, where the beam is tightly focused. Both cavities have already been used to demonstrate stable mode locking and the rather small beam size leads to efficient SHG. In fact, an intracavity frequency-doubled mode-locked AlGaInP-VECSEL was demonstrated [31] with the same V-shaped cavity which had been used to achieve picosecond pulses with 10 mW of average output power as described at the end of the previous section. Figure 10.9 shows the 3 mm BBO crystal which is placed directly in front of the SESAM. This leads to UV emission in one of the two beams, which becomes visible by the blue fluorescence on a sheet of paper after passing a filter to suppress the fundamental red emission. The mode-locked operation with 1.22 ps pulses is characterized at the fundamental wavelength and

Figure 10.9 Close-up view of the V-shaped mode locking setup with intracavity SHG. Frequency doubling is achieved by a BBO crystal in front of the SESAM and the emitted UV light becomes visible by the blue fluorescence on a sheet of paper. Source: Reprinted with permission from [31]. © The Optical Society.

no change is observed when the frequency doubling crystal is inserted in the cavity. The fundamental average output power at 650 nm still exceeds 10 mW per beam, while 0.5 mW of average output power is detected at 325 nm.

Since the direct measurement of ultrashort pulse durations with low output powers in the UV is rather challenging, the pulse duration is estimated by considering the temporal walk-off between the fundamental and the second harmonic pulse. This can be done by comparing the spectral width of the fundamental wave with the phase-matching bandwith of the frequency doubling crystal or, equivalently, by calculating the group velocity mismatch (GVM) with respect to the crystal length.

For the aforementioned results, the deconvolution of the optical spectrum at the fundamental wavelength with the response function of the spectrometer (see Figure 10.10) gives a value for the spectral FWHM which is close to the calculated [38] phase-matching bandwidth of 0.62 nm. This is in agreement with the estimation from the group velocity mismatch. The pulse-broadening effect due to GVM only becomes significant for a crystal length in the order of a characteristic length,

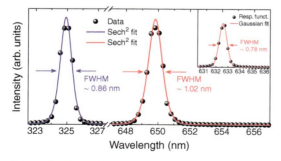

Figure 10.10 Optical spectrum of the pulsed laser showing the fundamental and the second harmonic emission (intensities are not to scale). Inset: The spectrum of a narrow-band helium-neon laser is measured, revealing the response function of the spectrometer with a FWHM of ∼0.8 nm. Source: Reprinted with permission from [31]. © The Optical Society.

which is given by the ratio of the fundamental pulse duration and the GVM [39]. The BBO length of 3 mm is below this characteristic length, indicating that the UV pulse duration is only slightly higher than the fundamental pulse duration of 1.22 ps.

However, the V-shaped setup with the frequency doubling crystal in front of the SESAM is not the best choice for intracavity SHG of a mode-locked VECSEL. Only half of the UV light generated in the crystal hits the outcoupling mirror, while the other half hits the absorber structure where most of it is absorbed. Besides the lowered efficiency, this can also lead to a faster degradation of the SESAM. These issues can be overcome with cavity configurations such as a Z-shaped cavity with two curved folding mirrors and the semiconductor samples as end mirrors. In this setup, the frequency doubling crystal can be included in the beam waist between the folding mirrors. By changing the arm lengths, the size of the beam waist can be adjusted for a desired conversion efficiency. With an increased efficiency, the same conversion is achieved with thinner crystals and pulse broadening by GVM is avoided also for shorter pulse durations in the femtosecond regime.

10.8 Summary and Outlook

SESAM mode-locked semiconductor disk lasers emitting in the red spectral range are compact laser sources providing ultrashort pulses with excellent beam profiles and repetition rates in the GHz regime. They are promising candidates for a variety of applications - not only, but also due to the limited availability of alternative lasers with equal emission properties. On the other hand, this makes the characterization of AlGaInP-SESAMs a challenging task, and most design considerations so far have been related to results from InGaAs-based SESAMs. Nevertheless, both GaInP-QWs and InP-QDs have been successfully used as active material in gain and absorber structures, achieving pulse durations ranging from several picoseconds to less than 250 fs. Furthermore, intracavity second harmonic generation of a SESAM mode-locked AlGaInP-VECSEL has been realized, showing the potential of these lasers for the generation of ultrashort UV-pulses.

A rather new and still not well understood method of mode locking dispenses with a semiconductor saturable absorber. The so-called "self-mode locking" described in Chapter 12 has been applied to a gain structure with GaInP-QWs in a simple linear cavity. The measurement results demonstrate pulsed operation with a repetition rate of 3.5 GHz and 22 ps-pulses with a complex substructure, which has not been reported for any mode-locked VECSELs before.

Further progress of the mode-locked systems such as increased output powers in a wide wavelength range is expected in the course of new developements of AlGaInP-based gain structures for vertical-emitting lasers. As described in Chapter 7.2.1, the poor charge carrier confinement compared to InGaAs-based gain structures requires a particularly effective thermal management as well as pumping schemes with a reduced amount of heat introduction. Whereas the latter can be realized with high efficiency by using an in-well pump with several passes through the active region, the heat extraction is optimized by a novel membrane approach.

The membrane external-cavity surface-emitting laser (MECSEL) presented in Chapter 7.3 enables the growth of structures emitting at wavelengths which cannot be accessed by the conventional disk laser approach using an active region on top of a DBR. The same holds for absorber membranes, which can then be used either bonded to dielectric mirrors or – just like the MECSEL – sandwiched between the heat spreaders. Together with the possibility of intracavity frequency conversion, these new concepts will open doors for compact pulsed semiconductor lasers at various new wavelengths.

References

1 Serbin, J., Egbert, A., Ostendorf, A. et al. (2003). Femtosecond laser-induced two-photon polymerization of inorganic organic hybrid materials for applications in photonics. *Optics Letters* 28 (5): 301–303. doi: 10.1364/OL.28.000301.
2 Helmchen, F. and Denk, W. (2005). Deep tissue two-photon microscopy. *Nature Methods* 2 (12): 932–940. doi: 10.1038/nmeth818.
3 Periasamy, A., Wodnicki, P., Wang, X.F. et al. (1996). Time-resolved fluorescence lifetime imaging microscopy using a picosecond pulsed tunable dye laser system. *Review of Scientific Instruments* 67 (10): 3722–3731. doi: 10.1063/1.1147139.
4 Dougherty, T.J., Gomer, C.J., Jori, G. et al. (1998). Photodynamic therapy. *Journal of the National Cancer Institute* 90 (12): 889–905. doi: 10.1093/jnci/90.12.889.
5 Okunaka, T., Kato, H., Konaka, C. et al. (1992). A comparison between argon-dye and excimer-dye laser for photodynamic effect in transplanted mouse tumor. *Japanese Journal of Cancer Research* 83 (2): 226–231. doi: 10.1111/j.1349-7006.1992.tb00090.x.
6 Kuznetsova, D.S., Shirmanova, M.V., Dudenkova, V.V. et al. (2015). Photobleaching and phototoxicity of killerred in tumor spheroids induced by continuous wave and pulsed laser illumination. *Journal of Biophotonics* 8 (11–12): 952–960. doi: 10.1002/jbio.201400130.
7 Roßbach, R., Schulz, W.M., Reischle, M. et al. (2007). Red to green photoluminescence of InP-quantum dots in InP. *Journal of Crystal Growth* 298: 595–598. doi: 10.1016/j.jcrysgro.2006.10.144.
8 Schlehahn, A., Gaafar, M., Vaupel, M. et al. (2015). Single-photon emission at a rate of 143 MHz from a deterministic quantum-dot microlens triggered by a mode-locked vertical-external-cavity surface-emitting laser. *Applied Physics Letters* 107 (4): 041 105. doi: 10.1063/1.4927429.
9 Michler, P. (2017). *Quantum Dots for Quantum Information Technologies*. Springer.
10 Gaponenko, M., Metz, P.W., Härkönen, A. et al. (2014). SESAM mode-locked red praseodymium laser. *Optics Letters* 39 (24): 6939–6941. doi: 10.1364/OL.39.006939.
11 Sarukura, N., Edamatsu, K., Semashko, V.V. et al. (1995). Ce_{3+}: $LuLiF^4$ as a broadband ultraviolet amplification medium. *Optics Letters* 20 (3): 294–296. doi: 10.1364/OL.20.000294.

12 Petrov, V., Ghotbi, M., Kokabee, O. et al. (2010). Femtosecond nonlinear frequency conversion based on BiB^3O^6. *Laser & Photonics Reviews* 4 (1): 53–98. doi: 10.1002/lpor.200810075.
13 Ranta, S., Härkönen, A., Leinonen, T. et al. (2013). Mode-locked VECSEL emitting 5 ps pulses at 675 nm. *Optics Letters* 38 (13): 2289–2291. doi: 10.1364/OL.38.002289.
14 Lederer, M.J., Kolev, V., Luther-Davies, B. et al. (2001). Ion-implanted ingaas single quantum well semiconductor saturable absorber mirrors for passive mode-locking. *Journal of Physics D: Applied Physics* 34 (16): 2455. doi: 10.1088/0022-3727/34/16/309.
15 Bek, R., Kahle, H., Schwarzbäck, T. et al. (2013). Mode-locked red-emitting semiconductor disk laser with sub-250 fs pulses. *Applied Physics Letters* 103 (24): 242101. doi: 10.1063/1.4835855.
16 Bek, R., Daghestani, N.S., Kahle, H. et al. (2014). Femtosecond mode-locked red AlGaInP-VECSEL. *Vertical External Cavity Surface Emitting Lasers (VECSELs) IV, Proceedings of SPIE*, San Francisco, CA, USA, [[vol]] 8966, p. 89660P. doi: 10.1117/12.2037828.
17 Smowton, P., Lutti, J., Lewis, G. et al. (2005). InP-GaInP quantum-dot lasers emitting between 690–750 nm. *IEEE Journal of Selected Topics in Quantum Electronics* 11 (5): 1035–1040.
18 Maas, D.J.H.C., Bellancourt, A.R., Hoffmann, M. et al. (2008). Growth parameter optimization for fast quantum dot sesams. *Optics Express* 16 (23): 18 646–18 656. doi: 10.1364/OE.16.018646.
19 Bek, R., Kersteen, G., Kahle, H. et al. (2014). All quantum dot mode-locked semiconductor disk laser emitting at 655 nm. *Applied Physics Letters* 105: 082 107. doi: 10.1063/1.4894182.
20 Schwarzbäck, T., Bek, R., Hargart, F. et al. (2013). High-power InP quantum dot based semiconductor disk laser exceeding 1.3 w. *Applied Physics Letters* 102 (9): 092101. doi: 10.1063/1.4793299.
21 Bek, R., Kersteen, G., Kahle, H. et al. (2015). Quantum dot based mode-locked AlGaInP-vecsel. *Vertical External Cavity Surface Emitting Lasers (VECSELs) V, Proceedings of SPIE*, San Francisco, CA, USA, [[vol]] 9349, p. 93490G. doi: 10.1117/12.2077164.
22 Schättiger, F., Bauer, D., Demsar, J. et al. (2012). Characterization of InGaAs and InGaAsN semiconductor saturable absorber mirrors for high-power mode-locked thin-disk lasers. *Applied Physics B: Lasers and Optics* 106: 605–612. doi: 10.1007/s00340-011-4697-7.
23 Paajaste, J., Suomalainen, S., Härkönen, A. et al. (2014). Absorption recovery dynamics in 2 μm GaSb-based sesams. *Journal of Physics D: Applied Physics* 47 (6): 065 102.
24 Grange, R., Haiml, M., Paschotta, R. et al. (2005). New regime of inverse saturable absorption for self-stabilizing passively mode-locked lasers. *Applied Physics B: Lasers and Optics* 80: 151–158. doi: 10.1007/s00340-004-1622-3.
25 Liau, Z.L. (2000). Semiconductor wafer bonding via liquid capillarity. *Applied Physics Letters* 77 (5): 651–653. doi: 10.1063/1.127074.

26 Hoogland, S., Dhanjal, S., Tropper, A.C. et al. (2000). Passively mode-locked diode-pumped surface-emitting semiconductor laser. *IEEE Photonics Technology Letters* 12 (9): 1135–1137. doi: 10.1109/68.874213.

27 Kuznetsov, M., Hakimi, F., Sprague, R., and Mooradian, A. (1997). High-power (> 0.5-W CW) diode-pumped vertical-external-cavity surface-emitting semiconductor lasers with circular TEM_{00} beams. *IEEE Photonics Technology Letters* 9 (8): 1063–1065. doi: 10.1109/68.605500.

28 Keller, U., Miller, D.A.B., Boyd, G.D. et al. (1992). Solid-state low-loss intracavity saturable absorber for nd:ylf lasers: an antiresonant semiconductor fabry–perot saturable absorber. *Optics Letters* 17 (7): 505–507. doi: 10.1364/OL.17.000505.

29 Brovelli, L., Jung, I., Kopf, D. et al. (1995). Self-starting soliton modelocked Ti-sapphire laser using a thin semiconductor saturable absorber. *Electronics Letters* 31 (4): 287–289. doi: 10.1049/el:19950184.

30 Keller, U., Weingarten, K.J., Kärtner, F.X. et al. (1996). Semiconductor saturable absorber mirrors (SESAM's) for femtosecond to nanosecond pulse generation in solid-state lasers. *IEEE Journal of Selected Topics in Quantum Electronics* 2 (3): 435–453. doi: 10.1109/2944.571743.

31 Bek, R., Baumgärtner, S., Sauter, F. et al. (2015). Intra-cavity frequency-doubled mode-locked semiconductor disk laser at 325 nm. *Optics Express* 23 (15): 19 947–19 953. doi: 10.1364/OE.23.019947.

32 Nicholson, J.W. and Rudolph, W. (2002). Noise sensitivity and accuracy of femtosecond pulse retrieval by phase and intensity from correlation and spectrum only (PICASO). *Journal of the Optical Society of America B* 19 (2): 330–339. doi: 10.1364/JOSAB.19.000330.

33 Klopp, P., Saas, F., Zorn, M. et al. (2008). 290-fs pulses from a semiconductor disk laser. *Optics Express* 16 (8): 5770–5775. doi: 10.1364/OE.16.005770.

34 Klopp, P., Griebner, U., Zorn, M. et al. (2009). Mode-locked InGaAs-AlGaAs disk laser generating sub-200-fs pulses, pulse picking and amplification by a tapered diode amplifier. *Optics Express* 17 (13): 10 820–10 834. doi: 10.1364/OE.17.010820.

35 Hanbury Brown, R. and Twiss, R.Q. (1956). The question of correlation between photons in coherent light rays. *Nature* 178: 1447–1448. doi: 10.1038/1781447a0.

36 Jung, I.D., Kärtner, F.X., Matuschek, N. et al. (1997). Semiconductor saturable absorber mirrors supporting sub-10-fs pulses. *Applied Physics B* 65: 137–150. doi: 10.1007/s003400050259.

37 Mateo, C.M.N., Brauch, U., Kahle, H. et al. (2016). 2.5 W continuous wave output at 665 nm from a multipass and quantum-well-pumped algainp vertical-external-cavity surface-emitting laser. *Optics Letters* 41 (6): 1245–1248. doi: 10.1364/OL.41.001245.

38 Meschede, D. (2007). *Optics, Light and Lasers – The Practical Approach to Modern Aspects of Photonics and Laser Physics*. Weinheim: Wiley-VCH Verlag GmbH & Co. KGaA.

39 Zheltikov, A., L'Huillier, A., and Krausz, F. (2007). *Springer Handbook of Lasers and Optics*. New York: Springer.

11

Colliding Pulse Mode-locked VECSEL

Alexandre Laurain

II–VI incorporated, Sherman, TX, USA

11.1 Introduction

Mode-locked VECSELs are emerging as the technology of choice for the generation of high brightness femtosecond pulses with multi-GHz repetition rates. The flexibility of the external cavity, combined with the advantages of a semiconductor gain medium, provides an excellent platform for stable mode locking with low intrinsic noise. The passive mode locking of VECSELs has been successfully demonstrated with various techniques and materials. The most commonly employed method is to incorporate a semiconductor saturable absorber mirror (SESAM) in one arm of the external cavity, in a V-shaped geometry, for example, [1, 2]. The SESAM typically consists of a single quantum well (QW) absorbing at the central wavelength of the desired pulse. It is grown on a semiconductor distributed Bragg reflector (DBR) to provide high reflectivity when the QW absorption is saturated at high intracavity field intensities. SESAMs need to be designed to have as fast a carrier recombination rate as possible to ensure short pulses. This can be achieved by low temperature growth of the QW creating defects, which cause rapid defect recombination [3], or by growing the single QW in close vicinity to the semiconductor surface (few nm) in order to facilitate fast recombination with surface states [4]. The saturable absorber can also be directly integrated into the gain structure, like the MIXSEL structure developed by the group of Ursula Keller [5]. This approach is ideally suited for very high repetition rate as the cavity can be extremely compact. However, it appears that this approach is less flexible and leaves less room for growth error, which has limited performance to lower power and longer pulses (>250 fs) than the external SESAM approach. The saturable absorber element is frequently a single QW, but it can also be replaced by a layer of quantum dots, or a more exotic material such as graphene [6] or carbon nanotubes [7] depending on the properties desired. Another mode locking technique that exploits the ultrafast-negative kerr effect has also been employed to generate a pulsed operation [8]; however, the mode locking performance and stability were not as good as with a SESAM. An alternative mode-locking technique called colliding pulse mode locking (CPM) was recently demonstrated with a VECSEL

Vertical External Cavity Surface Emitting Lasers: VECSEL Technology and Applications, First Edition.
Edited by Michael Jetter and Peter Michler.
© 2022 WILEY-VCH GmbH. Published 2022 by WILEY-VCH GmbH.

[9, 10], exhibiting very interesting properties in terms of stability and pulse characteristics. In this chapter, we will focus on this last technique. First, we present in detail the principle of operation of CPM and we discuss the design requirements to obtain CPM operation with a VECSEL. We will then present a typical design of a CPM VECSEL capable of producing ultrashort pulses. The experimental results obtained with such a device will be presented and analyzed, which will give us the opportunity to investigate the pulse interactions in the absorber. We will present a comprehensive model for the calculation of saturable losses in a CPM VECSEL, and we will show how the pulse interactions in the absorber lead to a reduction of the saturation fluence and a power balance between the counterpropagating beams, resulting in a more robust mode-locking regime. Finally, we will give an outlook of the potential applications of a CPM VECSEL.

11.2 Principle of Colliding Pulse Modelocking

The term CPM was first introduced in 1981 by Fork et al. [11]. It denotes the important role played by the interaction or "collision" between two oppositely directed pulses in a thin saturable absorber for the passive mode locking of a laser. Generally, a laser cavity without direction selective elements can support lasing for two counterpropagating beams, which may interact with each other at some locations along their pathway, provided that they cross in a medium with a nonzero susceptibility. When the cavity contains a saturable absorber, a pulse shaping mechanism may occur and provide a mode-locked state where the two counterpropagating pulses automatically synchronize their flight in the cavity to cross in the absorber. The absorber will indeed be saturated more effectively when the two counterpropagating pulses are colliding in the absorber, as it is saturated by two pulses instead of one, thus minimizing the losses seen by each pulse. Moreover, the coherent interaction of the pulses builds up a transient standing wave which locally increases the optical field intensity, further reducing the effective saturation fluence of the absorber. CPM can be obtained with numerous gain/absorber media and with various cavity geometries. A few examples of cavity geometries are shown in Figure 11.1.

When the absorber is used in transmission, the counterpropagating beams are collinear and the process of saturating the absorber by two colliding pulses bears some resemblance to the case of a standing-wave saturation by a single pulse when the absorber is attached to an end mirror. In this case, the field interference pattern is along the propagation axis (longitudinal) and creates a transient carrier grating in the population of the absorber, which will significantly reduce the saturation fluence, at the condition that the optical path in the saturable absorber is thinner than the desired pulse width. That means that for a pulse duration of 100 fs the optical path in the absorber needs to be less than 30 μm.

The first experimental demonstration of CPM was obtained in a ring cavity [11], with both the gain and the absorber used in transmission, a geometry similar to Figure 11.1a. They reported a passive mode locking of a Rhodamine 6-G dye laser with pulses as short as 90 fs and an average output power of 50 mW per output

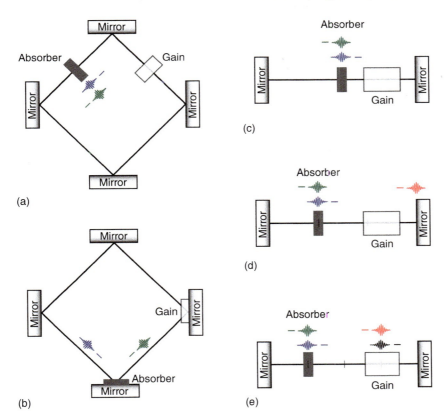

Figure 11.1 Cavity geometries and circulation of the pulses in a colliding pulse mode-locking laser. (a) Ring cavity geometry with the absorber used in transmission. The saturable absorber is placed at a distance equal to a quarter of the total length of the cavity from the gain medium, ensuring symmetrical amplification of the two pulses (b) Ring cavity with absorber and gain used in reflection. (c) Linear cavity supporting two intracavity pulses, absorber is placed at the center of the cavity. (d) Linear cavity supporting three intracavity pulses, the absorber is placed at one third of the cavity length from the end mirror. (e) Linear cavity supporting four intracavity pulses, the absorber is placed at one fourth of the cavity length from the end mirror.

beam. Interestingly, they also observed an increased mode-locking stability and an insensitivity of pulse width to a variation of pump power when compared to a more conventional mode-locking technique (linear cavity). We will see later that the improved stability is also observed when the technique is used for mode locking a VECSEL (see Section 11.5.1). The CPM technique was later used with an Nd:YAG laser [12] and a few years later with semiconductor lasers [13, 14], where the semiconductor absorber was placed in the middle of a linear cavity and surrounded by two gain regions, similar to the geometry of Figure 11.1c.

An alternative cavity configuration, where the absorber is used in reflection, is very interesting for a thin absorber like a SESAM. This time the counterpropagating beams are not collinear in the absorber, they collide at an angle θ which produces an interference pattern in the transverse direction instead. The field intensity is thus locally increased and has a maximum 4 times higher than the intensity of one beam alone. These interferences cause a transient transverse carrier grating in the absorber which will significantly reduce the effective saturation fluence of the absorber (see Section 11.6). This geometry is particularly well suited for a VECSEL and a SESAM and will be discussed in more details in the following sections.

11.3 Requirements for Stable Colliding Pulse Modelocking

To obtain stable CPM operation, we need to make sure that the counterpropagating pulses can actually overlap in time and space in the absorber and that the laser dynamic favors a CPM operation over a CW operation or unstable mode locking. The laser gain and absorber design for CPM is similar to what would be required for a standard mode-locking operation; however, the cavity design has a few specific requirements that are discussed below.

11.3.1 Pulse Timing

Evidently, for CPM to occur the counterpropagating pulses must be able to collide in the absorber periodically, without time delay over numerous round trips.

In the case of a linear cavity, this means that if n pulses oscillate in the cavity, then the absorber must be placed precisely at the submultiple n of the cavity length from one end mirror. The absorber placement for a linear cavity containing two, three, and four pulses is shown in Figure 11.1c–e. The precision on the placement of the absorber can be critical since the difference in arrival time of the two interfering pulses must be small compared to the pulse duration ($< 10\mu m$). This geometry is primarily used with monolithic semiconductor edge emitter lasers, where the gain and absorber sections can be precisely defined along the waveguide with standard lithography and chemical etching techniques [14].

On the other hand, in the case of a ring cavity geometry, the counterpropagating pulses will travel the same distance before they return to the absorber, regardless of the location of the absorber in the cavity. In most cases, the counterpropagating pulses will automatically synchronize their flight time to collide in the SESAM, provided that the gain dynamic allows them to coexist. It is generally the geometry of choice for laser technologies with an external cavity, such as a VECSEL.

11.3.2 Gain Recovery and Pumping Rate

When two or more counterpropagating pulses oscillate in the laser cavity, sharing the same gain medium, we must mitigate the competition for gain between them. If the gain seen by a particular pulse is higher than for the other pulses, it might "win" the competition for gain and CPM will cease. Indeed, if the difference of amplification between the pulses is greater than the reduction of absorption losses from

Figure 11.2 Illustration of the gain recovery dynamic after the passage of two successive counterpropagating pulses. (a) The absorber is placed at a quarter of the cavity length from the gain medium; amplification is symmetric. (b) The absorber is not optimally placed and the gain recovers slowly; the gain g_1 seen by pulse 1 has recovered over a time $\Delta t_1 > \Delta t_2$, leading to a stronger amplification. (c) The absorber is not optimally placed, but the gain recovers quickly; amplification is symmetric.

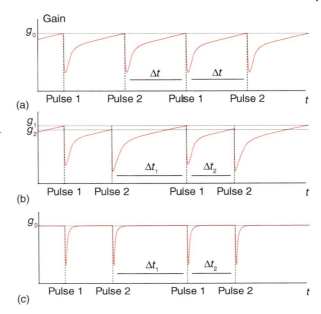

CPM operation, then the mode-locking will become unstable and possibly become unidirectional. This aspect is particularly important with a thin semiconductor gain media generating short pulses, where the gain will be strongly saturated by the passage of an intense pulse, thus reducing the available gain for the next pulse to hit the gain medium. To reduce the competition, the gain needs to recover to its unsaturated value or at least to a value similar to the one seen by the previous pulse. In a ring cavity, a gain symmetry can be achieved if the absorber is placed at one-fourth of the cavity length from the gain medium. But in this case, the precision of this placement is not dictated by the pulse duration like in a linear cavity, but by the gain recovery time and pumping rate. The influence of the absorber placement on the gain dynamic is illustrated in Figure 11.2.

Let us assume, for example, a total cavity length of 10 cm containing only two pulses (3 GHz repetition rate). If the absorber is placed 1 mm too close to the gain medium, one pulse will be amplified by a gain medium that has recovered for about 163 ps whereas the other pulse will see a gain that recovered for 170 ps. In the case of a semiconductor gain medium, this difference is negligible because a large fraction of the gain recovers on a shorter time scale. It has been shown that the kinetic hole burned in the carrier distribution by the passage of a short pulse will quickly recover due to kinetic hole filling from intraband scattering. A recent in situ measurement of the gain recovery dynamic revealed that the population inversion in a VECSEL gain structure can recover to more than 80% of its initial value in a time as short as 5 ps [15]. Therefore, for a semiconductor gain medium, the placement of the absorber in the ring cavity is not very stringent (Figure 11.2c). We should note, however, that for very short cavities, the absorber placement becomes increasingly important.

11.3.3 Polarization

The main advantage of CPM lies in the interference pattern created by the superposition of two pulses in the absorber. Evidently, the respective polarization of the two pulses will strongly affect the way they interact with each other. If the pulses are collinear (absorber used in transmission), the intensity will be fully modulated when the pulses have the same polarization. The contrast of the interference fringes will be maximal and the saturation of the absorber will be more effective. On the contrary, if the pulses' polarizations are orthogonal the interference pattern vanishes and the saturation of the absorber is less effective. The laser dynamic will thus tend to favor two pulses sharing the same polarization, minimizing the optical losses.

In the case of noncollinear pulses (absorber used in reflection), the situation is slightly different as the s- and p-polarization will also be discriminated. Similar to the collinear case, orthogonal polarizations will not produce any interference, but the s-polarization will be favored over the p-polarization for any non-null angle. Indeed, the amplitude of the transverse interference pattern from two s-polarized beams will be maximum regardless of the angle of incidence, whereas with two p-polarized beams the amplitude is maximum at $\theta = 0°$, decreases with the angle of incidence, and eventually vanishes for $\theta = 45°$, where the beams' polarizations become orthogonal [16]. The different polarization situations are illustrated in Figure 11.3. We should note that with a VECSEL, a very small absorption difference is generally sufficient to discriminate one polarization state over the other one. Consequently, the s-polarization state will be naturally selected by the nonlinear dynamics of the high finesse VECSEL cavity. It is thus best to optimize the gain element and the mirrors of the cavity for s-polarization, in particular the group delay dispersion (GDD) which will be different for each polarization state. In the following sections, we will assume that both beams are s-polarized (along the y axis).

11.3.4 Mode Waist and Saturation Fluence

It is well known that in order to obtain a stable mode-locking state, the absorber must saturate more strongly than the gain [17]. With a VECSEL cavity containing a gain medium and SESAM based on the same material system, the saturation energy of both the absorber and gain material will be similar. This usually imposes a design

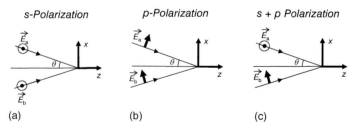

Figure 11.3 Example of polarization arrangements for two noncollinear beams colliding at z = 0. (a) s-Polarization, maximum interference contrast. (b) p-Polarization, partial interference contrast. (c) s + p Polarization, no interferences.

restriction for the mode area ratio between the VECSEL gain and the SESAM, which typically needs to be >10 to provide stable mode locking [18].

However, in a CPM VECSEL, the saturation fluence of the absorber will be reduced by about a factor of 3 due to the pulse interference (see Section 11.6.5). Therefore, the mode area ratio constraint is largely relaxed, without affecting the modulation depth of the absorber. This means that the cavity can be designed with a mode size on the absorber similar to the mode size on the gain, allowing for low-divergence beams. A larger mode size on the absorber also reduces the thermal impedance of the SESAM and increases the damage threshold, allowing a higher average and peak power. Another advantage is the possibility to operate far away from the stability limit of the cavity (see Section 11.4.1), providing a more stable and rugged oscillator, since a small cavity length fluctuation will hardly affect the cavity mode size.

11.4 Design of an Ultrafast CPM VECSEL

A common approach for the development of an ultrafast VECSEL is to start with the design and fabrication of a gain structure at the desired lasing wavelength, then design and fabricate a saturable absorber at this wavelength. These components are then tested and characterized individually before being assembled in a given cavity geometry to hopefully produce short pulses. However, if the required pulse characteristics are very demanding, like if the device has to generate sub-200 fs or even sub-100 fs pulses, it is better to consider the entire laser device when designing a specific element. For example, in a CPM VECSEL, the cavity geometry and the number of passes of a pulse on each element per round trip will be different than in a V-shaped cavity, and it is likely to affect the performances. In other words, a gain structure or SESAM that is designed and optimized for a standard cavity geometry might not be optimum when used in a ring cavity. It is thus better to first design the cavity where the gain and SESAM structure will be used and then design these elements according to this geometry.

11.4.1 The Optical Cavity

We already established that a ring cavity geometry, with the absorber placed at a quarter of the cavity length L from the gain is best suited for a CPM VECSEL. The choice of the total cavity length obviously depends on the repetition rate of the cavity that is desired, which for a ring cavity is it given by: $FSR = c/L$. We should note that the depletion of the carrier inversion will occur at twice this rate, which helps to avoid harmonic mode locking that might occur at high pump levels or at low repetition rate, if the time interval between consecutive pulses on the gain is much longer than the gain recovery time characteristics.

In order to keep the astigmatism and ellipticity of the laser beam to a minimum, the angle of incidence on the focusing mirror must be minimized. The more compact and simple way to accomplish this is to use a "folded ring cavity" geometry, as illustrated in Figure 11.4. To ensure that the beam waist is located on the SESAM,

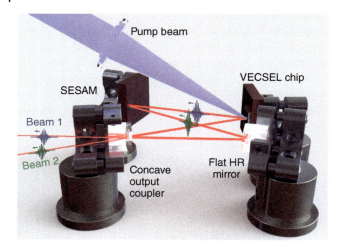

Figure 11.4 Schematic layout of a colliding pulse mode-locked VECSEL in a folded ring cavity configuration.

the focusing mirror must be placed at $L/2$ from the SESAM. This is, however, not strictly required, and the mode size on the SESAM can be increased by changing the relative distance of the concave mirror to avoid laser induced damage, for example. Another flat dielectric mirror is used to complete the cavity and is placed as close as possible to the gain structure to minimize the angle on the concave mirror. The distance to the flat mirror from the SESAM and from the concave mirror should total $L/2$. This cavity geometry will give two beams which can be chosen to output either from the concave mirror or from the flat mirror.

To ensure single transverse mode operation, the actual cavity length and the radius of curvature of the dielectric mirror must be selected such that the lasing mode waist matches the size of the pump beam. Or alternatively, the pump spot size and mirror can be adjusted to a given cavity length if a specific repetition rate is required. Figure 11.5 shows some cavity stability curves for four common values of radius of curvature of the focusing mirror with the SESAM set at $L/2$ from the curved mirror and the VECSEL set at $L/4$ from the SESAM. It is worth noting that the cavity mode size is evidently the same for both counterpropagating beams, and the radius of curvature of their wavefront is also equal but with opposite sign.

The mode propagation along each direction of the cavity for two counterpropagating beams synchronized on the SESAM is illustrated in Figure 11.6. It is clear from this graph that the gain recovery will be symmetric for the two counterpropagating beams and that the relative location of the flat mirror M_1 has no importance.

11.4.2 The Gain Structure

One of the most important aspects to consider when designing an ultrafast mode-locked VECSEL is to design a semiconductor gain structure that can support ultrashort pulses (<100 fs). The main strategy is to optimize a structure to provide

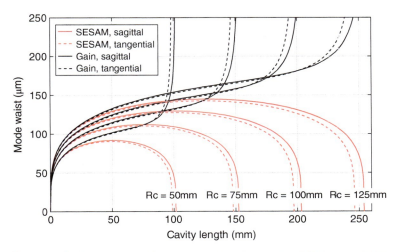

Figure 11.5 Calculation of the cavity mode width on the SESAM and VECSEL structure along the sagittal and tangential axis as a function of the total cavity length for four common values of concave mirror radius of curvature. An incidence angle of 10° and a wavelength of 1030 nm are assumed.

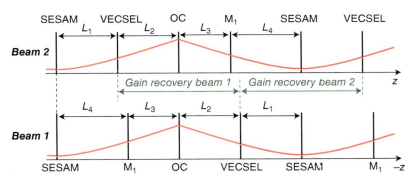

Figure 11.6 Beam width evolution along the cavity and illustration of the gain recovery symmetry for the two counterpropagating beams. OC, concave output coupler; M_1, flat mirror.

a very broad and spectrally flat gain, to ensure that the wide spectrum required for ultrashort pulses is not physically limited by the gain bandwidth. The type of gain material or system used in the structure, whether it is QWs or quantum dots, will provide a gain spectrum that is intrinsically curved, generally approximated to a parabolic function. This gain curvature is usually even more pronounced due to the field enhancement factor of the gain structure. Indeed, if the QWs or QDs are placed at the antinodes of the field, the gain will evidently be maximum, but it will also produce a sharp spectral resonance that further increases the curvature of the gain, which is not ideal for a short pulse generation. However, it is possible to mitigate this effect by a judicial placement of the QWs or QDs according to the standing wave in the structure. By distributing the QWs nonuniformly around the

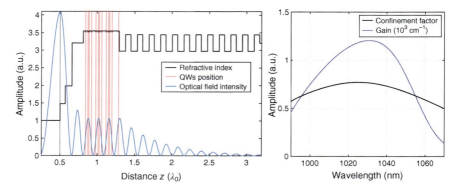

Figure 11.7 Left: Schematic layout of a VECSEL structure with nonuniform QW distribution around the antinodes. Right: simulated gain spectrum of an InGaAs QW used in the VECSEL structure for an emission wavelength of 1030 nm with a carrier sheet density of 5×10^{12} cm^{-2} and a temperature of 375 K, together with the calculated confinement factor of the VECSEL structure.

antinodes of the field, it is possible to flatten out the gain spectrum seen by a pulse which is more favorable for the generation of short pulses. The QWs position can be numerically optimized with a microscopic simulation of the pulse formation [19]. An example of a VECSEL structure optimized with such simulation is presented in Figure 11.7 with the corresponding enhancement factor.

The length of the microcavity (the active region between the DBR and the semiconductor top layer) is also an important factor since it directly affects the microcavity Q-factor, hence the spectral selectivity. To broaden this spectral filter, it is advantageous to keep the length of the microcavity to a minimum, but that also means that the pump-absorbing and strain-compensating barriers will be shortened. It is thus important to manage the reduced pump absorption by recycling whatever pump is not absorbed, for example, with a signal DBR transparent at the pump wavelength followed by an additional reflector like a metal (hybrid metal-semiconductor mirror) or a pump DBR (dual bandwidth DBR). By using a metallic mirror like gold or silver, which both have a very broad reflectivity spectrum, it is possible to enhance both the signal and the pump reflectivity, allowing one to reduce the number of DBR pairs significantly while keeping the same reflectivity level at the signal wavelength [20]. This has the advantage of reducing the thermal impedance of the device and providing a slightly wider bandwidth (∼10%). The DBR interference fringes outside the stopband will also be broadened, giving more flexibility for the pump wavelength and incidence angle acceptance. Figure 11.8 shows a reflectivity measurement of an hybrid DBR VECSEL structure designed at 980 nm with the same layout as Figure 11.7, realized with 12 DBR pairs and a pure gold reflector. A comparison with the same structure but with the gold reflector replaced by a titanium metalization is also given.

The strain management will also become more critical since, to keep an acceptable gain level (>2%), we have to stack more QWs in a limited space. If the local strain of the structure crosses the critical limit of the material grown, an array of

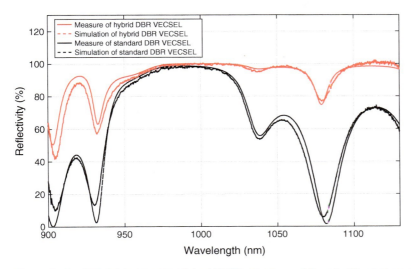

Figure 11.8 Reflectivity spectra of the VECSEL structure with and without the gold reflector measured at room temperature, together with the simulated spectra (dotted lines).

dislocations may form along the crystal axis and propagate vertically through the structure. These dislocation lines are to be avoided since they are centers of nonradiative recombination and drastically reduce the carriers' lifetime and the efficiency of the device. Since most QWs are not lattice matched to the host substrate, this limits the minimum spacing between them and the maximum number that can be stacked around an antinode and is increasingly challenging with heavily strained wells, i.e. at longer wavelengths for InGaAs QWs.

Another equally important design strategy is to manage the group delay dispersion (GDD) of the structure, to provide a broad and spectrally flat GDD with a value approaching 0 fs^2. When optimizing the GDD of the structure, it is important to consider the dispersion of the other elements that will be present in the cavity (SESAM, mirrors), and the respective number of passes per round trip on each of these elements, to ensure that the total cavity dispersion is optimized, not just the gain structure itself. One clear advantage of a CPM VECSEL in a ring cavity configuration is that the GDD contribution of the gain structure toward the total cavity GDD is minimized, since the pulses hit the structure only once per round trip instead of twice when it is placed as a folding mirror in a V-shaped cavity. Since the gain structure is usually the main contributor to the nonlinear GDD of the cavity, it is a notable advantage for the generation of ultrashort pulses.

The respective angle of incidence of the lasing field on each element should also be considered, as it significantly affects the GDD spectrum. A structure optimized for normal incidence will not necessarily be optimum for a non zero degree angle, as the reflectivity and GDD spectrum will blue shift with the angle and the third-order dispersion will also change. The simulation presented in Figure 11.9 shows the influence of the incidence angle on the GDD spectrum of a typical structure optimized for a flat GDD at 20° angle.

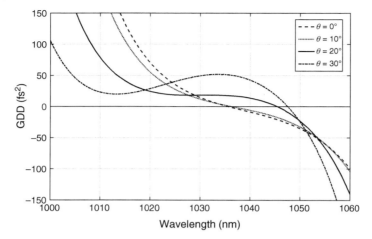

Figure 11.9 Influence of the incidence angle on the GDD spectrum of a typical gain structure optimized for a 20° angle at a central wavelength of 1030 nm.

The first step to optimize the GDD spectrum is to center the DBR stopband on the central pulse wavelength at the operating temperature and incidence angle. Second, the length of the active region is designed such that the field intensity has a node at the semiconductor surface, also at the operating angle and temperature. This kind of design is often referred as antiresonant, because it minimizes the field intensity in the structure. This approach will reduce the available gain, but will give a flatter gain and GDD spectrum, and will increase the gain saturation fluence. Finally, the structure should be completed by a precise antireflection coating. The simplest antireflection coating consists of one layer of a dielectric material transparent at the pump and signal wavelength, and having an index of refraction around $n_c = \sqrt{n_s}$, where n_s is the refractive index of the semiconductor [21]. Even though this provides a tremendous improvement of the GDD and is a satisfactory solution for the generation of sub-ps pulses, it is not sufficient for sub-100 fs pulses which require a broad spectrum with minimal third order dispersion. To further improve the GDD, it is possible to use a structure with a section made of several semiconductor layers and completed by a single layer of dielectric coating, generally SiO_2 [2]. This multilayer coating is usually engineered numerically to provide a flat GDD and to minimize the pump reflection. Another solution is to use a bilayer dielectric coating to gradually decrease the refractive index of the structure. In this case, the refractive index of the first material deposited must be significantly higher than the index of the second material. For example, this can be accomplished by a combination of Si_3N_4 or Ta_2O_5 with SiO_2. The thickness of each layer can be numerically optimized to obtain the GDD spectrum that will best compensate the other dispersive elements of the cavity and the self-phase modulation of the pulses. The influence of the type of coating on the gain and GDD is summarized in Figure 11.10. It shows that the third-order dispersion and gain curvature of a single-layer coating of Si_3N_4 is clearly higher and that even if one could find a material with an ideal refractive index n_c, a realistic bilayer coating would still be a better choice and should support shorter pulses.

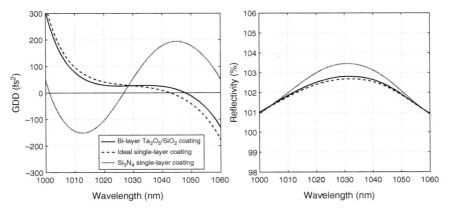

Figure 11.10 Influence of the type of coating on the gain and GDD spectrum. We simulate a structure designed for an emission wavelength of 1030 nm and assume a carrier sheet density of 5×10^{12} cm^{-2} and a temperature of 375 K.

Finally, for optimal thermal management the gain structure is usually grown as a bottom emitter and subsequently bonded to a diamond heatspreader. Another thermal management technique is to use a top emitter structure with an intracavity diamond heatspreader, but the spectral selectivity and dispersion caused by such element would drastically reduce the mode-locking performance and is usually reserved for CW operation or for substrate materials that cannot be easily removed by chemical etching.

11.4.3 The SESAM

The saturation fluence, the modulation depth, the GDD, and the recovery time of the absorber are the main parameters to consider when designing a SESAM. The saturation fluence and modulation depth will be mainly determined by the type of material system chosen for the absorber and by the field intensity enhancement of the structure. The recovery time is mainly affected by the density and proximity of non-radiative carrier recombination centers, whereas the GDD will be governed by the overall structural layout and dielectric coating. An example of a QW-based SESAM grown on a molecular beam epitaxy machine (MBE) is shown in Figure 11.11.

The different parts or "sections" of this SESAM can be divided as follow, in their growth order:

- A highly reflective DBR consisting of 24 quarter-wave layer pairs of AlAs/GaAs. The stop band center is aligned to the central wavelength of the pulse at operation temperature and angle of incidence.
- A 22 nm thick GaAs spacer layer separating the QW from the DBR, which tailors the field enhancement factor and thus the saturation fluence.
- An 8 nm thick InGaAs QW. The wavelength of the absorption peak is blue shifted by 10 nm from the lasing wavelength to provide minimal absorption at room temperature, facilitating alignment. The modulation depth can be adjusted while it is lasing by increasing the heat-sink temperature.

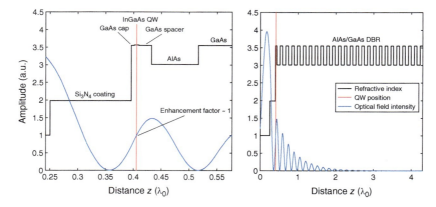

Figure 11.11 Schematic layout of a SESAM structure incorporating an InGaAs QW for an operating wavelength of 990 nm. On the left side is a magnification of the structure around the QW, whereas the right side shows the full structure.

- A 5 nm thick GaAs cap layer. This thin separation layer provides a high carrier recombination rate from the QW via tunneling to surface states, while protecting the QW from oxidation.
- A Si_3N_4 coating. Its thickness is not necessarily a quarter-wave like a regular AR coating but is engineered to minimize the total GDD of the cavity.

The GDD of the SESAM and gain structure, measured at room temperature and normal incidence, are shown in Figure 11.12. It shows that when the pulse hits the SESAM and gain structure only once per round trip like in a CPM VECSEL cavity, the total GDD is near zero and relatively flat at the designed wavelength of 990 nm.

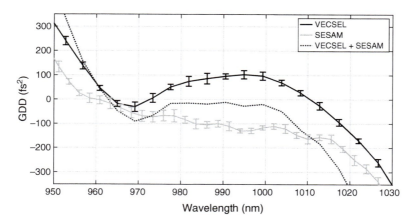

Figure 11.12 Measured group delay dispersion spectrum of the SESAM and VECSEL gain structure. The dotted line represents the sum of the VECSEL and SESAM GDD, and the vertical bars indicate the standard deviation over 20 measures.

11.5 Modelocking Results

The first experimental demonstrations of a CPM VECSEL were reported very recently [9, 10]; therefore, the amount of experimental data available to date is rather limited. Nevertheless, these first results, which we summarize in the following sections, have shown some very interesting pulse characteristics and performance in term of power, pulse duration, and stability of the mode-locking regime.

11.5.1 Robustness of the Modelocking Regime

Maybe the most remarkable feature observed with a CPM VECSEL was the robustness of the mode-locking regime with a variation of the pump power or with a change of the modulation depth of the absorber. For this demonstration, the VECSEL structure used consisted of two consecutive semiconductor Bragg mirrors containing, respectively, 23 and 13 pairs of $AlAs/Al_{0.12}Ga_{0.88}As$ to reflect the lasing and pump wavelength, followed by the active region, an InGaP cap layer and a single layer Si_3N_4 coating. The antiresonant active region contains eight InGaAs QWs placed in pairs on four antinodes of the field and is pumped in the GaAsP barriers with a 790 nm fiber-coupled pump diode. This gain structure and the SESAM were placed in a ring cavity according to the geometry shown in Figure 11.13, with a total cavity length of 136 mm giving a repetition rate of 2.2 GHz. The SESAM was placed 34 mm away from the gain ensuring an equal pumping duration and gain recovery for both pulses. The cavity was completed with a highly reflective concave mirror with a radius of curvature of 75 mm and a flat output coupler with a reflectivity of 99.2%. The angle of incidence on the SESAM and VECSEL was 7°. This geometry provides a mode radius of 152 µm on the gain and 85 µm on the SESAM.

Figure 11.14 shows the output power and pulse duration of one output beam measured at different pump powers and SESAM temperatures while the heat-sink temperature of the gain element was kept at 5 °C.

The highest average power was obtained with a low modulation depth of the absorber, at a SESAM temperature of 35 °C. An output power of 300 mW and a pulse duration of 250 fs were obtained at a central wavelength of 992 nm, giving a peak power of 480 W. At higher pump power, the higher gain and faster gain

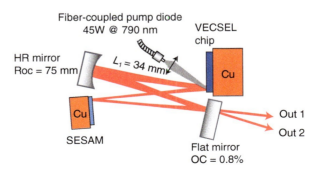

Figure 11.13 Schematic layout of the experimental CPM VECSEL device.

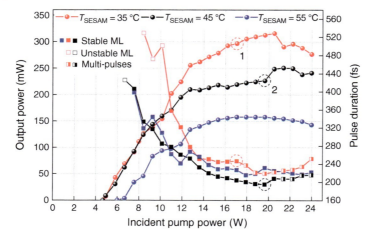

Figure 11.14 Output power from one output beam versus incident pump power at different SESAM temperatures. The corresponding pulse durations are also plotted (right axis). The empty, full, and half full squares represent, respectively, an unstable, a stable single pulse, and a stable but multipulsed mode-locking regime.

recovery enable a stable, but smaller, secondary pulse, lowering the main pulse peak power. When the SESAM temperature is increased to 45 °C, the maximum power in a single pulse regime drops to 225 mW and the pulse duration gets as short as 195 fs, giving a peak power of 460 W. The power, the spectrum, and the pulse duration of the two output beams have been measured simultaneously and are identical. Figure 11.14 shows the remarkable robustness of the mode-locking regime, which can be obtained over almost the entire range of pump power and for various modulation depths of the SESAM. As a comparison, when the same gain element was used as a folding mirror in a V-shaped cavity, with the same SESAM and a mode area ratio of 11, a similar pulse duration of 220 fs was obtained, but with only 50 mW of output power. Moreover, the mode-locking regime was only observed with a limited range of pump power, between 4.5 and 6 W and between 11 and 13 W, and only with a SESAM temperature restricted to the [45–53] °C range.

11.5.2 Cross Correlation of the Output Beams

The synchronization of the counterpropagating pulses on the SESAM, which is an explicit evidence of CPM operation, can be verified by a measurement of the cross correlation of the output beams. The concept of this measurement is illustrated in Figure 11.15. The output beams are focused and spatially overlapped in a thin nonlinear crystal, such as a beta barium borate (BBO) crystal, to generate a sum frequency signal (SFG). Since the incident beams are noncollinear, the signal can be spatially and spectrally filtered from the incident beams and measured with a simple photodiode. This signal will be maximum when the incoming pulses also coincide in time. By adjusting the delay of one of the beams to the maximum signal and

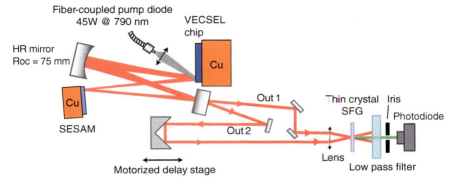

Figure 11.15 Measurement setup of the noncollinear sum frequency generation cross correlation of the two output beams.

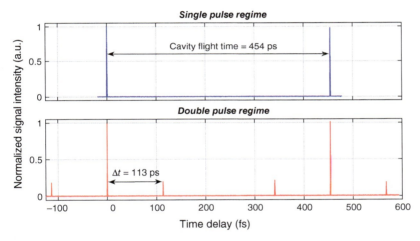

Figure 11.16 Measure of the noncollinear cross correlation of the two output beams in a single (top) and multipulse regime (bottom).

measuring the distance of each path, it is possible to determine where the pulses are synchronized in the cavity.

Figure 11.16 shows the cross-correlation traces obtained by scanning the time delay. It is calibrated such that the time zero corresponds to an equal path length from the crystal to the SESAM. The nonlinear crystal must be oriented to maximize the interaction of the two incoming s-polarized beams. We can see that the pulses are indeed synchronized on the SESAM (at $t = 0$), which confirms CPM operation.

A clean single pulse operation is verified by scanning the delay line by an entire cavity length. However, as mentioned before, at high pump power a smaller side pulse appears. In this case, the time delay between the main pulse and the side pulse is exactly a quarter of the cavity flight time ($\Delta t = 113$ ps). This corresponds to a symmetric gain recovery between successive pulses. From the gain dynamic point of view, it is equivalent to harmonic mode locking, however, the repetition

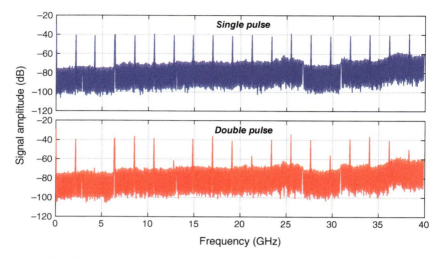

Figure 11.17 Measure of the microwave spectrum. In single pulse operation (top), all the harmonics are present and have the same amplitude whereas in double pulse operation (bottom) the harmonics amplitude are modulated at a frequency period of 4×FSR.

rate of the laser is not a multiple of the fundamental frequency. Instead, the microwave frequency spectrum has a "missing" harmonic at every frequency $f = \text{FSR} \times (4k + 2)$, where k is a positive integer. This peculiar behavior can be easily demonstrated mathematically by calculating the Fourier transform of the pulse train having a coherent side pulse delayed by 1/(4FSR). The experimental spectra shown in Figure 11.17 were recorded with a high speed photodiode and a high bandwidth heterodyne spectrum analyzer.

We should note that in multipulse operation, it is possible to determine the relative amplitude of the main and secondary pulse from the cross correlation measurement. Let us label the counterpropagating beams with the index letter "r" and "l" for the right and left-handed direction. If the beams consist of a main pulse of intensity I_1 and a secondary pulse of intensity I_2, the cross-correlation trace will have an amplitude proportional to $I_{1r}I_{1l} + I_{2r}I_{2l}$ when all pulses are synchronized on the crystal (maximum peak). Two smaller signal of amplitude $\propto I_{1r}I_{2l}$ and $\propto I_{2r}I_{1l}$ will appear when only two of the pulses are synchronized (side peaks). So if all pulses have the same intensity, the cross-correlation trace should consist of a main peak and two lateral peaks having half the amplitude of the main peak. In the measurement reported here, the secondary pulses have a smaller intensity than the main pulses by a factor ~4 which indicates that the gain dynamic favors the "initial" pulse.

11.5.3 Pulse Duration Optimization

It is pretty clear from Figure 11.14 that the pulse duration decreases when the pump power is increased. This is likely due to the pump-induced temperature rise that on one part increases the threshold carrier density (and thus broadens the gain bandwidth), and on another part shifts the optical spectrum to a wavelength more

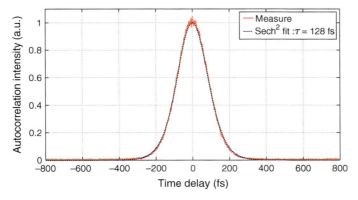

Figure 11.18 Measured noncollinear SHG autocorrelation of the single pulse operation output and simulated autocorrelation of a sech^2 pulse with a FWHM of 128 fs.

favorable for the GDD. It is, however, limited by the emergence of a "harmonic" mode-locking regime at high excitation. To avoid this regime, it is favorable to reduce the cavity length to limit the gain recovery time between successive pulses. And to further reduce the pulse duration, the gain structure can be optimized following the design guidelines given in Section 11.4.2.

These optimization steps were performed and presented in [10]. The VECSEL cavity was shortened from 136 to 92 mm, giving a repetition rate of 3.27 GHz. The double DBR of the gain structure was replaced by a short hybrid metal-semiconductor DBR consisting of 12 pairs of AlAs/AlGaAs quarter-wave layers completed by a pure gold reflector. The active region was shortened and contains 13 strain-compensated InGaAs QWs, placed nonuniformly around the antinodes of the field intensity. Finally, the single-layer coating was replaced by a bilayer dielectric coating of Ta_2O_5/SiO_2 providing a broader and flatter GDD that should support shorter pulses.

With this improved cavity and gain structure, the pulse duration was reduced to 128 fs. The gain element was kept at 25 °C and the SESAM at 55 °C, with an output power of 90 mW per output beam for a pump power of 22 W. The noncollinear SHG autocorrelation trace of the output shows an excellent agreement with a 128 fs sech^2 pulse shape (Figure 11.18). An autocorrelation scan over 20 ps was also performed to confirm the absence of side pulses.

The optical spectrum of the output beam is shown in Figure 11.19, together with the simulated spectrum of an ideal 128 fs sech^2 pulse. The spectrum is centered at 994 nm with a FWHM of 9.05 nm, about 1.12 times the bandwidth of an unchirped 128 fs pulse. The shape of the spectrum is relatively smooth and close to the ideal spectrum, without any spikes which would suggest a CW contribution.

The microwave spectrum of the output was also recorded using an ultrafast photodiode and a high bandwidth electrical spectrum analyzer. Figure 11.20 shows a scan from DC to 40 GHz with a resolution bandwidth (RBW) of 100 kHz and a zoom into the first harmonic with a span of 2 MHz and a RBW of 1 kHz. The repetition rate of the pulse is 3.724 GHz and the higher harmonics, up to the 12th, are clearly visible with a constant amplitude. The zoom into the first harmonic shows a resolution

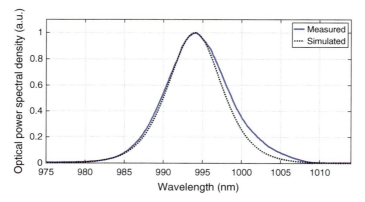

Figure 11.19 Measured optical spectrum of the output beam consisting of a single 128 fs pulse per round trip and simulated spectrum of an unchirped 128 fs sech^2 pulse.

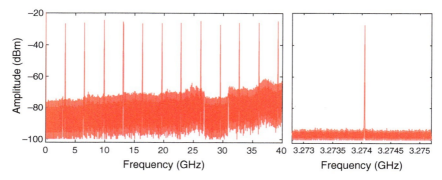

Figure 11.20 Microwave spectrum of the laser output, with a RBW of 100 kHz and a 40 GHz span (left), and with a RBW of 1 kHz and a span of 2 MHz (right).

limited linewidth with a signal to noise ratio of 68 dB. With this high dynamic range, a pulse train instability or strong timing jitter would typically result in side peaks or pedestal around the microwave signal, which is not observed here. These pulse characterizations indicate a clean and stable mode-locking regime.

11.5.4 Multipulse Regime

Once again, if the pump power is increased up to a certain threshold (35 W here), a secondary pulse is enabled. However, since the cavity was shortened, the secondary pulse is not delayed by a quarter of the cavity flight time but is generated shortly after the main pulse, benefiting from the residual absorber saturation of the main pulse and the relatively fast gain recovery. In this particular regime, an output power of 152 mW per output beam was observed. The autocorrelation trace reveals a side pulse containing 36% of the total energy. Figure 11.21 shows a perfect agreement between the measured autocorrelation and a simulated autocorrelation of a sech^2 pulse with a FWHM of 130 fs followed by a second 130 fs pulse delayed by 430 fs and a relative intensity of 56%.

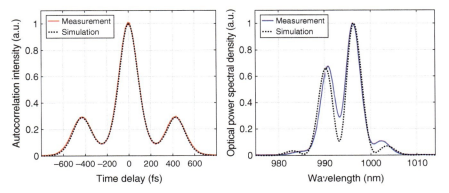

Figure 11.21 Left: Measurement of the noncollinear autocorrelation in multipulse operation and simulation of a sech² pulse with a FWHM of 130 fs followed by a 130 fs side pulse delayed by 430 fs with a relative intensity of 56%. Right: corresponding optical spectrum of the measured output beam and simulated pulses.

The optical spectrum of this regime is shown in Figure 11.21. The spectrum exhibits a strong intensity modulation, resulting from the interference of the main pulse with the side pulse. The simulated spectrum of a 130 fs sech² pulse followed by a second 130 fs pulse delayed by 430 fs, both centered at 994 nm, is also shown for reference. We should note that the interference pattern depends strongly on the phase and wavelength difference between the two pulses.

The peak power of 238 W reached by the main pulse in this regime is higher than the peak power of 189 W reached in the single pulse regime, which can be useful for applications based on nonlinear processes. However, for metrology applications [22, 23], a clean single pulse regime is usually desired. It is thus important to better understand the physical processes leading to the formation of side pulses. In the following sections, we investigate the pulse interactions in the absorber. This will allow us to calculate the optical losses seen by various pulses, the energy exchanged in the absorber, and the saturation fluence in a colliding pulse scheme.

11.6 Pulse Interactions in the Saturable Absorber

To investigate the pulse interactions in the absorber, it is necessary to evaluate the spatial distribution of the carrier density during the passage of the pulses. First, we need to express the evolution of the field intensity of colliding pulses, then we need to evaluate the absorption of this field to simulate the temporal evolution of the carrier density. Finally, the optical losses and energy transfer can be directly calculated from the carrier density dynamic.

11.6.1 Field Intensity Distribution

For this study, we consider the optical field in the slowly varying envelope approximation. We assume two counterpropagating beams with a sech² pulse envelope having the same central wavelength λ and duration $\tau_{fwhm} = 1.76\tau$, and a gaussian TEM_{00} transverse distribution. They propagate at a small angle θ to the z axis in the x, z plane and intersect within the saturable absorber located in the x, y plane at $z = 0$. We assume that both beams are linearly polarized along the y axis (s-polarization). They are focused at $z = 0$ with the same beam waist w_0. In the paraxial approximation, the "right handed" field E_r and the "left handed" field E_l may be written as [24]:

$$E_r(x, y, z, t) = E_{r_0} \frac{q_0}{q} e^{-j\frac{\pi}{\lambda}\left(\frac{(x-\theta z)^2 + y^2}{q} + 2\theta x - \theta^2 z\right)} \operatorname{sech}\left(\frac{t}{\tau}\right) \quad (11.1)$$

$$E_l(x, y, z, t) = E_{l_0} \frac{q_0}{q} e^{-j\frac{\pi}{\lambda}\left(\frac{(x+\theta z)^2 + y^2}{q} - 2\theta x - \theta^2 z\right)} \operatorname{sech}\left(\frac{t - t_d}{\tau}\right) \quad (11.2)$$

where E_{r_0} and E_{l_0} are the field amplitudes, t_d is the time delay between the pulses, $q_0 = j\frac{\pi w_0^2}{\lambda}$ and $q = q_0 + z$. The field intensity resulting from the collision of the two beams is given by:

$$I(x, y, z, t) = [E_l + E_r] \times [\overline{E_l} + \overline{E_r}] \quad (11.3)$$

The field intensity distribution of two synchronized beams colliding with an external incident angle of 7° is represented in Figure 11.22. Given the high refractive index of the SESAM material, this corresponds to an incidence angle of ∼ 2° in the absorber.

In the case of a thin absorber like a SESAM, i.e. where the interaction length with the absorber is negligible compared to the Rayleigh length, we can assume $q \simeq q_0$ and Eq. (11.3) can be simplified to:

$$I(x, y, t) = e^{-2\frac{x^2+y^2}{w_0^2}} \left[E_{r_0}^2 \operatorname{sech}^2\left(\frac{t}{\tau}\right) + E_{l_0}^2 \operatorname{sech}^2\left(\frac{t-t_d}{\tau}\right) + \cdots \right.$$
$$\left. 2E_{r_0} E_{l_0} \operatorname{sech}\left(\frac{t-t_d}{\tau}\right) \operatorname{sech}\left(\frac{t}{\tau}\right) \cos\left(\frac{\pi\theta x}{\lambda}\right) \right] \quad (11.4)$$

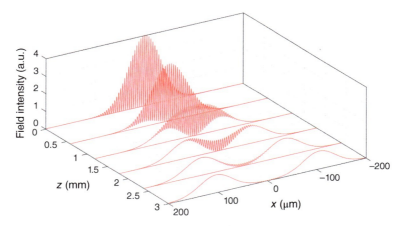

Figure 11.22 Field intensity distribution of two coherent Gaussian beams colliding at $z = 0$.

It is clear that the interference fringes created by the collision of the two beams will be maximum when they are synchronized (i.e. $t_d = 0$), giving a maximum intensity 4 times higher than the intensity of a beam alone. The absorption of such a field distribution in the absorber will create a carrier grating, which is analyzed in the following sections.

11.6.2 Saturable Absorption Model

A very common way to model the saturation effects of the absorber is to introduce an empiric parameter referred as the saturation fluence, which is defined as the pulse energy per unit area necessary to reduce the absorption by $1/e$ ($\approx$37%) of its initial value. This factor is usually determined experimentally via nonlinear reflectivity measurements [25]. This macroscopic parameter can be sufficient to describe the dynamic behavior of the absorber in the case of a clean single pulse operation [26]; however, it is not adequate for the simulation of multiple pulses or for colliding pulses. It also fails to describe the dependance of the saturation fluence on the pulse duration. In order to describe the interaction of the pulses in the absorber, we need to account for the carriers generated by the previous pulses, which might not have fully recombined and will certainly affect the absorption seen by the delayed or side pulses. For synchronized colliding pulses, one would have to evaluate the saturation fluence with the same field distribution as in the laser cavity to account for the interference effects, which would be extremely challenging to measure.

For our simulations, we evaluate the SESAM absorption α as a function of the carrier density N, instead of the pulse intensity I. First, the QW absorption spectra are computed using a microscopic approach based on the semiconductor Bloch equations, which are described in Ref. [27] and the references therein. Figure 11.23a shows the computed linear absorption of an InGaAs QW at a temperature of 350 K for various carrier densities.

To account for the field enhancement of the structure, the full reflectivity spectra of the SESAM are also simulated using conventional transfer-matrix methods, giving

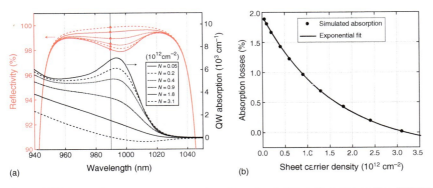

Figure 11.23 (a) Microscopically calculated QW absorption and corresponding SESAM reflectivity spectra for various carrier density N. (b) Simulated absorption losses as a function of the carrier density N of the SESAM.

a realistic absorption evaluation (Figure 11.23a). It has been shown recently that for a better recovery time and less pulse distortion, an excitation energy higher than the excitonic bandgap of the QW is preferable [28]. Therefore, it is best to evaluate the absorption below the excitonic peak, where the central pulse wavelength is supposed to be optimum. Figure 11.23b shows the simulated absorption at 990 nm, 4 nm below the excitonic peak, as a function of the carrier density. We can see that the simulation data can be fitted accurately with a decreasing exponential function, which in this case is given by:

$$\alpha(N) = 2.244 \times \exp\left(-\frac{N}{1.585 \times 10^{12}}\right) - 0.3 \tag{11.5}$$

where N is expressed in cm^{-2}. This equation is only valid for the specific SESAM structure used here, as the type of absorber or the distance between the DBR and the QW will change the field strength and absorption. However, the pulse interaction effects and the conclusions drawn in the following sections could be generalized for other types of absorbers with a scaling of the fluence.

We should note that with a finite spectral bandwidth around 990 nm, the spectral component of the pulse that is closer to the excitonic peak will saturate the absorption with a lower carrier density than the spectral component away from the excitonic peak. This is due to the higher density of energy states that needs to be filled at higher photon energy. The evaluation of the absorption at the central wavelength will average these effects with good accuracy as long as the absorption dispersion can be linearly approximated. However, for pulses with a very wide spectrum (>10 nm), or when the SESAM is operated near its band gap edge, a spectral integration of the absorption will become necessary.

In order to model the pulse interaction in the SESAM, it is necessary to determine the carrier recovery dynamics of the absorber. This is usually accomplished by measuring the transient reflectivity change of the SESAM in a pump and probe setup. In the following example, the pump pulses are focused onto the sample with a fluence of about $25\,\mu J\,cm^{-2}$ under a 5° incidence angle and the probe pulses are focused onto the same spot with a fluence below $100\,nJ\,cm^{-2}$ and under a different angle for spatial separation of the beams. Figure 11.24 shows the measured transient

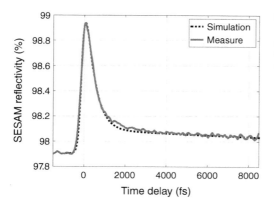

Figure 11.24 Experimentally measured and simulated reflectivity change in the SESAM after excitation with a 80 fs pulse at 990 nm with a fluence of $25\,\mu J\,cm^{-2}$.

reflectivity of the SESAM used in the experimental setup. The recovery dynamic can usually be fitted by a double exponential function with a slow and fast time constant as followed:

$$R(t) = R_0 + \Delta R \times \left(A \cdot e^{-t/\tau_{\text{slow}}} + (1-A) \cdot e^{-t/\tau_{\text{fast}}}\right) \qquad (11.6)$$

where $R_0 = 97.9\%$ is the linear reflectivity, $\Delta R = 1.04\%$ is the reflectivity change at $t = 0$, and $A = 0.085$ is the amplitude of the slow component. The fast decay rate $\tau_{\text{fast}} = 400$ fs is governed by intraband electron-electron and electron-phonon scatterings of the spectrally localized carrier distribution created by the pulse and subsequently through phonon emission [28]. The second relaxation process has a decay rate $\tau_{\text{slow}} = 20$ ps and is governed by the tunneling of carriers into surface states. In the following, we will assume that the relaxation time constants and ratio A are independent of the carrier density.

11.6.3 Dynamics of the Carrier Density Distribution

To compute the carrier density evolution, we have to solve the carrier rate equations, taking into account the field distribution and the absorption properties described previously. For this simulation, we assume that the optical power remains constant, which is a reasonable assumption considering the low level of losses in the SESAM (<2%). We will also neglect the carrier diffusion, as transverse diffusion will remain negligible while the pulses are present. Diffusion will also show a strong saturation behavior and is much suppressed for high local intensities [29], which is the case in this study. In this simplified analysis, the governing rate equations are

$$N(x,y,t) = N_{\text{gen}}(x,y,t) + AN_{\text{slow}}(x,y,t) + (1-A)N_{\text{fast}}(x,y,t) \qquad (11.7)$$

$$\frac{\partial N_{\text{gen}}}{\partial t} = \frac{\eta}{h\nu} \times \alpha(N) \times I(x,y,t) \qquad (11.8)$$

$$\frac{\partial N_{\text{slow}}}{\partial t} = -\frac{N_{\text{gen}}(x,y,t) + N_{\text{slow}}(x,y,t)}{\tau_{\text{slow}}} \qquad (11.9)$$

$$\frac{\partial N_{\text{fast}}}{\partial t} = -\frac{N_{\text{gen}}(x,y,t) + N_{\text{fast}}(x,y,t)}{\tau_{\text{fast}}} \qquad (11.10)$$

Where N_{gen} is the carrier density generated from the absorption of the pulses, $h\nu$ is the photon energy and η is the photon conversion efficiency. N_{slow} and N_{fast} represent, respectively, the carriers relaxing with a slow and fast time constant. Here, we ignore ambipolar diffusion effects due to the high local intensities [29], and we will assume that every photon absorbed excites a carrier, i.e. $\eta = 1$.

We are now left with a system of coupled equations that can be solved numerically, using a Runge-Kutta method for example. Figure 11.25 shows a snapshot at $t = 0$ of the spatial distribution of carrier density in the case of the synchronous collision of two pulses having an equal energy of 50 nJ cm^{-2} and 50μJ cm^{-2}. To obtain a realistic simulation result, we used the experimental pulse duration (128 fs), incident angle (7°), and beam waist radius (67μm).

In the case of weak pulses, the carrier density distribution follows the interference pattern of the field intensity, but with strong pulses the carrier density at the

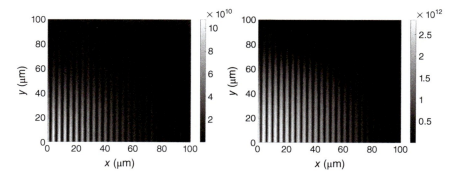

Figure 11.25 Carrier density in the SESAM QW generated by the colliding of two pulses of 50 nJ cm^{-2} (a), and 50 µJ cm^{-2} (b). The incident beams are centered at $(x = 0, y = 0)$ and are axially symmetric.

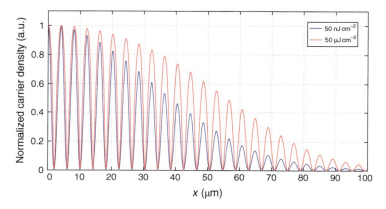

Figure 11.26 Normalized carrier density distribution in the SESAM QW along the x axis at $y = 0$ generated by the colliding of two pulses of 50 nJ cm^{-2} and two pulses of 50 µJ cm^{-2}.

field maximum becomes saturated, increasing the relative amplitude of the carrier density near the nodes and the tails of the field intensity distribution. This effect is clearly visible on the 2D images of Figure 11.25 and on the normalized cross-section plots of Figure 11.26.

The evolution over time of the spatial distribution of carrier density is then calculated, starting at a time prior to the pulses' arrival in the SESAM until they are completely gone. The carrier density dynamic at the center of the beam ($x = y = 0$) is plotted in Figure 11.27. For this simulation, we assumed a single 128 fs pulse with an energy of 51.8 µJ cm^{-2}, corresponding to the experimental output power of 90 mW presented previously. We also plotted the experimental case of a side pulse operation, with two 130 fs pulses delayed by 430 fs and an average output power of 152 mW.

This result clearly shows the effect of the fast relaxation of carriers during the presence of the pulse, which makes the saturation fluence strongly dependent on the pulse duration. In the situation of a delayed pulse, it is also clear that it will benefit from the carriers generated by the earlier pulse, resulting in absorption losses decreasing with a short delay.

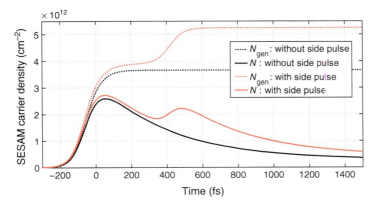

Figure 11.27 Simulated carrier density in the SESAM at the center of the beam ($x = y = 0$), from a single 128 fs pulse of 51.8 µJ cm^{-2} (90 mW average output power) and a double 130 fs pulse of 87.4 µJ cm^{-2} (152 mW output power) separated by 430 fs. N_{gen} represents the generated carrier density, not accounting for the carriers relaxation.

We should note that if the side pulse has a different central wavelength than the first pulse, it will only partially benefit from the saturation created by the earlier pulse. Indeed, if the pulses have a different spectral content, the first pulse will burn a kinetic hole in the carrier distribution that will partly overlap with the second pulse's spectrum. However, after a delay of 430 fs, the initial kinetic holes will be significantly filled and smoothed out via intra-subband scattering [30, 31], which will reduce the spectral dependence of the side pulse interaction. This model does not account for such phenomena as it would require a spatially resolved microscopic analysis of nonequilibrium dynamics in the semiconductor and would be numerically too expensive.

11.6.4 Absorption Losses and Pulse Shaping

The optical power lost in the absorber, P_{lost}, during the passage of the pulses is directly related to the amount of carriers generated by the absorption of these pulses, N_{gen}. It is given by the spatial integration of the carrier density generated as followed:

$$P_{lost}(t) = \frac{h\nu}{\eta} \times \frac{\partial}{\partial t} \left(\iint_{xy} N_{gen}(x, y, t) \, dx \, dy \right) \tag{11.11}$$

We should note that this expression does not account for other kind of losses such as two-photon absorption (TPA), free-carrier absorption (FCA), scattering from rough surfaces, and transmission losses. Scattering and transmission losses can usually be kept relatively small and remain constant regardless of the pulse duration and intensity. However, TPA will become increasingly important at high fluence and short pulse duration as its probability of occurrence increases with the square of the photon number and can thus dominate over linear absorption at high field intensities [32]. In the scope of this book, we want to investigate the effect of CPM on the linear absorption only and will thus neglect these additional loss mechanisms.

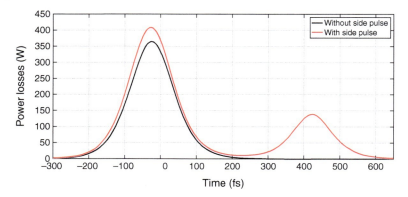

Figure 11.28 Simulated absorption losses from a single pulse of 90 mW and a dual pulse of 152 mW separated by 430 fs.

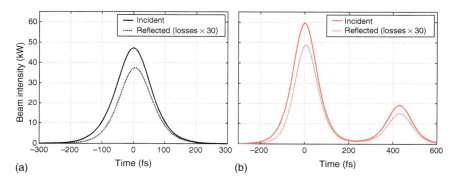

Figure 11.29 Simulated pulse intensity before and after reflection on the SESAM for CPM without side pulse (a) and for CPM with a side pulse (b). The SESAM losses are artificially increased by a factor 30 for illustration purposes.

Figure 11.28 shows the power lost in the SESAM for the two experimental cases presented previously. We can see that the power lost reaches a maximum located 25 fs before the main pulse maximum, which is centered at $t = 0$ s. The more intense the pulse, the more this maximum is shifted toward the leading edge of the pulse, as the saturation will occur earlier. This effect is obviously responsible for the pulse shortening occurring in the absorber, with a leading edge significantly more attenuated, as it can be seen in Figure 11.29. For a better visualization of the pulse shaping, we plotted a pulse before and after reflection on the SESAM, with the losses artificially scaled by a factor of 30.

The total energy lost by the pulses is given by the temporal integration of the power lost as:

$$\mathcal{E}_{\text{lost}} = \int P_{\text{lost}}(t)\, dt \tag{11.12}$$

In the experimental case with a side pulse operation, the main pulse loses 0.89% of its energy, whereas the side pulse only loses 0.49%. The main pulse has thus

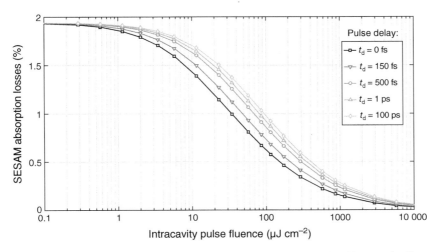

Figure 11.30 Absorption losses from the SESAM as a function of the intracavity fluence for different delays of the two counter-propagating pulses.

indirectly "transferred" some of its energy to the side pulse through the interaction in the absorber. This energy transfer will obviously be balanced by the lower gain seen by the side pulse, since a similar gain recovery dynamic will occur in the gain medium.

11.6.5 Saturation Fluence of the Absorber

To evaluate the saturation fluence of the absorber, we must calculate the total absorption losses $\mathcal{E}_{lost}$ as a function of the fluence of the pulses. For this study, we assume a collision of two identical single pulses, with a pulse duration of 128 fs and the geometrical parameters described in the experimental section. We varied the delay between the pulses to assess the effect of the synchronization on the saturation fluence. Figure 11.30 shows a clear reduction of the saturation fluence when the pulses are delayed by less than 1 ps.

For perfectly synchronized pulses, the saturation fluence is reduced by a factor of 2.9, from $F_{sat1} = 50\,\mu J\,cm^{-2}$ to $F_{sat2} = 17.2\,\mu J\,cm^{-2}$. With a pulse fluence of $51.8\,\mu J\,cm^{-2}$, corresponding to the 90 mW output power result, the absorption losses are reduced from 1.23 to 0.85% when the pulses are synchronized. The mode locking with synchronized colliding pulses is thus clearly favored. We should note that the interference pattern is also present in a CW regime, which helps to transition from the CW to the mode-locked regime since the absorber is saturated more easily. This could explain the robustness of the mode-locking regime observed in the experimental setting, when compared to a standard V-shaped cavity.

In the case of the side pulse regime investigated, the saturation fluence is initially higher at $F_{sat1} = 65.1\,\mu J\,cm^{-2}$ with a single beam. This is due to the longer time allowed for carrier relaxation while the pulse and side pulse are present. The saturation fluence is however reduced by a factor of 2.8 to $F_{sat2} = 23.2\,\mu J\,cm^{-2}$ when two

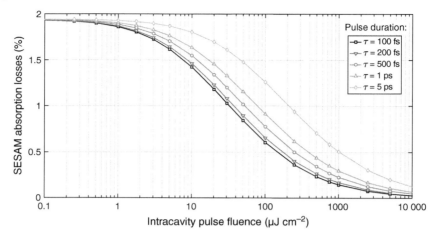

Figure 11.31 Absorption losses from the SESAM as a function of the intracavity fluence for different pulses durations.

beams are colliding. With the experimental pulse energy of 87.4 µJ cm^{-2}, the losses are reduced from 1.12 to 0.75% when the pulses are synchronized.

A possible strategy to avoid the emergence of side pulses at high pump power is to decrease the fast carrier recombination time constant of the absorber. It could be realized either by exciting the SESAM at a higher energy, where intraband electron-electron and electron-phonon scattering are faster, or by using a graphene-based absorber which has a very fast carrier relaxation time due to the strong Coulomb and phonon interaction [28]. Another strategy would be to increase the recovery time of the gain medium. However, that would be very challenging with an intense short pulse since it would cause a deep spectral hole burning in the carrier distribution, which will recover quickly via intraband scattering [33].

We should note that for longer pulses, in the 400 fs to 5 ps range, the saturation fluence will be significantly higher due to the extensive carrier relaxation occurring while the pulses are present, enabling much higher output powers without side pulses [34, 35]. To illustrate this effect, we simulated the loss saturation for different pulse durations of two synchronized pulses (Figure 11.31). This graph shows that the saturation fluence is about a factor of 5 higher for a pulse duration of 5 ps than for a duration of 100 fs. This factor obviously depends on the recovery dynamic of the SESAM and will decrease with longer time characteristics. This result also illustrates the shortcomings of a model based solely on the energy of the pulse for the calculation of the absorption (E_{sat} parameter), especially if one wants to model the pulse formation dynamic [31] where the pulse duration will change significantly from the initial conditions to the final steady state.

11.6.6 Power Balance in CPM Operation

The interaction of the pulses in the absorber not only leads to a lower saturation fluence but it can also reduce the power imbalance between the two

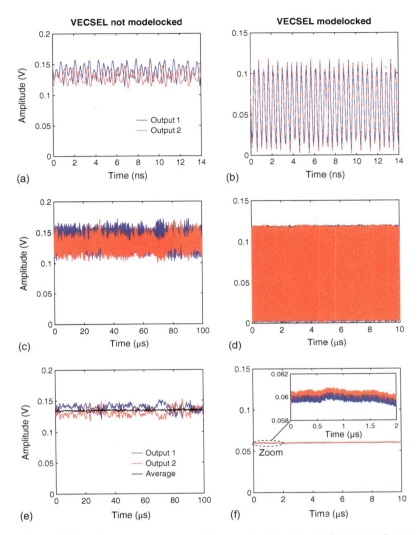

Figure 11.32 Recorded intensities of the two counter-propagating output beams: (a) Short time scale in CW regime, (b) Short time scale in CPM regime, (c) Long time scale in CW regime, (d) Long time scale in CPM regime, (e) Long time scale with low pass filter in CW regime, (f) Long time scale with low pass filter in CPM regime.

counter-propagating beams. Indeed, if one beam has more power than the other one, it will saturate the absorber faster, resulting in more absorption losses. Over the numerous round trips in the high-Q VECSEL cavity, the average power of each beam will naturally balance and the mode competition frequently observed in a ring cavity [36] will be strongly reduced.

Figure 11.32 shows an example of the mode competition and power imbalance observed in a ring cavity while in the CW regime, with a comparison to a mode-locked regime with the same cavity. The two output beams of a ring cavity VECSEL were recorded simultaneously at the same distance from the SESAM with two identical fast photodiodes and a 6 GHz bandwith oscilloscope. To obtain the CW operation regime, the SESAM is simply cooled to 20°C to blue-shift the absorption spectrum.

Figure 11.32a and b shows a time trace on a short time scale, revealing a relatively "noisy" CW regime where the fluctuations are not synchronized, while in the mode-locked regime the outputs are nearly identical and perfectly synchronized. The small intensity variation from pulse to pulse is likely due to an aliasing with the oscilloscope. Figure 11.32c and d represents the same outputs but over a longer time window, confirming a "noisy" CW operation and perfectly stable and synchronized pulse trains when mode locked. In Figure 11.32e and f, we applied a digital low pass filter to the signals. It reveals a CW operation with a clear competition between the two beam directions while the total average power remains constant, whereas the mode-locked outputs reveal a very stable output power for each direction. The zoom in the inset of Figure 11.32f shows the two beams fluctuating together rather than competing. This is clear evidence of the energy exchange occurring in a colliding pulse mode-locked regime. This energy exchange could also compensate for a small difference of gain between the pulses, due to an imperfect position of the absorber in the cavity, for example (see Section 11.3.2), and is key for the stability of the mode-locking regime.

11.7 Summary and Outlook

The constant improvement of VECSEL technology and better understanding of the pulse formation mechanisms in a passive mode-locking regime have enabled the generation of ultrashort pulse at high repetition rates, showing tremendous potential for numerous applications. These sources have become not only a viable alternative to other laser technologies due to their flexibility, compactness, and performance, but they are emerging as the technology of choice for the generation of ultrashort pulse in the 1-20 GHz repetition rate regime, where other ultrafast systems such as Ti:Sa lasers, microresonators, or fiber lasers are more challenging to operate. Recent progress has led to the demonstration of average powers >100 mW with pulse durations close to 100 fs, which is shorter than any other semiconductor laser without external pulse compression. To further improve the pulse duration and output power, it is important to identify and analyze the physical limitations, either numerically or experimentally, and find new ways to overcome them. It has been recently shown that the maximum peak power achievable with sub-200 fs pulse durations is mostly limited by the stability of the fundamental mode-locking regime as side pulses or harmonic mode locking emerge at high pump power, leading to a trade-off between the pulse duration and the average power achievable. A key parameter for high-power mode-locked VECSELs is the saturation fluence of the absorber, which must be kept low to start and stabilize passive mode locking and minimize the losses. The CPM of a VECSEL takes advantage of the pulse interference to artificially decrease the saturation fluence of the absorber. The numerical calculation of the absorber losses presented here shows a reduction of the saturation fluence by about a factor of three. This new technique has been employed to demonstrate state-of-the-art performance in term of power and pulse duration combination and, more importantly, has shown great potential in terms of stability and robustness of the mode-locking regime. These qualities are paramount for real-world applications where the laser might be submitted to temperature and pump fluctuations or mechanical vibrations. The more stable and rugged oscillators of CPM VECSELs minimize the sensitivity to mechanical vibration, while the saturation dynamics of the absorber strongly favor a mode-locking regime for a wide range of pump powers and temperatures, increasing the reliability of the device. The ring cavity geometry of CPM VECSEL could also be adapted to realize an optical gyroscope using the Sagnac effect.

These ultrafast sources could potentially replace conventional ion-doped solid-state lasers used in numerous applications such as multiphoton and biomedical imaging, optical coherence tomography, high-resolution time domain terahertz spectroscopy with asynchronous optical sampling, etc. The excellent noise performance exhibited by VECSEL sources [37, 38] is particularly promising for the generation of ultrastable frequency combs. The most widespread method to stabilize a frequency comb relies on the generation of a coherent, octave spanning supercontinuum that is launched into an f-to-2f interferometer to detect and stabilize the carrier envelope offset frequency. The power level and pulse duration reached by these VECSELs could enable a direct generation of such a supercontinuum in a photonic crystal fiber or nonlinear waveguide without preamplification, which would be a major step to simplify and reduce the cost of existing technologies. Frequency combs are extremely precise instruments that can be used in an ever growing field of applications such as optical frequency metrology, chemical analysis, molecular spectroscopy, astronomical spectrograph calibration, optical atomic clocks, environmental monitoring, dielectric laser accelerator, medical diagnostics, etc.

Acknowledgments

This material is based upon work supported by the Air Force Office of Scientific Research under award number FA9550-14-1-0062 and FA9550-17-1-0246. I also would like to acknowledge my colleagues Robert Rockmore, Caleb Baker, Hsiu-Ting Chan, Jorg Hader, Jason Jones, Stephan Koch, and Jerry Moloney at the university of Arizona, as well as Declan Marah and John McInerney at the university of Cork Ireland, who contributed to the development of the first CPM VECSEL. I am also grateful to Ganesh Balakrishnan and Sadhvikas Addamane at the University of New Mexico for providing the semiconductor saturable absorbers, and to Antje Ruiz Perez and Wolfgang Stolz at the university of Marburg for growing the VECSEL structures. Finally, I would like to thank Michael Jetter for putting this book collection together.

References

1 Klopp, P., Griebner, U., Zorn, M. et al. (2009). Mode-locked InGaAs-AlGaAs disk laser generating sub-200-fs pulses, pulse picking and amplification by a tapered diode amplifier. *Optics Express* 17 (13): 10 820–10 834.

2 Waldburger, D., Link, S.M., Mangold, M. et al. (2016). High-power 100 fs semiconductor disk lasers. *Optica* 3 (8): 844–852.

3 Siegner, U., Fluck, R., Zhang, G., and Keller, U. (1996). Ultrafast high-intensity nonlinear absorption dynamics in low-temperature grown gallium arsenide. *Applied Physics Letters* 69 (17): 2566–2568.

4 Garnache, A., Hoogland, S., Tropper, A.C. et al. (2002). Sub-500-fs soliton-like pulse in a passively mode-locked broadband surface-emitting laser with 100 mW average power. *Applied Physics Letters* 80 (21): 3892–3894.

5 Mangold, M., Golling, M., Gini, E. et al. (2015). Sub-300-femtosecond operation from a MIXSEL. *Optics Express* 23 (17): 22 043–22 059.

6 Zaugg, C.A., Sun, Z., Wittwer, V.J. et al. (2013). Ultrafast and widely tuneable vertical-external-cavity surface-emitting laser, mode-locked by a graphene-integrated distributed Bragg reflector. *Optics Express* 21 (25): 31 548–31 559.

7 Seger, K., Meiser, N., Choi, S.Y. et al. (2013). Carbon nanotube mode-locked optically-pumped semiconductor disk laser. *Optics Express* 21 (15): 17 806–17 813.

8 Albrecht, A.R., Wang, Y., Ghasemkhani, M. et al. (2013). Exploring ultrafast negative Kerr effect for mode-locking vertical external-cavity surface-emitting lasers. *Optics Express* 21 (23): 28 801–28 808.

9 Laurain, A., Marah, D., Rockmore, R. et al. (2016). Colliding pulse mode locking of vertical-external-cavity surface-emitting laser. *Optica* 3 (7): 781–784.

10 Laurain, A., Rockmore, R., Chan, H.T. et al. (2017). Pulse interactions in a colliding pulse mode-locked vertical external cavity surface emitting laser. *Journal of the Optical Society of America B* 34 (2): 329–337.

11 Fork, R.L., Greene, B.I., and Shank, C.V. (1981). Generation of optical pulses shorter than 0.1 psec by colliding pulse mode locking. *Applied Physics Letters* 38 (9): 671–672.

12 Vanherzeele, H., Eck, J.L.V., and Siegman, A.E. (1981). Colliding pulse mode locking of a Nd:YAG laser with an antiresonant ring structure. *Applied Optics* 20 (20): 3484–3486.

13 Vasil'ev, P., Morozov, V., Popov, Y., and Sergeev, A. (1986). Subpicosecond pulse generation by a tandem-type AlGaAs DH laser with colliding pulse mode locking. *IEEE Journal of Quantum Electronics* 22 (1): 149–152.

14 Chen, Y.K., Wu, M.C., Tanbun-Ek, T. et al. (1991). Subpicosecond monolithic colliding-pulse mode-locked multiple quantum well lasers. *Applied Physics Letters* 58 (12): 1253–1255.

15 Baker, C., Scheller, M., Koch, S.W. et al. (2015). In situ probing of mode-locked vertical-external-cavity-surface-emitting lasers. *Optics Letters* 40 (23): 5459–5462.

16 Eichler, H.J. (1985). *Laser-Induced Dynamic Gratings*, vol. 50. Springer Series in Optical Sciences.

17 Haus, H.A. (2000). Mode-locking of lasers. *IEEE Journal of Selected Topics in Quantum Electronics* 6 (6): 1173–1185.

18 Lorenser, D., Unold, H., Maas, D. et al. (2004). Towards wafer-scale integration of high repetition rate passively mode-locked surface-emitting semiconductor lasers. *Applied Physics B* 79 (8): 927–932.

19 Kilen, I., Koch, S.W., Hader, J., and Moloney, J.V. (2017). Non-equilibrium ultra-short pulse generation strategies in VECSELs. *Optica* 4 (4): 412–417.
20 Gbele, K., Laurain, A., Hader, J. et al. (2016). Design and fabrication of hybrid metal semiconductor mirror for high-power VECSEL. *IEEE Photonics Technology Letters* 28 (7): 732–735.
21 Orfanidis, S.J. (2013). Electromagnetic Waves and Antennas. New Brunswick, NJ: Rutgers University New Brunswick.
22 Good, J.T., Holland, D.B., Finneran, I.A. et al. (2015). A decade-spanning high-resolution asynchronous optical sampling terahertz time-domain and frequency comb spectrometer. *Review of Scientific Instruments* 86 (10): 103107.
23 Zaugg, C.A., Klenner, A., Mangold, M. et al. (2014). Gigahertz self-referenceable frequency comb from a semiconductor disk laser. *Optics Express* 22 (13): 16 445–16 455.
24 Siegman, A.E. (1977). Bragg diffraction of a Gaussian beam by a crossed-Gaussian volume grating. *Journal of the Optical Society of America* 67 (4): 545–550.
25 Maas, D.J.H.C., Rudin, B., Bellancourt, A.R. et al. (2008). High precision optical characterization of semiconductor saturable absorber mirrors. *Optics Express* 16 (10): 7571–7579.
26 Sieber, O., Hoffmann, M., Wittwer, V. et al. (2013). Experimentally verified pulse formation model for high-power femtosecond VECSELs. *Applied Physics B* 113: 113–145.
27 Hader, J., Koch, S., and Moloney, J. (2003). Microscopic theory of gain and spontaneous emission in GaInNAs laser material. *Solid-State Electronics* 47 (3): 513–521.
28 Hader, J., Yang, H.J., Scheller, M. et al. (2016). Microscopic analysis of saturable absorbers: semiconductor saturable absorber mirrors versus graphene. *Journal of Applied Physics* 119 (5): 053102.
29 Haug, H. and Koch, S.W. (1989). Semiconductor laser theory with many-body effects. *Physical Review A* 39 (4): 1887–1898.
30 Hader, J., Moloney, J.V., and Koch, S.W. (2014). Microscopic analysis of non-equilibrium dynamics in the semiconductor-laser gain medium. *Applied Physics Letters* 104 (15): 151111.
31 Hader, J., Scheller, M., Laurain, A. et al. (2017). Ultrafast non-equilibrium carrier dynamics in semiconductor laser mode-locking. *Semiconductor Science and Technology* 32 (1): 013 002.
32 Alfieri, C.G.E., Diebold, A., Emaury, F. et al. (2016). Improved SESAMs for femtosecond pulse generation approaching the kW average power regime. *Optics Express* 24 (24): 27 587–27 599.
33 Kilen, I., Hader, J., Moloney, J.V., and Koch, S.W. (2014). Ultrafast nonequilibrium carrier dynamics in semiconductor laser mode locking. *Optica* 1 (4): 192–197.
34 Wilcox, K.G., Tropper, A.C., Beere, H.E. et al. (2013). 4.35 kW peak power femtosecond pulse mode-locked VECSEL for supercontinuum generation. *Optics Express* 21 (2): 1599–1605.

35 Baker, C.W., Scheller, M., Laurain, A. et al. (2017). Multi-angle VECSEL cavities for dispersion control and peak-power scaling. *IEEE Photonics Technology Letters* 29 (3): 326–329.

36 Mignot, A., Feugnet, G., Schwartz, S. et al. (2009). Single-frequency external-cavity semiconductor ring-laser gyroscope. *Optics Letters* 34 (1): 97–99.

37 Myara, M., Sellahi, M., Laurain, A. et al. (2013). Noise properties of NIR and MIR VECSELs. *Proceedings of SPIE* 8606: 86 060Q–86 060Q–13.

38 Laurain, A., Mart, C., Hader, J. et al. (2014). Optical noise of stabilized high-power single frequency optically pumped semiconductor laser. *Optics Letters* 39 (6): 1573–1576.

12

Self-Mode-Locked Semiconductor Disk Lasers
Arash Rahimi-Iman

Department of Physics and Materials Sciences Center, Philipps-Universität Marburg, Marburg, Germany

12.1 Introduction

Mode-locked (ML) semiconductor disk lasers (SDLs), also known as ML VECSELs, have attracted considerable attention from science and industry beyond the VECSEL community owing to their remarkable and promising features [1–3]. As is evidenced in previous chapters of this book dealing with mode-locking (also abbreviated ML), ultrafast optical pulse generation is still one of the hot topics of the laser community and has opened the door for a wide range of different applications in biology, medicine, manufacturing, and metrology, where ML SDLs can be employed [4–8]. Particularly, the shortening of pulse durations with advanced ML lasers have enabled various measurement techniques with high resolution in the time-domain [9, 10] – some even used to characterize VECSELs [11, 12] – and have given rise to remarkable spatial precision in the field of material processing [7] as well as high peak powers which are required for multiphoton-absorption-based technologies [6].

Yet, saturable-absorber-mirror-based ML VECSELs have a dominant share in the development of efficient and versatile high-power, ultrashort-pulse SDL devices [13–16], but a new generation of saturable-absorber-free ML VECSELs is already on the verge of catching up with its established counterpart [17–20]. With comparable features such as sub-ps pulse durations, high peak powers, and high beam quality, they are promising candidates for the realization and spread of cost-effective, robust, and compact ML SDLs. Given the independence from an additional intracavity element for pulse formation, saturable-absorber-free ML VECSELs are also referred to as self-mode-locked (SML) VECSELs/SDLs [21], in analogy to the terminology used in the case of Kerr lens ML solid-state lasers [22].

Owing to the noticeable interest in the achievements and the developments in the field of SML VECSELs, this chapter is dedicated to such devices, their history, and the obtained highlights. In order to understand the differences between different device concepts, a brief summary on the existing ML techniques used for optically pumped SDLs is provided in the following. Then, light is shed on the rise of SML optically pumped VECSELs, and the performance similarities to semiconductor-saturable-absorber-mirror (SESAM)-operated ML VECSELs are

Vertical External Cavity Surface Emitting Lasers: VECSEL Technology and Applications, First Edition.
Edited by Michael Jetter and Peter Michler.
© 2022 WILEY-VCH GmbH. Published 2022 by WILEY-VCH GmbH.

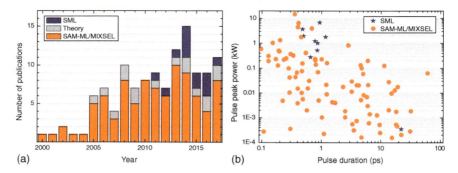

Figure 12.1 (a) Overview on the annual publication output concerning ML VECSELs and (b) on the achieved peak powers with respect to pulse durations. Results for SML VECSELs are highlighted in this direct comparison to their (SE)SAM-ML counterparts. The data are extracted from the literature to the best of the author's knowledge at the time of surveying. The shortest unambiguous pulse from a ML VECSEL has been demonstrated to be as short as 100 fs. Given the comparison of peak powers, similar ultrashort pulsing results in the domain of SML can be expected for correspondingly optimized devices. For (a), all studies related to SML operation have been included as "SML."

emphasized in an overview on SESAM-free ML achievements (Figure 12.1). Hereby, also other SESAM-free concepts for mode-locking of SDLs will be highlighted. Although SML VECSELs are believed to be an equal match to their SESAM ML counterparts, they are far less understood when it comes to the mechanisms that allow pulse formation and a stable ML regime. This is briefly addressed at the end of this chapter and an outlook on applications as well as development directions is provided.

12.2 Mode-Locking Techniques for Optically Pumped SDLs at a Glance

A variety of passively mode-locked VECSELs has been demonstrated to date, having been triggered by the pioneering work of Hoogland et al. published in the year 2000 [23]. The very first ML VECSEL delivered pulses in the ps range and was operated with the help of a quantum well (QW) SESAM. In fact, QW SESAMs had been the essential intracavity element in a VECSEL resonator which enabled stable pulsing, before additional saturable absorber structures were introduced to the VECSEL world. These structures include quantum dot (QD)-based SESAMs [23–29], graphene saturable absorber mirrors (GSAMs) [30, 31], and carbon nanotubes (CNTs) intracavity saturable absorbers [32]. This allowed the demonstration of mode-locking for various resonator geometries, as shown in Figure 12.2, and led to a wide range of pulsing results with some highlights summarized in Table 12.1. Undoubtedly, saturable absorbers served well with regard to mode-locking of SDLs, with pulsing achieved for different gain/absorber combinations i.e. QW-gain/QW-absorber, QW-gain/QD-absorber, QD-gain/QW-absorber, and QD-gain/QD-absorber [3].

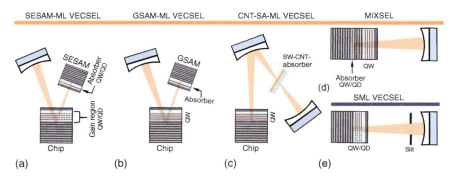

Figure 12.2 Different mode-locking techniques for VECSELs. (a) SESAM ML device, (b) graphene, (c) carbon-nanotube-based saturable absorber ML devices, (d) MIXSEL as well as (e) self-mode-locked VECSEL. Source: Sketches drawn after Ref. [3].

Table 12.1 Highlights of SESAM mode-locking with QW-SDLs

	Record value	No. QWs	Wavelength (nm)	Year	Ref.
Peak power	4.35 kW	10	1013	2013	[5]
Average power	5.1 W	10	1030	2012	[33]
Pulse duration	107 fs	4	1030	2011	[13]
	107 fs (compressed to 97 fs)	10	1033	2016	[14]
	95 fs (colliding-pulse ML)	10	1033	2016	[15]
	253 fs (MIXSEL)	10	1030	2015	[16]
Repetition rate	Highest: 50 GHz	7	958	2006	[24]
	Lowest: 85.7 MHz	16	989	2013	[25]

One simple resonator configuration for SESAM-ML VECSELs is depicted schematically in Figure 12.2a, with the VECSEL acting as the folding mirror, and the SESAM as one of the cavity's end mirrors. In this example, both VECSEL and SESAM can contain one or more QW or QD layers as optically active/absorptive medium, respectively. Usually, these layers are grown by means of epitaxy, for instance using vapor-phase or molecular beam epitaxy, on top of a highly reflective distributed Bragg reflector (DBR) to form either the gain chip or the SESAM, often capped by some additional layer to suppress surface recombination of charge carriers and to control dispersion and the absorption behavior [1]. However, it is the SESAM's absorbance which adds further optical losses to a VECSEL and implies that additional heat dissipation has to be coped with.

Yet, the employment of SESAMs in VECSELs sets some limitations to the design and the compactness of the resonator, in which they have to be integrated as additional optical element (thus referred to as resonator-integrated SESAMs, cf. Figure 12.2a–c). The desire to make the resonator more compact and production of devices more cost-efficient has triggered the conception of chip-integrated

SESAMs. Very comparable mode-locking results have been achieved with this approach, referred to as mode-locked integrated external cavity surface-emitting laser (MIXSEL) configuration [34], which also promises repetition rate scaling due to its design principles [2]. Introduced in 2007 [34], MIXSELs even feature ultrashort pulses in the 200-fs range, as has been recently demonstrated [16]. A schematic of the simple linear cavity achievable with MIXSELs, which can make use of both QW and QD gain or absorber regions, is shown in Figure 12.2d.

Although SESAMs are widely established in the field of ML SDLs, they not only impose resonator design limitations but also restrict wavelength flexibility and cost-effectiveness. Interestingly, GSAMs and CNT-based absorbers introduced to the community try to tackle this issue. With a series of different gain chips in combination with a single-layer GSAM, ML was demonstrated over a wavelength range of 46 nm from 935 to 981 nm with sub-500-fs pulses and power levels in the mW range [30]. In contrast, another experiment showed 10 W of output power of a 353-fs-pulsed GSAM-ML VECSEL [31]. Similarly, single-walled CNTs were successfully employed on a transparent carrier for insertion into a VECSEL's open cavity for mode-locking [32]. Both, GSAM and CNT schemes are sketched in Figure 12.2b and c, respectively.

Recently, the family of ML VECSELs has been expanded by the introduction of the SML VECSEL scheme, which complements the pool of previously widely used active and passive mode-locking techniques. With their already comparable performance with respect to their SESAM counterparts, SML VECSELs promise efficient, compact, and cost-effective pulsed light sources based on SDL technology [20]. Furthermore, SML VECSELs can circumvent limitations naturally imposed on the device's performance by SESAMs. Similar to the achievements made with solid-state lasers more than two decades ago by the introduction of Kerr lens mode-locking [22, 35], this scheme simply relies on the insertion of an aperture into the resonator in order to give preference to the ML regime by modulating losses for the cavity mode. While the aperture can be either a soft or hard one – depending on the VECSEL configuration – the presumed effect behind SML operation is considered to be primarily an intensity-dependent Kerr lensing effect in the SDL. Based on the first assumptions regarding the working principles [17, 18, 36], different examples of SML VECSELs have been demonstrated up to date [19, 21, 37–39] and preliminary studies regarding the mechanism behind SML kicked off [40–43]. A schematic representation of an SML VECSEL in its simplest configuration is shown in Figure 12.2e, employing a hard aperture in a straight cavity in front of the resonator out-coupling mirror.

12.3 History of Saturable-Absorber-Free Pulsed VECSELs

12.3.1 Self-Mode-Locked Optically Pumped VECSELs

In the following, an overview on the history of SML VECSELs is presented and the developments are summarized which have been undergone toward a broader acceptance and availability of SML results. Based on these results, record-high peak powers of up to 6.8 kW [18] and sub-500-fs pulses [37] were reported individually. Furthermore, recent activities regarding SML operation involving both quantum-well [19, 21] and quantum-dot-based [38] gain media are highlighted. In fact, it took some time until unambiguous SML results were presented to the VECSEL community: the clean demonstration and thorough characterization of stable SML operation in a VECSEL can be attributed to the works published in 2014 [19, 21].

12.3.1.1 Once Upon a Time – Beyond Magic

The first SESAM-free mode-locked VECSEL was reported by Chen et al. in 2011 [17] at a time when SESAM-ML was very well established and no reason was actually given to change the status quo in terms of how pulse formation in VECSELs should be achieved. Rather an anomaly, which magically took place in the classical straight cavity configuration that Chen et al. used [17], triggered a burst of unforeseen SML reports in the VECSEL domain. In their work, the authors experienced pulsing of their VECSEL spontaneously formed and intuitively suggested that mode-locking in their SESAM-free device results from saturable absorption in some of the QWs of the gain chip which are not fully pumped to inversion, similar to the behavior obtained due to extra layers incorporated in a MIXSEL for chip-integrated saturable absorbance. Remarkably, pulses as short as 654 fs had been presented by Chen et al. with an average output power of 0.45 W at a repetition rate of 2.17 GHz. The authors claimed to have almost no CW background and that the pump-to-mode size ratio should be between approximately 0.7 and 1.3 in their linear cavity in order to reach stable spontaneous mode-locking. A common characterization approach for such a VECSEL is displayed in Figure 12.3.

In the following year, an alternative version of SML operation of a VECSEL was shortly after also demonstrated by Kornaszewski et al. [18]. In contrast to

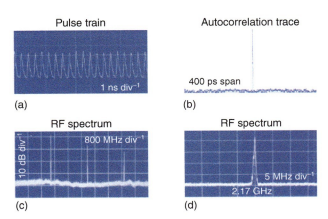

Figure 12.3 Pulse characterization as shown for the first spontaneously mode-locked VECSEL. (a) Pulse train measurement, (b) Long time span intensity autocorrelation trace, and (c) and (d) RF spectra with different frequency window, respectively. Source: Adapted with permission. [17] Copyright 2011, Optical Society of America.

the previous report, a folded six-mirror VECSEL cavity configuration was used. Moreover, the origin of mode-locking was attributed to an intensity-dependent Kerr lens experienced by the circulating intracavity light field in the semiconductor gain medium. This effect was expected to lead to an intensity-dependent modulation of the beam waist in front of the cavity outcoupling mirror, where a slit would act as an aperture to the intracavity beam waist and expose the CW mode to losses. As commonly used in mode-locked titanium-sapphire (Ti:Sa) lasers, the slit would give preference to the pulsed regime, which would experience stronger lensing owing to its high peak intensities compared to CW intensities and, thus, less losses by the well-placed slit than any low-intensity light field. An explanation of this principle is for instance given in [44] (also see Figure 12.4), which shed light on the phenomenon referred to in its early days as magic mode-locking – since it was hardly imaginable to achieve mode-locking without a saturable absorber inside the cavity until SML operation of a Ti:Sa oscillator was demonstrated [22, 35]. According to Eqs. (12.1) and (12.2), the intensity-dependent refractive index is comprised of the common intensity-independent refractive index n_0 and the nonlinear refractive index term $n_2 I$ which scales with the square of the electrical field $I = |E|^2$:

$$n(I) = n_0 + n_2 I \tag{12.1}$$

$$n_2 = (2\pi/n_0)^2 \chi^{(3)} \tag{12.2}$$

with n_2 related to the third-order nonlinear susceptibility $\chi^{(3)}$.

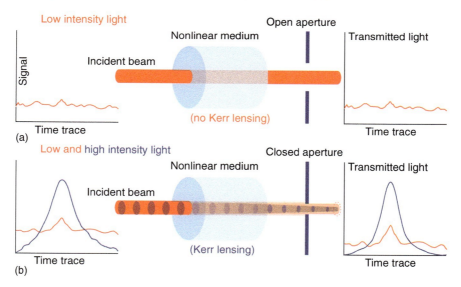

Figure 12.4 Schematic representation of the Kerr lensing effect for high-intensity light, which experiences less losses than low-intensity light when an aperture behind the nonlinear medium is closed. (a) Sketch of the transmitted low-intensity light without Kerr lensing and beam waist truncation, compared to the case (b) with intensity-dependent lensing and closed aperture. The survival of high-intensity light components in a resonator with a Kerr medium and slit in the optical path gives rise to the formation of pulses due to the naturally higher transmitted intensities for pulse peaks than for continuous-wave signal.

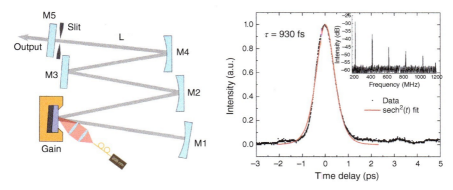

Figure 12.5 Schematic representation of an SML-VECSEL's cavity configuration (left). As part of the emission's characterization, an intensity autocorrelation trace is commonly presented, although not sufficient evidence for stable ML operation (right). Inset: RF spectrum. Source: Reproduced with permission. [18] Copyright 2012, Wiley-VCH.

In the aforementioned work by Kornaszewski et al., two different configurations were reported to give rise to mode-locked operation. A scheme of the used folded cavity configuration is depicted in Figure 12.5. On the one hand, stable mode-locking was obtained when the cavity operated near its stability limit without the influence of any intracavity aperture. In that case, single pulses with pulse duration of about 1.5 ps were achieved at an average output power level of 700 mW at a repetition rate of 200 MHz, and pulse formation was attributed to soft-aperture-assisted Kerr lensing. On the other hand, operating the cavity in its stability region and inserting a hard aperture near the outcoupling mirror enabled mode-locked operation with pulses of 930 fs duration (cf. Figure 12.5) at a repetition rate of 210 MHz and an average output power level of 1.5 W at 985 nm. From these parameters, a record-high 6.8 kW peak power was deduced and reported [18].

12.3.1.2 Mode Competition – A Struggle for Acceptance

As previously stated in the reviews by Gaafar et al. [3] and Rahimi-Iman [45], the early achievements were not only met by enthusiasm and belief, but also by critical remarks and strong skepticism. Obviously, the first reports on SML operation in SDLs gave rise to some controversy, otherwise there would be little reason for the correspondence between different actors of the ML VECSEL community on this issue and the following desire to put things straight: most prominently, the discussion on the validity of SML results presented in Refs. [36, 46] is remembered by many involved groups, which was triggered by the second report on SML VECSELs [18], although the criticism was also partly directed to the previous publication [17]. Yet, the impact on later work has not been unnoticed. Furthermore, it also shows how constructive criticism can be turned into positive outcomes.

One simple reason for this struggle for acceptance of SML achievements was given by the lack of complete or unambiguous characterization of the laser; another one was given by the fact that the mechanism responsible for SML operation had not been clear, so far. While the latter is still subject of ongoing investigations and not well understood even today, the former problem has been tackled by work performed in the years after the first publications on SML VECSELs.

The authors of Ref. [46] frankly argued that additional experiments need to be performed in order to have clear evidence for stable and true mode-locked operation. Such experimental proof would require clean (long time span) intensity autocorrelation as well as pulse train measurements, a high signal-to-noise ratio in radio frequency (RF) signal measurements, and no side bands and no considerable intensity drop off between the harmonics in the RF signal up to the point where the measurement equipment chokes the signal off – to cut a long story short. While the optical spectrum and autocorrelation trace cannot be used reliably to claim ML operation of a laser, care has to be taken when relying on RF and oscilloscope signal as well. Moreover, evidence for the high peak powers typical for pulsed light was appreciated, for instance by the demonstration of nonlinear effects such as supercontinuum generation or second-harmonic generation outside the VECSEL resonator. In fact, these demands were carefully taken into account in later reports on SML VECSEL, such as in the work of Gaafar et al. [19] that provides a thorough characterization in order to support the authors' claim of a stable SML regime.

However, no clear attribution of the cause for SML to one (or more) of the expected effects responsible for a stable mode-locked operation regime has been made so far. While SESAM-free mode-locking might rely on different principles based on the VECSEL design, for instance involving other types of saturable absorber structures [30–32] or extra Kerr media [47], particularly the SML case with no additional intracavity element remains somehow "magic" (in analogy to the term used in the early days of Kerr lens mode-locking): although many ML-VECSEL groups suspect a Kerr lensing effect to be involved and design their cavities in assumption of negative or positive nonlinear lensing, other reasons for mode-locking are still imaginable. This will be briefly discussed in the last section of this chapter.

12.3.1.3 More Than a Flash in the Pan – Triggered Wave of Results

After the first reports of SML, other groups carried out similar experiments and observed the SML phenomenon as well. Both soft and hard apertures inside the optical cavity were used to achieve Kerr lens mode-locking as well, such as in the work reported in 2013 by Albrecht et al. [37]. In their work, the authors analyzed the possibility of Kerr lensing in a typical VECSEL cavity – in this case a V-shaped one. They further verified their predictions that Kerr lens action can give rise to SML operation in such cavity configuration by the demonstration of pulsing at a repetition rate of 1 GHz and with pulse durations between 2 ps and 500 fs. However, a residual background or pedestal was observed in their pump-power-dependent investigations which can be attributed to a CW background. Furthermore, their RF spectrum indicated instable operation similar to previous SML reports, while the pulse train fluctuated over a timescale of some microseconds.

The study by Albrecht et al. showed that at low pump powers, the achieved pulses were longer than 1 ps, which had been reduced to sub-ps levels at high powers [37]. In addition, by insertion of an intracavity-fused silica glass Brewster plate at a fixed

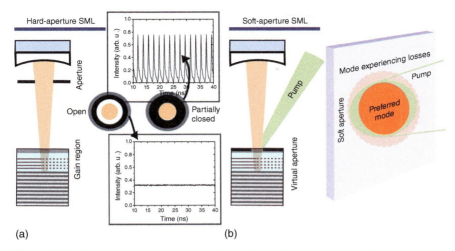

Figure 12.6 Schematic representation of (a) hard- and (b) soft-aperture Kerr lens ML.

pump power, the pulse length was controllable to a certain extent. This means of compensation (i.e. using a glass plate) of, both, negative group velocity dispersion (GVD) and negative self-phase-modulation (SPM) in the gain crystal ($n_2 < 0$) showed that for increasing positive GVD the pulse length could be minimized. However, this variation of pulse length had only been effective at pump power ranges, for which the authors assumed some balance between GVD and SPM.

In a two-mirror plane-concave resonator configuration, commonly referred to as linear cavity VECSEL, Albrecht et al. also achieved similar SML operation without any hard aperture. According to their modeling work for negative Kerr lensing, this configuration could only enable soft-aperture SML operation. In contrast to the hard-aperture case (see Figure 12.6a), in soft-aperture configuration, the pump area on the chip can determine a soft aperture for the resonator mode, which can add some losses to the CW cavity beam by truncating the beam waist of low-intensity light in the VECSEL at the position of the gain medium. This is schematically shown in Figure 12.6b.

On the other hand, the observation of self-starting pulses in a SESAM-free cavity was seen in connection with the occurrence of high-order transverse modes by Liang et al. [39]. Experiments concerning pump-to-mode-size ratios were performed and the conclusion drawn from experiments and numerical analysis, that the critical pump power, at which the transition to SML operation takes place, coincides with the pump threshold for exciting the TEM$_{10}$ mode. Hence, it was believed that SML is assisted by high-order transverse modes. Interestingly, in the first paper on Kerr lens mode-locking, a similar explanation was given [22]. However, the work of Keller et al. in 1991 and 1992 showed that higher-order transverse modes are not required in order to start and sustain mode-locking, as it is unclear how higher-order modes should give rise to self-amplitude modulation, which is understood to stabilize the mode-locked operation [35, 48].

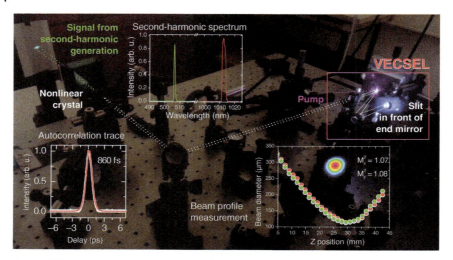

Figure 12.7 Photo of the actual setup used for external SHG, using the original laser output of the SML VECSEL. (Colored) dotted lines indicate the beam path inside and outside the VECSEL cavity. Top inset: optical spectrum of SHG signal and the original laser light. Bottom left and right insets: autocorrelation trace and beam quality measurement data, respectively. Source: Graphical elements reused with permission. [3] Copyright 2016, Optical Society of America.

The observation of an excellent beam quality in the SML operation regime with fundamental transverse mode profile by Gaafar et al. in their work on harmonic SML of VECSELs [21] had already ruled out a dependence of SML on higher-order transverse modes, as was emphasized in their follow-up publication [19]. Beam quality (M^2) measurements confirmed fundamental transverse mode operation in the SML regime with M^2 values of less than 1.1 for both axes (see Figure 12.7). Triggered by the discussion on the validity of ML claims in 2013 [36, 46], Gaafar et al. also focused on a detailed characterization with the aim to verify mode-locked operation in their SML-VECSEL configuration [19]. This included a measurement with an external nonlinear crystal – here, beta barium borate (BBO) – for second harmonic generation (SHG). Frequency doubling outside the VECSEL resonator was only achievable due to high peak powers of the SML-VECSEL pulses, converting the infrared pulsed laser light (1014 nm) into green pulses (507 nm). In contrast, no SHG signal was obtained for the CW emission mode of the same VECSEL.

12.3.2 Harmonic Self-Mode-Locking

Due to the early results on SML, the teams exploring this operation regime quickly gained confidence and more publications on this topic appeared. With it, also the acceptance of SML-VECSEL results increased more and more in the ML-VECSEL community and beyond.

In 2014, Gaafar et al. contributed to another important SML result be successfully demonstrating harmonic SML with a QW-based VECSEL up to the third harmonic

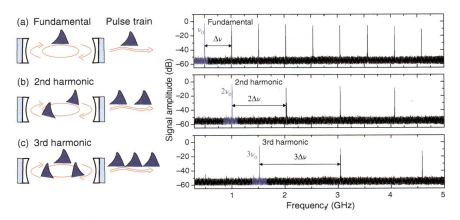

Figure 12.8 (a)–(c) Number of pulses in the cavity (sketch) and corresponding RF spectra for the SML VECSEL in the regimes of fundamental, second-harmonic and third-harmonic mode-locking, respectively.

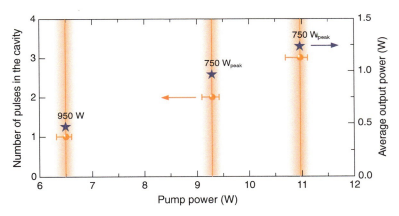

Figure 12.9 Number of pulses in the cavity (orange balls, left axis) and the average output power (blue stars) as a function of the optical pump power of the device.

[21] (cf. Figure 12.8). The authors achieved SML operation in a Z-shaped cavity with 300-mm resonator length and observed stable pulsing for each harmonic only for nearly discrete pump power levels in repeated investigations. This feature is seen in Figure 12.9. Here, the number of pulses, which were circulating in the cavity, for each harmonic at the respective pump power level is displayed together with the corresponding average output power. For slightly lower or higher powers around the proper pump power for each harmonic, instable mode-locking was evidenced. Thereby, horizontal error bars were derived in order to indicate islands of stability for SML operation.

Fundamental mode-locking with a repetition rate of 500 MHz was accomplished with sub-ps pulses of about 860 fs at the lowest pump level. Yet, the pulse lengths for the higher harmonics are of the same order. The authors attributed the obtained and almost equivalent time bandwidth products of 0.69 (first), 0.73 (second), and 0.72

(third harmonic), which were not transform-limited, to the remaining group-delay dispersion (GDD) caused by the VECSEL chip.

The resonator with intracavity hard-aperture (slit) was designed upon the assumption that negative Kerr lensing would reduce the beam waist in front of the plane high-reflectivity end mirror of the device, while light was coupled out at the opposite end mirror with about 1 % transmission rate. For increased pump levels, a stable SML regime was only achieved for a doubled and tripled repetition rate, corresponding to two and three pulses circulating in the cavity, respectively [21]. In this device, SML was initiated by moving the slit or by narrowing its gap. Remarkably, for all repetition rates, i.e. 500 MHz, 1.0 and 1.5 GHz, peak powers almost reached the 1 kW level, with about 950 W for the first, 750 W for the second, and 750 W for the third harmonic, respectively. According to the authors, the nearly constant peak power for the different power levels indicates that a certain intracavity power is needed for the underlying ML mechanism.

Shortly after, Gaafar et al. also presented a detailed characterization of the VECSEL emission from the very same device, which was shown to be free from double pulses and a CW component [19]. Furthermore, they highlighted the excellent beam quality and further bolstered their claim of stable SML operation with additional checks. This included long time span autocorrelation trace as well as pulse train measurements and RF signal analysis.

12.3.3 Self-Mode-Locking Quantum-Dot VECSEL

In order to demonstrate that the SML phenomenon is not restricted to a certain chip or gain medium type, Gaafar et al. demonstrated the first passively SML QD VECSEL, which was built in a classical linear cavity configuration and pumped optically from the side [38]. Such a device is schematically depicted in Figure 12.10. The QD-based SML VECSEL employed a chip with 35 layers of Stranski–Krastanov grown InGaAs QDs, which are organized in five stacks of seven QD layers in the active region, and a GaAs/AlGaAs-based DBR (see Figure 12.10a).

QD-based VECSELs have long been attractive candidates for ML operation owing to their shorter carrier lifetimes than QW structures (also valid for SESAMs) and promise, both, higher repetition rates that are beneficial for optical transmission schemes and shorter pulses – a feature desired for a plethora of applications. Particularly, their inhomogeneously broadened emission spectrum serves well with respect to wavelength tunability and the generation of shorter pulses in correspondence to the time-bandwidth principle. In addition, a stronger temperature dependence of the emission can be used for a better variation of pulse durations for an operational ML device. For more details on the benefits of QD structures and their use in lasers, the interested reader is referred to Refs. [6, 49, 50].

In the presented SML QD VECSEL, temperature controllable sub-picosecond pulse durations with average output powers between 300 and 750 mW have been demonstrated. Due to the short linear cavity configuration, pulses were delivered at a repetition rate of 1.5 GHz. The measured autocorrelation trace revealed pulse lengths between 830 and 950 fs for heat sink temperatures between 38 and 5 °C,

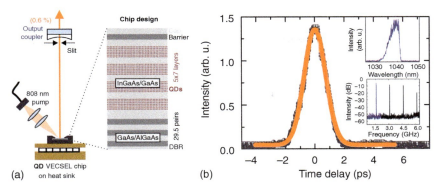

Figure 12.10 (a) Schematic drawing of the optically pumped SML QD-VECSEL setup with structure of the QD-VECSEL chip. (b) Autocorrelation trace of the SML QD VECSEL. Black dotted: experimental data. Red line: fit curve assuming a sech² pulse. Insets: Optical spectrum (top) and RF spectrum measured over a span of 6.5 GHz and a resolution bandwidth of 100 kHz, showing the first four harmonics (bottom).

respectively. Figure 12.10b summarizes these results. For this device, the optical spectrum was centered at 1038 nm and the RF spectrum indicated the quality of ML operation.

12.3.4 SML Cavity Configurations

As is evident from multiple reports on SML VECSELs, obtaining (stable) spontaneously formed pulsing is not restricted to a certain cavity geometry, provided that the correct assumptions were made in the design of the resonator. Both QW and QD VECSELs have shown SML operation in a straight cavity [17, 20, 37–39]. For QW VECSELs, various resonator shapes have led to the desired effect based on the assumption of nonlinear lensing in the chip [18, 19, 21, 37]. The early results in the SML domain by Kornaszewski et al. also show that even long cavities with multiple folding elements can be employed to obtain SML (see Figure 12.5).

Interestingly, a linear cavity fulfills its job with regard to SML operation, although it is still not clear what exactly gives rise to the observation of the desired behavior. The assumption of soft-aperture SML based on negative Kerr lensing in a linear cavity is a reasonable one, given the fact that most SML cavities were successfully operated under the assumption of a negative n_2 of the chip structure and preliminary Z-scan investigations supporting this assumption (see Section 12.5.1). In the work by Albrecht et al. in 2013, the authors had already modeled the influence of a negative intracavity lens on the beam waist modulation. Alternatively, according to resonator beam waist considerations as a function of a lensing effect at the position of the chip, a linear resonator also supports the use of a hard aperture in front of the out-coupling mirror, provided that n_2 is positive [20]. Such hard-aperture configuration was used to achieve SML operation for a QD VECSEL [38]. This is schematically presented in Figure 12.11a.

As summarized in [20], the assumed Kerr lens' focal length can be described by

$$f_{\text{Kerr}} = \pi\omega^4/n_2 LP, \tag{12.3}$$

whereas ω is the beam radius at the lens position, L is the thickness of the Kerr medium, and P is the power of the laser beam. The only variable in this equation that can be negative is the nonlinear refractive index n_2. In contrast to hard-aperture SML in a linear cavity, hard-aperture SML in a V-shaped or Z-shaped cavity relies on a negative n_2. These folded cavity geometries are commonly used in VECSEL studies and have their specific advantages as they give, for instance, access to slower repetition rates. Here, the calculated beam radius as a function of a Kerr lens at the chip position is shown in Figure 12.11b. Thereby, it is shown that a reduction of the beam radius in front of the end mirror can occur as a consequence of a negative

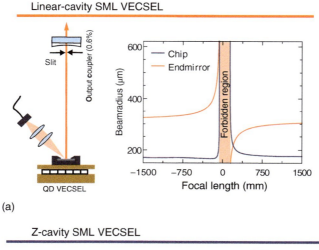

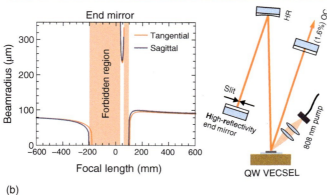

Figure 12.11 (a) Linear-cavity geometry as used for an SML QD VECSEL. (b) Z-cavity configuration of a harmonic SML QW VECSEL with 500 MHz repetition rate. For both example configurations, the corresponding calculated evolution of the intracavity beam radius at crucial intracavity positions as a function of a Kerr lens' focus length is shown.

Table 12.2 Developments concerning SML VECSELs; reported results (extrema in bold)

Year	Emission wavelength (nm)	Gain region (layer no.)	Pulse duration (fs)	Peak power (W)	Average power (W)	Repetition rate (GHz)	Ref.
2011	**1064**	QW (multi)	778	—	2.35	2.17	[17]
			654	—	0.45		
			1170	—	**5.1**		
2012	985	QW (10×)	930	**6800**	1.5	**0.21**	[18]
2013	1026	QW (multi)	1320	—	—	1	[37]
			1010	—	—		
			758	—	—		
			482	—	—		
2014	1038	**QD (35×)**	>830	<460	<0.75	1.5	[38]
	1014	QW (10×)	860	948	—	0.504	[21]
			1120	752	—	1.008	
			950	754	—	1.512	
2015	1059	QW (multi)	2350	—	—	1.567	[40]
2016	1014	QW (10×)	3500	55	0.32	1.56	[20]
2017	**666**	QW (20×)	**22 000**	0.3	0.03	**3.52**	[51]

Kerr lensing effect. In fact, these considerations are not perfect and detailed, but they can give a reasonable hint at the expected behavior in the respective resonator geometry, until more sophisticated modeling work and experimental confirmation of the obtained effects are available.

12.3.5 SML VECSEL at Other Wavelengths

So far, all SML VECSELs described in this chapter are based on the InGaAs material system with an emission wavelength around 1 μm. Only recently, the first SESAM-free ML VECSEL emitting in the visible range was realized [51]. With an AlGaInP-based gain chip including 20 GaInP QWs and a wedged diamond heat spreader on top, an average output power of 30 mW was achieved in a linear cavity with a 0.3% outcoupling mirror (see Figure 12.12). The rather short cavity length of ~42 mm results in a repetition rate of 3.5 GHz, which is the highest value for an SML VECSEL reported to date (see Table 12.2). In this experiment, mode-locking is obtained by carefully adjusting the spot of the 532 nm pump laser on the gain chip to a size slightly larger than the laser mode. For a certain adjustment, stable fundamental mode-locking at a wavelength of 666 nm is indicated by the oscilloscope trace and the RF spectrum (see Figure 12.13, left).

A typical autocorrelation (AC) trace measured in a non-collinear setup is shown in the right part of Figure 12.13. In contrast to previously reported ML VECSELs,

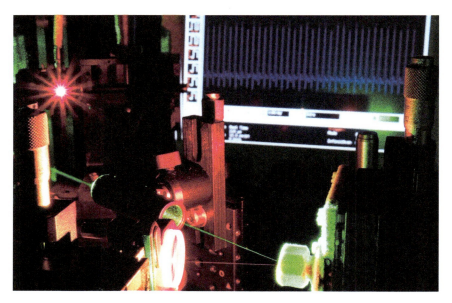

Figure 12.12 Photo of the SML AlGaInP VECSEL with characterization of the pulse train. In a linear cavity, the gain chip is pumped by a green laser and the red emission is focused onto a photodiode (bright red spot on the left). The temporal signal measured by an oscilloscope is displayed as a blue curve in the background of the picture. Source: Courtesy of Marius Grossmann, Institut für Halbleiter und Funktionelle Grenzfläche, Universität Stuttgart.

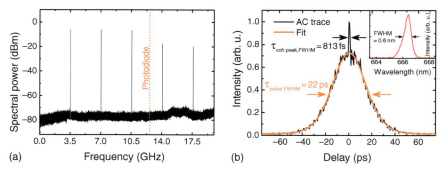

Figure 12.13 Red SML VECSEL results. Source: Adapted with permission. [51] Copyright 2017, AIP Publishing. (a) RF spectrum with peaks of equal intensities within the photodiode bandwidth of 12.5 GHz. (b) Autocorrelation (AC) trace showing a 22 ps sech^2 pulse with an additional coherence peak on top due to a complex pulse structure. The 813 fs FWHM of the shortest feature is obtained from a high-resolution AC measurements with 5 ps scan range. Inset: Optical spectrum with a FWHM of 0.6 nm.

a coherence peak appears on top of a 22 ps sech² pulse. The authors assume that it results from a complex pulse structure and represents the shortest temporal feature of this substructure. High-resolution AC traces of the coherence peak reveal a FWHM of 813 fs for this short feature. Together with the spectral bandwidth of 0.6 nm shown in the inset of Figure 12.13, it gives a time-bandwidth product of 0.33 which is close to the Fourier limit. The complex pulse structure revealed by the coherence peak is an indication for partial mode-locking as described theoretically [52] and is often observed for ML fiber lasers with "noise-like" or "double-scale" pulses [53, 54].

A Gaussian beam profile and measured M^2 values below 1.1 demonstrate the absence of higher order modes and confirm the results of Ref. [19]. Kerr lensing with soft aperture – as supposed in several reports on SML VECSELs before – could be the mechanism responsible for the mode-locked operation if a negative nonlinear refractive index is assumed in the active region. However, without reliable n_2 values for the AlGaInP gain region and with a positive n_2 contribution from the diamond, these findings cannot be explained by the Kerr lens effect without any further investigations.

12.4 Overview on SESAM-Free Mode-Locking Achievements

In this section, a brief overview on SML highlights is given and put in relation with general achievements of ML VECSEL development. Furthermore, some alternative SESAM-free ML techniques other than SML operation are summarized.

12.4.1 Spotlight on SML VECSELs

An important feature of ultrashort pulse lasers is that they can combine high peak powers, very short pulse durations, and high repetition rates. Most of the time, applications demand these features from a device simultaneously, but often, it is sufficient to tailor one's device to be at least optimal with respect to two of these characteristics. In addition, many applications require an excellent beam quality and robust operation, not to mention cost-efficiency. The results highlighted in Table 12.2 are a good indicator that SML VECSELs can be regarded as promising candidates for a wide range of applications and demonstrate that they show similar characteristics as their SESAM counterparts, which had experienced a decade of optimization efforts before the appearance of SML devices. Naturally, there are synergy effects arising from the optimization work on any ML VECSEL, and yet one can expect even better performance values for SML VECSELs once they have become more mature devices. In the following, the aforementioned important features are briefly reviewed.

12.4.1.1 Pulse Duration
Shortening pulse durations generally enables the achievement of higher peak powers at a given output rate, but the pulse length itself is also of great importance to applications such as spectroscopy and material processing, as mentioned earlier. The pulse length determines, for instance, the shortest events that can be resolved in a

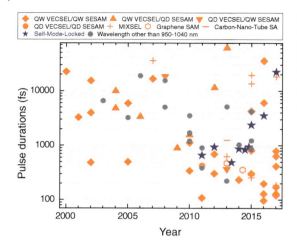

Figure 12.14 Pulse durations of mode-locked optically pumped VECSELs. Source: After Ref. [3]. Big symbols: Devices emitting around a wavelength of 1 μm (950–1050 nm). Small dots: Devices with other wavelengths than 1 μm.

pump-probe experiment. On the other hand, it reduces the impact of thermal effects in laser cutting applications. Additionally, due to high peak powers obtained from ultrashort pulses, nonlinear experiments such as two-photon absorption, supercontinuum generation, and frequency conversion become more efficient.

The pulse durations of VECSELs have been consecutively reduced to below the few-100-fs regime [13–15] at sub-Watt-level output powers over the years (see Figure 12.14), demonstrating the great potential of ML VECSELs. Herein, SML VECSELs show a similar performance with respect to pulse lengths obtainable; however, the record values are not as good as for optimized SESAM-based ML VECSELs. Currently, reported pulse durations range from 500 fs [37] to 22 ps [51]. Harmonic SML operation of a QW VECSEL and fundamental SML of a QD VECSEL were achieved with about 800–900 fs at repetition rates between 500 and 1500 MHz [21, 38], while other SML results reported so far are not far off in either direction.

12.4.1.2 Peak Power

The general trend of ML-VECSELs' peak powers can be well learned from recent reviews, such as Ref. [3] (see figure 5 therein). In Section 12.1, a chart has linked pulse durations and peak powers in order to show the span of accessible peak powers (see Figure 12.1), while the chart does not cover the whole range but only the main window of achieved pulse durations. Peak powers in the kW range have been demonstrated in recent years for both SESAM-based as well as SESAM-free ML-VECSELs: The assumed 6.8 kW peak power of the SML VECSEL demonstrated in 2012 by Kornaszewski et al. [18] compares well with a similar peak power obtained for a SESAM ML device presented in 2013 by Wilcox et al., who demonstrated incoherent supercontinuum generation with their 400-fs pulses [5]. The peak powers obtained from harmonic SML operation nearly reached the 1 kW level for each harmonic [21], while the SML QD VECSEL featured peak powers up to about 0.5 kW [38]. Based on these achievements, it is reasonable to assume that optimized SML devices will be capable of delivering multi-kW-level peak powers in the near future.

12.4.1.3 Repetition Rate

As the repetition rate of pulsed lasers matters for various applications, it is worth taking a look at the current situation for SML VECSELs. For instance, biomedical applications on the one hand need low repetition rates when it comes to long fluorescence lifetimes of their employed or investigated molecules [55]; on the other hand, multiphoton fluorescence imaging microscopy also benefits from high repetition rates when it comes to scanning applications. Beyond classical telecom applications of high-repetition-rate lasers, quantum communication is also a candidate for the employment of ultrafast ML VECSELs in order to achieve record-high-speed single-photon-based transmission schemes [56]. In the lower limit of repetition rates, one clearly finds the SML VECSEL of Kornaszewski et al., who presented their long-cavity device with 200 MHz repetition rate [18], followed by the 500-MHz VECSEL by Gaafar et al. [21]. On the other side, short linear cavities enable SML operation at repetition rates of a few GHz [17, 38, 51].

12.4.2 SESAM-Free Alternatives to SML VECSEL

12.4.2.1 Graphene or Carbon Nanotube Saturable Absorber Mode-Locked VECSELs

An alternative to SESAM ML VECSELs is not only given by SML VECSELs but also by devices which employ graphene or CNT saturable absorbers [3]: For instance, the wide bandwidth of GSAM can be advantageous when it comes to wavelength flexibility, whereas the ultrafast charge-carrier dynamics of graphene can help to achieve very short pulses [30]. While the output power of the first GSAM ML VECSEL was limited due to the GSAM's high non-saturable losses and low damage threshold, an antiresonant GSAM with low non-saturable losses and high saturation fluence showed high-power operation of a GSAM ML VECSEL with up to 10 W average-output power [31]. Both GSAM ML devices delivered pulses of a few-hundred femtoseconds around 1 μm emission wavelength, although at different power levels. In addition to GSAMs, single-walled carbon nanotubes (SWCNTs) were successfully employed as ultrafast intracavity saturable absorber (SA) element for the demonstration of SWCNT-SA ML VECSELs [32]. The low loss element with broadband absorption is a promising alternative for widely tunable SESAM-free ML VECSELs as well, having brought up ps pulses at a repetition rate of around 600 MHz.

12.4.2.2 SESAM-Free VECSEL Design with Intracavity Kerr Medium

In fact, many SML VECSEL results were attributed to Kerr lens ML of SESAM-free VECSELs, with the assumed Kerr medium being the chip itself. Nonetheless, an example of a VECSEL design with an intracavity Kerr medium employed for ML operation was presented by Moloney et al. in 2014 [47] (see Figure 12.15). The authors claimed that an extra Kerr medium – in that case an yttrium orthovanadate (YVO4) crystal – inside the cavity enhanced the Kerr lensing effect in such a system, which was meant to facilitate the observation of a stable Kerr-lens-assisted ML regime. As YVO4 exhibits a nonlinear refractive index three times higher than that of titanium–sapphire, it was expected that a strong effect is seen. However, first

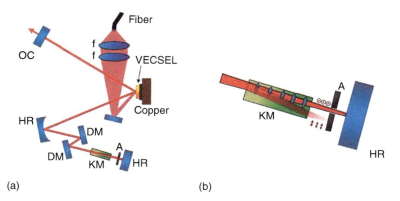

Figure 12.15 (a) Schematics of the laser cavity used for SESAM-free mode-locking, in which the Kerr medium (KM) was placed close to the end mirror. (b) The YVO4 crystal provides polarization control: the ordinary laser beam passes the crystal directly, while the orthogonal polarization is displaced and is blocked by the aperture (A). Source: Reproduced with permission. [47] Copyright 2014, Optical Society of America.

results with this scheme only yielded instable pulsing with measured autocorrelation traces, from which a pulse duration of 850 fs was derived, showing a peak that sits on a high background pedestal – an indicator for a strong quasi-CW component in the emission.

12.5 Investigations into the Mechanisms and Outlook

12.5.1 First Studies Concerning the Mechanisms Behind SML

While Chen et al. supposed the gain region itself to act as a saturable absorber [17], both Kornaszewski et al. and Albrech et al. took Kerr lensing as their working hypothesis [18, 37]. The first investigation into nonlinear lensing in a VECSEL was directly performed by Albrecht et al. in their first publication on SML VECSELs, which modeled the Kerr-lens-induced beam radius change inside a linear (can be folded in a V-shape) cavity for different conditions. Thereby, a variation of the Kerr lens position with respect to one of the end mirrors was taken into account. It was shown that a negative n_2 value (negative nonlinear lens) will enable soft-aperture mode-locking and also hard-aperture mode-locking if the chip acts as a folding mirror in a V-shaped cavity. However, in a linear cavity, hard-aperture mode-locking is not possible for negative n_2 as no beam narrowing will occur anywhere between the Kerr medium and the end mirror. Yet, a positive nonlinear lens shall enable SML operation with a slit inside the resonator, as the signs in the modeled results would reverse. For more details, the interested reader is referred to the Ref. [37]. So far, the assumption is made that Kerr lensing in the chip structure is the mechanism which gives rise to SML operation in VECSELs. In fact, this has become the primary working hypothesis behind many of the following cavity configurations [20, 51]. Although preliminary studies with a direct aim to reveal the nonlinear refractive index of the gain mirror have concluded that the effect can be the cause for a sufficiently high perturbation in the cavity for SML operation to set in [41], a direct attribution to Kerr lensing is overdue. In fact, thorough investigations will be essential for the improvement of SML VECSELs and there is still a lack in this regard. To ramp up the performance and to tailor the pulse durations of SML devices, it is important to understand which effects cause pulse formation and to which extent they are involved and determine the characteristics of the achieved ML operation.

12.5.2 Z-Scan Measurements of the Nonlinear Refractive Index in a VECSEL Chip

The nonlinear refractive index n_2 of a medium is defined by (see Eq. (12.1)

$$n(I_0) = n_0 + n_2 I_0. \tag{12.4}$$

Here, $n(I_0)$ is the total refractive index as a function of incident radiation intensity (irradiance) and n_0 is the intensity-independent refractive index. For ultrashort pulses with Gaussian beam profile, the strength of the nonlinearity is set in relation to the on-axis peak power density of the pulse, represented by the intensity value I_0. This equation is only exact for an instantaneous Kerr medium. For processes with a finite response time with respect to the optical pulse length (e.g. carrier effects), this is an approximation which must be validated afterward through a linear dependency of the nonlinear index change on the irradiance. In this case, the nonlinear refractive index has to be understood as an effective one. This means that the temporally integrated response function of the refractive index over the pulse depends linearly on the peak intensity I_0.

Z-scan measurements allow one to directly measure the nonlinear refractive index changes in a material. Thus, the Z-scan technique has become an effective and sensitive tool to probe intensity-dependent nonlinear lensing, as well as nonlinear absorption, in a medium [57].

The Z-scan technique, measuring the above-defined figure n_2, cannot discriminate between noninstantaneous contributions and instantaneous contributions, i.e. those induced by free-carrier-related nonlinearities (FCN) [58] and bound electronic Kerr effect (BEKE) [59], respectively. However, the relative contributions from FCN and BEKE to the total nonlinear refractive index could be unraveled with the help of time-resolved investigations, as the timescales of these contributing effects differ. It can be anticipated that a precise knowledge of the different timescales involved in the effective third-order refractive nonlinearity n_2 will ultimately facilitate the explanation and modeling of the mechanisms behind nonlinear-lens-induced SML, with important implications for future chip designs [60].

Analogous to nonlinear refraction, nonlinear absorption can be defined by an intensity-dependent absorption term. The nonlinear absorption coefficient β is defined by

$$\alpha(I_0) = \alpha_0 + \beta I_0, \tag{12.5}$$

whereas α_0 is the linear absorption coefficient and thereby $\alpha(I_0)$ the total intensity-dependent absorption coefficient. Z-scan measurements readily deliver information about the nonlinear absorption coefficient β and reveal to which extent intensity-dependent absorption can act as possible loss mechanism. While contributions to the effective nonlinear refractive index n_2 of a VECSEL chip can take place on different timescales, β only comprises ultrafast $\chi^{(3)}$ effects. Regarding the design of SML-optimized VECSELs, the Z-scan technique can provide a comparative measure about how strongly nonlinear refraction dominates over nonlinear absorption which is commonly expressed in the figure of merit (FOM), $\frac{2kn_2}{\beta}$, with k being the wave number. The FOM of the investigations in Ref. [60] exceeded 2 for a chip structure not specifically optimized for SML.

While several preliminary investigations of the nonlinear refractive index in VECSEL gain structures [41–43] reported first refractive index data, they are not providing a definite proof that SML operation is indeed caused by Kerr lensing. Nonetheless, the demonstrated results have already indicated the possibility of nonlinear lensing being a factor in SML operation and have motivated a more detailed understanding of the nonlinear lensing properties in VECSELs. The introductory studies mentioned above were exclusively performed at arbitrarily selected single wavelengths.

Consequently, experiments prior to those of Kriso et al. [60] were not taking into account the microcavity resonance as well as the strong dispersion of the nonlinear refractive index around the excitation band edge. This edge is characteristic for contributions to the nonlinear refractive index, both, from the FCN existing above that edge [58] and from the BEKE existing for optical frequencies below the band gap [59]. While the effect of the microcavity resonance is discussed in Ref. [60], the underlying origin of the nonlinear lensing (BEKE and/or FCN) is still unclear and must be investigated in time-resolved experiments.

With the wavelength-dependent Z-scan measurements of Ref. [60], it has been indeed reassured that nonlinear lensing could be the effect behind SML owing to the sufficiently high negative n_2 that can provide a considerable intensity-dependent defocusing effect. Particularly for wavelengths resonant to the microcavity structure, the wavelength-dependent n_2 can lead to a strong perturbation of the intracavity beam profile [60]. Furthermore, it has been unambiguously and uniquely demonstrated that nonlinear refraction and nonlinear absorption are strongly shaped by the microcavity resonance and, thus, are angle-tunable, according to wavelength- as well as incidence-angle-dependent Z-scan characterization (see Figure 12.16). However, it needs to be considered that VECSEL chips optimized for ultrashort pulse generation are usually designed in an antiresonant manner to reduce dispersion and to minimize its impact on the pulses. This might also reduce the strength of the nonlinear refractive index and needs to be taken into account when designing a VECSEL optimized for mode-locking by nonlinear lensing.

Although self-mode-locking behavior in VECSELs has been attributed to Kerr lens mode-locking, the role of a nonlinear refractive index in the VECSEL chip has not been clearly unraveled from all other possible contributions to an overall pulse forming mechanism, such as for example the gain dynamics. This remains more challenging than simply performing measurements of a nonlinear refractive index change in the cavity for relevant intensities. Nevertheless, an important step toward the

Figure 12.16 Wavelength and angle-dependent nonlinear absorption (a) and nonlinear refraction (b): the behavior of β and n_2 measured for the incidence angles of 10°, 20°, and 30° are plotted as a function of the wavelength (left axes), with guides to the eyes. The photoluminescence (PL) from the quantum-well gain medium and the different angle-dependent longitudinal confinement factor (LCF) peaks are indicated by arrows on top of the diagram. Here, the trends are directly compared to the corresponding surface PL at given angle of incidence (shaded plots) as well as reflectivity spectra (line plots), both normalized to 1 and with respect to the right axes. A schematic inset displays the probe geometry with respect to the VECSEL chip. For further details and explanations see Ref. [60]. Source: Reproduced with permission. [60] Copyright 2019, Optical Society of America.

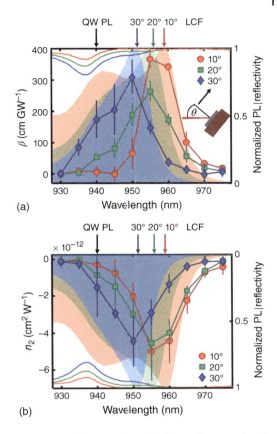

characterization of SML behavior is given by the wavelength-dependent study of nonlinear lensing in the gain mirror structure. Further studies with the same structure of Ref. [60] showed that optical pumping has only little influence on both nonlinear absorption and refraction, as only a small change of nonlinear optical properties with optical pump density up to the rollover point were obtained. However, as Kerr mode-locking is known to be quite sensitive to changes in intracavity power and resulting changes of the nonlinear lens, this might have to be considered in the cavity design of a nonlinear lens mode-locked VECSEL.

Recent findings may ultimately lead to novel chip concepts with regard to tailored SML behavior achievable by peculiar Kerr lens chip designs for cost-effective, robust, and compact fs-pulsed semiconductor lasers. In fact, beyond Z-scan measurements, the direct observation of a lensing behavior for the intracavity beam or for a synchronized probe beam incident on the chip being defocused from the mode's spot as a consequence of a spatial nonlinearity remains a desirable goal. In addition, pulse characterization will further help to identify whether Kerr lens mode-locking or quasi-soliton mode-locking is the dominating mechanism in SML VECSELs due to their different characteristic phase profiles. Similarly, these studies can be further supported by experiments for investigations of self-phase-modulation and gain dynamics. Eventually, one cannot rule out that one effect initiates pulse formation, while the other maintains mode-locking, particularly in connection with an intracavity aperture for the introduction of mode-locking.

12.5.3 Applications and Expected Advances

ML VECSELs are attractive sources of pulsed light for both scientific and industrial applications which promise to provide high peak powers, high repetition rates, and an excellent beam quality. Thus, it is natural to predict that SML VECSELs will address similar applications as their SESAM ML counterparts. Although SML operation of VECSELs is a quite young technique, their performance compares well with that of other ML VECSELs. Key to improvements in their performance is a better understanding of the mechanisms that give rise to ML in such a device and the corresponding optimization of the chip structure. If operated under optimum conditions, high peak powers for ultrashort pulses and high operation stability can be expected for SML devices, which can be designed for various applications. While a number of interesting applications and examples for the use of ML VECSELs can be found in recent reviews [2, 3, 45], this section shall briefly emphasize few of these.

Typical applications for fs-pulsed lasers are micromachining [7] and nonlinear optical microscopy [6], such as of biological organisms [55]. Here, (S)ML-VECSELs are promising devices due to their cost-effectiveness, compactness, and wavelength versatility. Other possible biophotonics applications were highlighted in the literature as well, such as optical coherence tomography [6]. Particularly, QD-based SML VECSELs [38] could be beneficial due to the broader tunability of QD devices [50]. In addition to micromachining, surface texturing with ultrafast VECSELs can serve as a means of preparation of microfluidic devices and lab-on-a-chip settings [61].

An exotic use of SML VECSELs can be found as an ultrafast optical pump for high-repetition-rate single-photon sources, similar to the example using a SESAM-ML device in Ref. [56]. Currently, Ti:sapphire lasers are commonly employed as a source of ultrashort laser pulses to drive various nonclassical light sources for quantum optical experiments which could be replaced by (S)ML VECSELs specially designed for these applications in the future. Typically, not peak power values are in the focus for pump lasers of quantum light sources, but repetition rate optimization and wavelength flexibility for ideal excitation conditions. Furthermore, if the spectral bandwidth of the pulsed light matters, for instance in resonant excitation schemes, even the adjustment of pulse durations may become necessary. Overall, pulsed light from SDLs shows promise in these directions.

Next, it shall be noted that the construction of optical frequency combs, which may require supercontinuum generation for self-referencing, has been a major motivation in the peak power and pulse length scaling efforts of VECSELs. Supercontinuum generation usually requires high optical intensities contained in ultrashort pulses and a highly nonlinear medium. Since high-power sub-ps-pulsed VECSELs with high peak powers up to 4.4 kW have been successfully employed to demonstrate direct supercontinuum generation [5] in photonic crystal fibers, the use of SML VECSELs for the same purpose is naturally implied. Supercontinuum light generation by SML devices would represent a key proof of their capabilities. If initially the peak power of SML VECSELs is not sufficient for direct supercontinuum generation, one could use a master-oscillator power-amplifier setting to post-amplify the SML-VECSEL output to desired power levels – something desirable also for laser cutting applications.

Ultimately, SML technology allows one to think wavelength-independent, since SESAM technology is not well developed for every wavelength range and generally adds up an additional design as well as cost factor in the production chain for SESAM-ML VECSELs. Additionally, SML operation helps overcoming degradation-related lifetime problems of SESAMs which have been recently discussed for ultrashort-pulsed VECSELs with intense intracavity light [62]. Such degradation phenomenon is not completely understood yet and can considerably affect long-term operation of ultrafast SESAM-ML VECSELs. Yet, the challenge remains to spread SML operation further in the VECSEL community and to make the outcomes of SML endeavors highly predictable which naturally is linked to a continuously improved understanding of the whole subject.

Acknowledgments

The author thanks the editors for the generous invitation to contribute to this exciting book project with a chapter on self-mode-locking of VECSELs. Moreover, the author is grateful to (1) W. Stolz, S. W. Koch, and M. Koch, for their sincere support and the fruitful collaboration on VECSELs in Marburg, Germany, (2) M. Jetter and P. Michler and their team in Stuttgart, Germany, for joint efforts on the demonstration of a red SML VECSEL and the characterization of the nonlinear refractive index in VECSEL chips, and (3) E. U. Rafailov and K. A. Fedorova from the Aston University, UK, for invaluable discussions and the fruitful collaboration on quantum-dot VECSELs.

Furthermore, special thanks are devoted to the former PhD students and the author's team members working on VECSELs in Marburg, particularly M. Wichmann, M. Gaafar, D. Al-Nakdali, C. Möller, and F. Zhang who helped paving the way for this book chapter to happen, as well as current PhD student and the author's project staff member C. Kriso for his great efforts and contributions to the Z-scan investigations and follow-up experiments. The author is also grateful for the help of all other students involved in the author's VECSEL projects and in the partners' teams, particularly R. Bek, M. Vaupel, S. Kress, S. Kefer, M. Grossmann, M. Alvi, and T. Munshi who supported the author's endeavors in the field of SML VECSELs, and also J. Quante, S. Wang, H. Guoyu, O. Mohiuddin, A. Barua, M. Gao as part of the author's VECSEL team.

The provided information in this chapter is summarized to the best of the author's knowledge, and may reflect mainly the developments at the time of writing/surveying. No warranty, expressed or implied, is given with respect to the material contained in this chapter or for any errors or omissions that may have been made.

The German Research Foundation DFG (GRK 1782, SFB 1083 and RA 2841/1-1, as well as RA 2841/1-3) is thanked for financial support over the past several years.

References

1 Keller, U. and Tropper, A.C. (2006). Passively modelocked surface-emitting semiconductor lasers. *Physics Reports* 429: 67–120.
2 Tilma, B.W., Mangold, M., Zaugg, C.A. et al. (2015). Mode-locked semiconductor disk lasers. *Light: Science and Applications* 4: e310.
3 Gaafar, M.A., Rahimi-Iman, A., Fedorova, K.A. et al. (2016). Mode-locked semiconductor disk lasers. *Advances in Optics and Photonics* 8: 370.
4 Schulz, N., Hopkins, J.M., Rattunde, M. et al. (2008). High-brightness long-wavelength semiconductor disk lasers. *Laser & Photonics Reviews* 2: 160–181.
5 Wilcox, K.G., Tropper, A.C., Beere, H.E. et al. (2013). 4.35 kW peak power femtosecond pulse mode-locked VECSEL for supercontinuum generation. *Optics Express* 21: 1 599–1 605.
6 Rafailov, E.U. (2014). *The Physics and Engineering of Compact Quantum Dot-based Lasers for Biophotonics*. Wiley-VCH.
7 Markovic, V., Rohrbacher, A., Hofmann, P. et al. (2015). *160 W 800 fs Laser System without CPA for High Speed Surface Texturing*, CA-9.1. Munich: CLEO-Europe.
8 Newbury, N.R. (2011). Searching for applications with a fine-tooth comb. *Nature Photonics* 5: 186–188.
9 Zewail, A.H. (1988). Laser femtochemistry. *Science* 242: 1 645–1 653.
10 Kaindl, R., Carnahan, M.A., Hägele, D. et al. (2003). Ultrafast terahertz probes of transient conducting and insulating phases in an electron–hole gas. *Nature* 423: 734–738.
11 Baker, C., Scheller, M., Koch, S.W. et al. (2015). In situ probing of mode-locked vertical-external-cavity-surface-emitting lasers. *Optics Letters* 40: 5 459–5462.
12 Lammers, C., Stein, M., Berger, C. et al. (2016). Gain spectroscopy of a type-II VECSEL chip. *Applied Physics Letters* 109: 232 107.
13 Klopp, P., Griebner, U., Zorn, M. et al. (2011). Pulse repetition rate up to 92 GHz or pulse duration shorter than 110 fs from a mode-locked semiconductor disk laser. *Applied Physics Letters* 98: 071103.
14 Waldburger, D., Link, S.M., Mangold, M. et al. (2016). High-power 100 fs semiconductor disk lasers *Optica* 3: 844–852.
15 Laurain, A., Kilen, I., Hader, J. et al. (2018). Modeling and experimental realization of modelocked VECSEL producing high power sub-100 fs pulses. *Applied Physics Letters* 113 (12): 121 113.
16 Mangold, M.,Golling, M., Gini, E. et al. (2015). Sub-300-femtosecond operation from a MIXSEL. *Optics Express* 23: 22 043–22 059.
17 Chen, Y.F., Lee, Y.C., Liang, H.C. et al. (2011). Femtosecond high-power spontaneous mode-locked operation in vertical-external cavity surface-emitting laser with gigahertz oscillation. *Optics Letters* 36: 4 581–4 583.
18 Kornaszewski, L., Maker, G., Malcolm, G.P.A. et al. (2012) SESAM-free mode-locked semiconductor disk laser. *Laser & Photonics Reviews* 6: L20–L23.

19 Gaafar, M., Richter, P., Keskin, H. et al. (2014). Self-mode-locking semiconductor disk laser. *Optics Express* 22 (23): 28 390–28 399.
20 Rahimi-Iman, A., Gaafar, M., Möller, C. et al. (2016). Self-mode-locked vertical-external-cavity surface-emitting laser. *Proceedings of SPIE* 9734: 97340M.
21 Gaafar, M., Möller, C., Wichmann, M. et al. (2014). Harmonic self-mode-locking of optically pumped semiconductor disc laser. *Electronics Letters* 50: 542–543.
22 Spence, D.E., Kean, P.N., and Sibbett, W. (1991). 60-fsec pulse generation from a self-mode-locked Ti: sapphire laser. *Optics Letters* 16: 42–44.
23 Hoogland, S., Dhanjal, S., Tropper, A.C. et al. (2000). Passively mode-locked diode-pumped surface-emitting semiconductor laser. *IEEE Photonics Technology Letters* 12: 1 135–1 137.
24 Lorenser, D., Maas, D.J.H.C., Unold, H.J. et al. (2006). 50-GHz passively mode-locked surface-emitting semiconductor laser with 100-mW average output power. *IEEE Journal of Quantum Electronics* 42: 838–847.
25 Butkus, M., Viktorov, E.A., Erneux, T. et al. (2013). 85.7 MHz repetition rate mode-locked semiconductor disk laser: fundamental and soliton bound states. *Optics Express* 21: 25 526–25 531.
26 Lorenser, D., Unold, H.J., Maas, D.J.H.C. et al. (2004). Towards wafer-scale integration of high repetition rate passively mode-locked surface-emitting semiconductor lasers. *Applied Physics B* 42: 927–932.
27 Wilcox, K.G., Butkus, M., Farrer, I. et al. (2009). Subpicosecond quantum dot saturable absorber mode-locked semiconductor disk laser. *Applied Physics Letters* 94: 251 105.
28 Hoffmann, M., Sieber, O.D., Maas, D.J.H.C. et al. (2010). Experimental verification of soliton-like pulse-shaping mechanisms in passively mode-locked VECSELs. *Optics Express* 18: 10 143–10 153.
29 Zaugg, C.A., Hoffmann, M., Pallmann, W.P. et al. (2012). Low repetition rate SESAM modelocked VECSEL using an extendable active multipass-cavity approach. *Optics Express* 20: 27 915–27 921.
30 Zaugg, C.A., Sun, Z., Wittwer, V.J. et al. (2014). Ultrafast and widely tuneable vertical-external-cavity surface-emitting laser, mode-locked by a graphene-integrated distributed Bragg reflector. *Optics Express* 21: 31 548–31 559.
31 Husaini, S. and Bedford, R.G. (2014). Graphene saturable absorber for high power semiconductor disk laser mode-locking. *Applied Physics Letters* 104: 161 107.
32 Seger, K., Meiser, N., Choi, S.Y. et al. (2013). Carbon nanotube mode-locked optically-pumped semiconductor disk laser. *Optics Express* 21: 17 806–17 813.
33 Scheller, M., Wang, T.L., Kunert, B. et al. (2012). Passively modelocked VECSEL emitting 682 fs pulses with 5.1 W of average output power. *Electronics Letters* 48: 588–589.
34 Maas, D.J.H.C., Bellancourt, A.-R., Rudin, B. et al. (2007). Vertical integration of ultrafast semiconductor lasers. *Applied Physics B* 88: 493–497.
35 Keller, U., 'tHooft, G.W., Knox, W.H., and Cunningham, J.E. (1991). Femtosecond pulses from a continuously self-starting passively mode-locked Ti:sapphire laser. *Optics Letters* 16: 1 022–1 024.

36 Kornaszewski, L., Maker, G., Malcolm, G.P.A. et al. (2013). Reply to comment on SESAM-free mode-locked semiconductor disk laser. *Laser & Photonics Reviews* 7: 555–556.

37 Albrecht, A.R., Wang, Y., Ghasemkhani, M. et al. (2013). Exploring ultrafast negative Kerr effect for mode-locking vertical external-cavity surface-emitting lasers. *Optics Express* 21: 28 801–28 808.

38 Gaafar, M., Al Nakdali, D., Möller, C. et al. (2014). Self-mode-locked quantum-dot vertical-external-cavity surface-emitting laser. *Optics Letters* 39: 4 623–4 626.

39 Liang, H.C., Tsou, C.H., Lee, Y.C. et al. (2014). Observation of self-mode-locking assisted by high-order transverse modes in optically pumped semiconductor lasers. *Laser Physics Letters* 11: 105 803.

40 Tsou, C.H., Liang, H.C., Wen, C.P. et al. (2015). Exploring the influence of high order transverse modes on the temporal dynamics in an optically pumped mode-locked semiconductor disk laser. *Optics Express* 23: 16 339–16 347.

41 Quarterman, A.H., Tyrk, M.A., and Wilcox, K.G. (2015). Z-scan measurements of the nonlinear refractive index of a pumped semiconductor disk laser gain medium. *Applied Physics Letters* 106 (1): 011 105.

42 Shaw, E.A., Quarterman, A.H., Turnbull, A.P. et al. (2016). Nonlinear lensing in an unpumped antiresonant semiconductor disk laser gain structure. *IEEE Photonics Technology Letters* 28 (13): 1 395–1 398.

43 Quarterman, A.H., Mirkhanov, S., Smyth, C.J.C., and Wilcox, K.G. (2016). Measurements of nonlinear lensing in a semiconductor disk laser gain sample under optical pumping and using a resonant femtosecond probe laser. *Applied Physics Letters* 109 (12): 121 113.

44 Keller, U. (2003). Recent developments in compact ultrafast lasers. *Nature* 424: 831–838.

45 Rahimi-Iman, A. (2016). Recent advances in VECSELs. *Journal of Optics* 18: 093 003.

46 Wilcox, K.G. and Tropper, A.C. (2013). Comment on SESAM-free mode-locked semiconductor disk laser. *Laser & Photonics Reviews* 7: 422–423.

47 Moloney, J.V., Kilen, I., Bäumner, A. et al. (2014). Nonequilibrium and thermal effects in mode-locked VECSELs. *Optics Express* 22: 6 422–6 427.

48 Keller, U., Knox, W.H., and W'tHooft, G. (1992). Ultrafast solid-state mode-locked lasers using resonant nonlinearities. *IEEE Journal of Quantum Electronics* 28: 2 123.

49 Rafailov, E.U., Cataluna, M.A., and Sibbett, W. (2007). Mode-locked quantum-dot lasers. *Nature Photonics* 1: 395–401.

50 Rafailov, E.U., Cataluna, M.A., and Avrutin, E.A. (2014). *Ultrafast Lasers Based on Quantum Dot Structures: Physics and Devices*. Wiley-VCH.

51 Bek, R., Großmann, M., Kahle, H. et al. (2017). Self-mode-locked AlGaInP-VECSEL. *Applied Physics Letters* 111: 182 105.

52 Shapiro, S.L. (1984). *Ultrashort light pulses*. Berlin, Heidelberg: Springer-Verlag.

53 Kobtsev, S., Kukarin, S., Smirnov, S. et al. (2009). Generation of double-scale femto/pico-second optical lumps in mode-locked fiber lasers. *Optics Express* 17 (23): 20 707–20 713.
54 Wang, Q., Chen, T., Li, M. et al. (2013). All-fiber ultrafast thulium-doped fiber ring laser with dissipative soliton and noise-like output in normal dispersion by single-wall carbon nanotubes. *Applied Physics Letters* 103 (1): 011 103.
55 Aviles-Espinosa, R., Filippidis, G., Hamilton, C. et al. (2011). Compact ultrafast semiconductor disk laser: targeting GFP based nonlinear applications in living organisms. *Biomedical Optics Express* 2: 739–747.
56 Schlehahn, A., Gaafar, M., Vaupel, M. et al. (2015). Single-photon emission at a rate of 143 MHz from a deterministic quantum-dot microlens triggered by a mode-locked vertical-external-cavity surface-emitting laser. *Applied Physics Letters* 107: 041 105.
57 Sheik-Bahae, M., Said, A.A., Wei, T.H. et al. (1990). Sensitive measurement of optical nonlinearities using a single beam. *IEEE Journal of Quantum Electronics* 26: 760–769.
58 Lee, Y.H., Chavez-Pirson, A., Koch, S.W. et al. (1986). Room-temperature optical nonlinearities in GaAs. *Physical Review Letters* 57: 2 446–2 449.
59 Sheik-Bahae, M., Hutchings, D.C.D., Hagan, D.J., and Van Stryland, E.W. (1991). Dispersion of bound electron nonlinear refraction in solids. *IEEE Journal of Quantum Electronics* 27: 1296–1309.
60 Kriso, C., Kress, S., Munshi, T. et al. (2019). Microcavity-enhanced Kerr nonlinearity in a vertical-external-cavity surface-emitting laser. *Optics Express* 27: 11914–11929.
61 Coffey, V.C. (2014). High-energy lasers: new advances in defense applications. *Optics and Photonics News* 25: 28–35.
62 Addamane, S., Shima, D., Laurain, A. et al. (2018). Degradation mechanism of SESAMs under intense ultrashort pulses in modelocked VECSELs. *Proceedings of SPIE* 10515: 105150T.

Index

a
active and passive mode-locking
 techniques 360
autocorrelation (AC) 291, 292, 310, 312, 313, 314, 339–341, 361, 363, 364, 366, 368, 369, 371, 372, 376

b
beta barium borate (BBO) 248, 315, 336, 366
biological imaging 252–253
birefringent filter (BRF) 43, 80, 187, 211, 213, 216, 218–220

c
"class B" dynamics 136
colliding pulse mode-locked VECSEL
 modelocking results
  cross correlation, of output beams 336–338
  modelocking regime robustness 335–336
  multipulse regime 340–341
  pulse duration optimization 338–340
 principle of 322–324
 pulse interactions, in saturable absorber
  absorption losses and pulse shaping 347–349
  carrier density distribution 345–347
  field intensity distribution 341–343
  power balance in CPM operation 350–352
  saturable absorption model 343–345
  saturation fluence of absorber 349–350
 stable
  gain recovery and pumping rate 324–325
  mode waist and saturation fluence 326–327
  polarization 326
  pulse timing 324
 ultrafast CPM VECSEL design
  gain structure 328–333
  optical cavity 327–328
  SESAM 333–334
colliding pulse modelocking (CPM) 322–327
commercial femtosecond lasers 232
continuous tunability 110, 118, 122, 125, 129, 139
continuous wave (CW) operation 7, 28, 197
Coulomb potential 274

d
DBR-free optically pumped
 semiconductor disk lasers
  broadband tunability 180–182
  conventional dielectric mirrors 176
  device fabrication 182–184
  heat spreader and heatsink 175
  implementation
   broad tunability 187–189

Vertical External Cavity Surface Emitting Lasers: VECSEL Technology and Applications, First Edition.
Edited by Michael Jetter and Peter Michler.
© 2022 WILEY-VCH GmbH. Published 2022 by WILEY-VCH GmbH.

DBR-free optically pumped semiconductor disk lasers (*contd.*)
  high power operation 185–187
  wafer-scale processing 189
  longitudinal mode structure 180–182
  opportunities and advantages 177–178
  thermal analysis 178–180
  TIR-based monolithic ring 190
dielectric and metamorphic DBRs 33–34
diode-pumped solid-state lasers (DPSSLs) 22, 121
diode-pumped VECSEL 210
direct red-emitting AlGaInP VECSELs
  AlGaInP material system 199–201
  GaInP quantum well VECSELs. *see* GaInP quantum well VECSELs
  power scaling via quantum well and multi-pass pumping 208–211
  second harmonic generation, UV-A spectral range 211–212
direct wafer bonding 39–44
distributed Bragg reflectors (DBRs) 7, 66, 124, 175, 199, 234, 268, 271, 277, 306, 321, 359
distributed feedback (DFB) cavities 147
distributed feedback fiber laser 121
dots-in-a-well (DWELL) structure 31
double heterostructure (DHS) 176
double-metal waveguide 146
double-plasmon waveguide 146
double sided heatspreader (DSH) 73, 77–78, 100
Drude model parameters 150

e

edge-emitting diode laser 84, 233
edge-emitting InSb laser 5
effective value or RMS 132
electrically pumped EP-VECSELs 235
epitaxial layer design, AlGaInP-SESAM structures
  cavity designs 309–310
  characterization methods 310–311
  mode locking. *see* mode locking
  quantum dot SESAMs 307
  quantum well SESAMs 306–307
  second harmonic generation, UV spectral range 315–317
  temporal response of AlGaInP SESAMs 307–309
Epitaxial Products International (EPI) 10
erbium-doped fiber amplifier 253
external quantum efficiency (EQE) 72

f

field intensity distribution 341–343, 346
flashlamp pumped solid-state ruby laser 4
focusing metasurface 162–165, 170
folding cavity (F-cavity) 273
frequency discriminators 136, 137
frequency noise spectrum 137–139
full-width-half-maximum (FWHM) 112, 121, 272

g

GaAs-based gain mirror technologies
  GaAsSb QWs 31–32
  GaInNAs QWs 28–30
  InAs QDs 30–31
  InGaAs QWs 28
GaAsSb QWs 31–32
gain-embedded meta-mirror (GEMM) 191
gain filtering 235, 238–239
GaInNAs QWs 28–30, 199
GaInP quantum well VECSELs
  characterization results 204
  experimental setup 203–204
  internal efficiency 204–208
  semiconductor structures architecture 202–203
GaSb-based VECSEL 63, 64, 68, 69, 72–74, 76, 80, 81, 85, 92, 96, 100
graphene saturable-absorber mirrors (GSAMs) 282, 358
group delay dispersion (GDD) 236, 307, 326, 331, 334, 368

h

harmonic self-mode-locking 366–368
Hermite–Gauss 111, 113, 119
heterodyne 63, 86, 88, 90, 138, 338
highly coherent single-frequency tunable VeCSELs
    characteristics 127–129
    coherence properties 118
    high coherence 118–121
    ideal laser 111–113
    laser applications 109–111
    limits and solutions 125–127
    single-mode operation 113–118
    spatial coherence 131
    time domain coherence and noise
        intensity noise of 135–136
        phase noise, frequency noise and linewidth 136–139
        photonics 131–135
    ultrahigh-purity single-mode operation 129–131
high power multi-segmented semiconductor lasers 11, 75, 77, 91
high-reflectivity (HR) backside mirror 111
high-resistivity (HR) silicon window 160

i

ideal laser 91, 110–113
III-Sb material system 66–68
InAs QDs 30–31, 42
InGaAs/GaAs laser 13
InGaAs QWs 28, 29, 31, 331, 335, 339
InP-based gain mirror technologies
    dielectric and metamorphic DBRs 33–34
    gain structures in transmission 47–50
    monolithic InP-based DBRs 32–33
    semiconductor-dielectric-metal compound mirrors 34–37
    wafer-bonded GaAs-based DBRs. *see* wafer-bonded GaAs-based DBRs
interband cascade lasers (ICLs) 64

intraband-scattering 283, 285, 286, 325, 350
intracavity heatspreader (ICH) 41, 65, 73–76, 81–91, 94–99, 179, 180, 185, 186, 247
intra-cryostat cavity configuration 165
inverted quantum well 282–286, 288
in-well pumping 49, 69–70, 188, 197

k

Kerr-lensing 360, 362–365, 368, 369, 371, 373, 375–378
Kerr lens mode-locking 256, 360, 364, 365, 378, 379
kinetic hole filling 269, 274–277, 284–286, 293, 294, 325

l

Laguerre–Gauss 111, 114, 119
lasing-spaser 170
light detection and ranging (LIDAR) 63
linear field regime 283
LQD-VECSEL 71, 72, 76, 77, 79, 100
long-wavelength infrared VECSELs
    GaAs-based gain mirror technologies 28–32
    InP-based gain mirror technologies. *see* InP-based gain mirror technologies
low quantum deficit barrier pumping 70–72
lumped model 268

m

Maxwell's equations 268–270, 273
Maxwell-semiconductor Bloch equations 268
molecular beam epitaxy (MBE) 10, 42, 199, 306, 333, 359
membrane external-cavity surface-emitting laser (MECSEL)
    characterization
        beam profile and beam quality factor 218

membrane external-cavity surface-emitting laser (MECSEL) (contd.)
   output power measurements 216–218
   spectra 218–220
   DBR 212–213
   semiconductor active region membrane 213–215
   setup 215–216
metal–metal waveguide 146–151, 155, 157, 166, 169
metal-organic chemical vapor deposition/vapor phase epitaxy (MOCVD/MOVPE) 10
metasurface
   design 149–152
   QC-VECSEL 152–159
Micracor
   core technologies 8
   DARPA STTR program 18
   development 11–20
   Epitaxial Products International (EPI) 10
   GaAsP layers 16
   OPS laser power 19
   SBIR program 9
   semiconductor technology 10
   VECSELs 20–22
ML-VECSELs 231, 233, 239, 241–244, 246–249, 253–255, 357–360, 371, 374, 375, 380
modelocked integrated external-cavity surface-emitting lasers (MIXSELs) 21, 231, 233–235, 241–246, 248, 252, 254, 255, 273, 321, 359–361
monolithic InP-based DBRs 32–33, 37
modal uniformity 155
mode-locked laser 232, 233, 235, 239–241, 244, 246, 251, 307, 310, 313, 315
mode locking
   for optically-pumped SDLs 358–360
   quantum dot mode-locked AlGaInP VECSELs 314–315
   quantum well mode-locked AlGaInP VECSELs 311–314
monochromatic wave 112
mounting technologies
   double sided heatspreader (DSH) 77–78
   intracavity heatspreader 74–76
   thin device 76–77
multiphoton-excited fluorescence microscopy 252
multiple-quantum-well (MQW) 16, 278

n

noise 251–252
   buildup to pulse 281–282
   frequency 136–139
   intensity 135–136
   phase 136–139
   in photonics 131
   time domain coherence and 131–135
noncollinear pulses 326
nonlinear coupling drives 252
nonuniformity modal 154, 155
Novalux Extended Cavity Surface-Emitting Laser (NECSEL) 20

o

OPS CW output power 17, 18
optical fiber sensors (OFS) 34
optically pumped semiconductor lasers (OPSL)
   concept and history 4–8
   development 11–20
   Micracor 8–10, 20–22
optical sampling by cavity tuning (OSCAT) 247, 255

p

passively mode-locked semiconductor lasers
   mode-locked VECSELs 248–249
   noise 251–252
   pulse duration 244–246
   pulse repetition rate 246–248
   simulation and modelling 249–251

passively mode-locked VECSELs   358
passive metasurface reflectance   151
passive mode-locking   231, 232, 234, 306, 321, 322, 352, 360
photonic crystal fiber (PCF)   253, 353, 380
polarization coherence   112
polarization control, QC-VECSELs   166–169
Pound-Drever-Hall (PDH) stabilization technique   88
power scaling
   multi-pass pumping   208–211
   quantum well pumping   208–210
pulsed laser power   15
pulse molecule   267, 291–295
pulse propagation   250, 269–273, 290
pulse repetition rate   6, 233, 246–248, 252, 254, 255
pulse timing   251, 324

q

QC-VECSEL model
   confinement factor   156–157
   metasurface and cavity optimization   157–159
   modal uniformity   155
   QC-laser active material   152
quantum cascade lasers (QCL)   64, 254
quantum optics   63, 78, 253, 305
quantum wells (QWs)   7, 9, 11–13, 15–17, 27, 64, 117, 118, 122–126, 145, 146, 148, 175, 197, 199–201, 208–210, 233, 236–238, 244, 246, 248, 249, 267, 268, 273, 278, 282–286, 288, 306–307, 311–314, 321, 361, 379
quasi-Fermi Dirac distribution   276

r

Rabi-flopping   283, 284, 288
Rabi frequency   274
radiating mirror   5, 6, 176
resonant periodic gain (RPG)   12, 14, 31, 43, 68, 84, 176, 180, 182, 197, 237, 272

resonator out-coupling mirror   360
rigorous coupled wave analysis (RCWA)   191

s

saturable-absorber-free ML VECSELs   357
saturable-absorber-free pulsed VECSELs, history of
   harmonic self-mode-locking   366–368
   self-mode-locked optically-pumped. *see* self-mode-locked optically-pumped VECSELs
   self-mode-locking quantum-dot VECSEL   368–369
   SML cavity configurations   369–371
   SML VECSEL at other wavelengths   371–373
saturable-absorber-mirror based ML VECSELs   357
Schawlow–Townes limit   121
self-mode-locked optically-pumped VECSELs
   high-order transverse modes   365
   Kerr lens action   364
   magic mode-locking   362
   mode competition   363–364
self-mode-locked semiconductor disk lasers
   ML solid-state lasers   357
   optically-pumped VECSELs   360–366
self-mode-locking quantum-dot VECSEL   368–369
self-phase-modulation (SPM)   365, 379
semiconductor Bloch equations (SBE)   268, 269, 273, 343
semiconductor-based lasers   64, 65, 199, 212
semiconductor-dielectric-metal compound mirrors   34–37
semiconductor laser mode-locking   267–300
semiconductor lasers for the MID-IR range   64–66

semiconductor-saturable-absorber-mirror (SESAM)
  domain setup/modeling 277–282
  microscopic theory 273–277
  numerical results 282–299
  pulse propagation 269–273
SESAM-free mode-locking achievements
  SESAM-free alternatives to SML VECSEL
    GSAM 375
    intracavity Kerr medium 375–376
  SML VECSELs
    peak power 374
    pulse duration 373–374
    repetition rate 375
SESAM mode-locked VECSEL 234, 255, 309, 311, 312
SESAM mode-locking 235, 309, 359
side-of-fringe method 88
silicon carbide (SiC) 69, 177, 179, 185
single-crystalline diamond 69, 185, 186, 189, 192
single-frequency operation (SFO)
  key parameters for 79–81
  intracavity heatspreader with
    active stabilization and influence of sampling time 88–90
    emission linewidth 85–87
    laser cavity setup 82–83
    wavelength tuning 83–85
  intracavity heatspreader without 94–99
  microcavity VECSELs 92–94
  wedged heatspreader 91–92
single-walled carbon nanotubes (SWCNTs) 243, 375
single-bounce OPS laser configuration 12, 13
soft-aperture mode-locking 376
spatial coherence 111, 119, 131, 132, 139
spatial hole burning (SHB) 116, 117, 154, 155
spectral hole burning 117, 350
stable colliding pulse modelocking
  gain recovery and pumping rate 324–325

mode waist and saturation fluence 326–327
polarization 326
pulse timing 324
stable mode locked pulses 267, 278, 280, 281, 291, 295
stable mode-locking 240, 315, 326, 340, 363
stable pulsed operation mode 367
Stranski–Krastanow (S–K) growth 30
submonolayer (SML) 30
supercontinuum generation and frequency combs 253–254
surface-emitting gain chip design 235–238
surface-emitting semiconductor laser
  gain chip design 235–238
  gain filtering 238–239
  gain saturation and recovery 239–241
  saturable absorbers for ML-VECSELs and MIXSELs 241–244

t

terahertz imaging and spectroscopy 254–255
THz QC-lasers waveguides 146–147
THz QC-VECSEL performance
  external cavity configuration 160
  focusing metasurface VECSEL 162–165
  intra-cryostat cavity QC-VECSEL 165–166
  metasurface on spectrum effect 160–161
  output coupler effect 161–162
total internal reflection (TIR) 177, 191
traditional VECSELs 175–182, 186, 192
two-mirror plan-concave resonator configuration 365
2–3 µm VECSELs
  in-well pumping 69–70
  low quantum deficit barrier pumping 70–72
  SFO. *see* single-frequency operation (SFO)
  standard barrier pumped structures 68–69

u

ultrafast lasers   232–233, 235, 243, 246, 252, 305–307
ultrafast nonequilibrium carrier dynamics
  domain setup/modeling
    gain region   278–280
    noise buildup to pulse   281–282
    relaxation rates and the round trip time   280–281
    VECSEL cavity   277–278
  microscopic theory   273–277
  mode-locked VECSELs
    gain, absorption, and dispersion   288–290
    influence of loss   294–296
    mode-locked pulse family   291–294
    pulse and hysteresis effect   296–299
    pulse buildup and initial conditions   290
    self phase modulation from QWs   290–291
  pulse propagation   269–273
  second Born Markov approximation
    inverted quantum well   282–285
    saturable absorber   285–288
ultrafast optical pulse generation   357
ultrafast pulse formation, in surface-emitting semiconductor laser
  gain filtering   238–239
  gain saturation and recovery   239–241
  saturable absorbers for ML-VECSELs and MIXSELs   241–244
  surface-emitting gain chip design   235–238
ultrafast semiconductor lasers   233–235, 240, 243
ultrahigh purity single-mode operation   129–131
ultrashort pulses   34, 175, 232, 236, 237, 244, 248, 256, 277, 279, 300, 305, 306, 316, 317, 322, 328, 329, 331, 352, 357, 360, 373, 374, 377, 378, 380, 381
uniformity efficiency factor   155

v

van der Waals bonding   39, 44, 179, 182, 189
V-cavity configurations   42, 272

w

wafer-bonded GaAs-based DBRs
  direct wafer bonding   39–44
  generic procedure   38
  low temperature bonding   44–47
wafer-fused VECSELs   21, 39, 41–43
wafer-scale processing   189, 192

y

ytterbium-doped power amplifiers (YDFAs)   253

z

zinc germanium phosphide (ZGP)   63
Z-scan measurements   377–380